£61.00

Spectral Techniques in Digital Logic

Spectral Techniques in Digital Logic

S. L. HURST
University of Bath
Bath, England

D. M. MILLER
University of Manitoba
Winnipeg, Manitoba

J. C. MUZIO
University of Victoria
Victoria, British Columbia

1985

ACADEMIC PRESS
(Harcourt Brace Jovanovich, Publishers)

London Orlando San Diego New York
Toronto Montreal Sydney Tokyo

ACADEMIC PRESS INC. (LONDON) LTD.
24–28 Oval Road
LONDON NW1 7DX

United States Edition published by
ACADEMIC PRESS, INC.
Orlando, Florida 32887

British Library Cataloguing in Publication Data

Hurst, S. L.
Spectral techniques in digital logic.
1. Electronic digital computers 2. Logic design
I. Title II. Miller, D. M. III. Muzio, J. C.
621.3819'582 TK7888.3

Library of Congress Cataloging in Publication Data

Hurst, S. L. (Stanley Leonard)
Spectral techniques in digital logic.

Bibliography: p.
Includes index.
1. Logic design. I. Miller, D. M. II. Muzio, Jon C.
III. Title.
TK7868.L6H88 1985 621.3819'5835 83-83427
ISBN 0–12–362680–3 (alk. paper)

PRINTED IN THE UNITED STATES OF AMERICA

85 86 87 88 9 8 7 6 5 4 3 2 1

Contents

Preface

During the past two decades a new field of digital logic theory has been born and its boundaries continuously extended. Two motives for this work may be discerned: first, the purely academic one of applying existing mathematical techniques to pastures new, and second, the increasing appreciation of the limitations of existing algebraic and geometric methods in handling digital data for logic network design purposes.

Digital logic design is a peculiar discipline in that the logic design for a given requirement, which may be in total a very large system, is frequently made without the use of any sophisticated design procedure. Very great sophistication, however, may be present in the logical verification of the complete assembly and in its translation into a microelectronic realisation, but the basic logic formulation may remain largely a hand-assembled synthesis, albeit aided by the experience of the logic designer who may be very proficient in intuitively recognizing patterns or symmetries in the network being designed. Indeed, many industrial designers will state that this is a perfectly acceptable situation, but this may be because of more demanding pressures from the existing areas of sophistication. Equally, the increasing capabilities of LSI and VLSI fabrication have so far kept pace with the designers' need, and no strong pressure for "better" design techniques has arisen.

However, the new field of digital logic theory represents a modern approach to expressing conventional digital data, one which can provide various insights into the structure of the data which are absent from classical Boolean algebra and truth-table formats. The pioneering work, particularly that of Karpovsky, originally in Russia, and Lechner, in the United States of America, forms the basis of this approach. Others have contributed and amplified the basic concepts and have translated the underlying mathematics into engineering tools which may be more acceptable to a digital logic designer.

The first chapters of this book will attempt to introduce the underlying theory of this area, that of orthogonal transforms and resulting spectral data. We assume that the reader will be conversant with conventional digital logic theory, such as is contained in any graduate textbook. Subsequent chapters will be concerned with the application of spectral data to Boolean function classification, logic network synthesis, fault diagnosis, and other aspects relevant to digital logic design. While the underlying theory of this area is applicable to any-valued logic, and not exclusively to the binary case, we will generally confine our discussions herewith to two-valued digital networks. Nevertheless, it is particularly significant that should future technology adopt a higher-valued logic, say quaternary, then the basic design techniques such as those discussed here will be applicable for such developments.

We hope that this book will be of interest to all working in computer engineering and digital system design, as well as to academic and research establishments. Our final hope is that it may maintain and increase interest in this area, leading to yet further developments which must surely be forthcoming.

S.L.H.
D.M.M
J.C.M.

Acknowledgements

The authors would like to express their thanks and appreciation for the work of others who have initiated and contributed to the present state of knowledge and application of spectral methods in digital logic design. Many have been referenced in the several chapters of this text, and we hope that this will be taken as our thanks to them for their work and inspiration.

The authors must also acknowledge financial support received from many sources in furthering their own work in this area and that of the research students with whom they have been associated. Among these benefactors we thank the United Kingdom Science and Engineering Research Council (SERC) for research studentships at Bath and the Natural Sciences and Engineering Research Council of Canada (NSERC) for their support of the work done in Canada.

In addition the SERC, the NSERC, the British Council, and particularly NATO have provided means for the authors to co-operate across the Atlantic for a number of years.

Finally, our thanks go to many colleagues in several countries for their personal comments and guidance in this still-evolving area of digital logic design.

S.L.H.
D.M.M.
J.C.M.

List of Symbols

a	Integer threshold-gate input weight
b	Entries of **B** (see below)
d	Integer threshold-gate output discrimination (threshold) value
e	Parameter used in fault detection procedures only
f	Function of all the input variables
g, h	Decompositions of a function
i	Independent input variable subscripts, $i = 1$ to n
j	Row subscripts of a matrix, $j = 0$ to $2^n - 1$
k	Column subscripts of a matrix, $k = 0$ to $2^n - 1$
m	Minterm identification with decimal notation subscript, see tabulation, also used in Fig. 3.8

m	x_n	x_{n-1}	$\cdots$	x_2	x_1
m_0	0	0	$\cdots$	0	0
m_1	0	0	$\cdots$	0	1
m_2	0	0	$\cdots$	1	0
$\vdots$					
m_j					
$\vdots$					
m_{2^n-1}	1	1	$\cdots$	1	1

Note that x_n = most-significant digit and x_1 = least-significant digit; see also Appendix A, Section 1.

n	Number of independent x_i input variables
p	Number of multi-valued logic levels, $p \geq 2$, $p = 2$ in binary
r	Entries of **R**
s	Entries of **S**
t	Entries of **T**, or as a superscript the transpose of a vector or matrix
u	Sub-function identifier
x	Independent input variable x_i, $i = 1$ to n, x_1 least significant
y	Entries of **Y**, except locally in Figs. 3.1 and 3.8
z	Entries of **Z**
X	$\{x_1, \ldots, x_n\}$
$f(X)$	$f(x_1, \ldots, x_n)$
B	Vector of $n + 1$ modified Chow parameters
N	2^n
Z	Truth-table column vector for function $f(x)$, entries $\in\langle 0, 1\rangle$, in truth-table order $m_0, m_1, \ldots, m_{2^n-1}$ unless otherwise stated

$\mathbf{Y}$ Truth-table column vector for function $f(x)$, entries $\in\langle +1, -1\rangle$, in truth-table order $m_0, m_1, \ldots, m_{2^n-1}$ unless otherwise stated

$\mathbf{T}$ Any transform

$\mathbf{T}^n$ $2^n \times 2^n$ transform

(Note that all $\mathbf{T}^n$ transforms will be in Hadamard ordering unless otherwise stated.)

$\mathbf{T}_p^n$ $p^n \times p^n$ transform matrix (used principally in Section 2.7)

$\mathbf{T}_{j*}^n$ jth row of $\mathbf{T}^n$ (row vector), $j = 0$ to $2^n - 1$

$\mathbf{T}_{*k}^n$ kth column of $\mathbf{T}^n$ (column vector), $k = 0$ to $2^n - 1$

t_{jk} jth row, kth column entry of $\mathbf{T}^n$

$\mathbf{R}$ Resultant spectrum of function $\mathbf{Z}$, $\mathbf{Z}\in\langle 0, 1\rangle$

$\mathbf{S}$ Resultant spectrum of function $\mathbf{Y}$, $\mathbf{Y}\in\langle +1, -1\rangle$

r_j jth entry of column vector $\mathbf{R}$, $j = 0$ to $2^n - 1$

s_j jth entry of column vector $\mathbf{S}$, $j = 0$ to $2^n - 1$

$\mathbf{I}^n$ $2^n \times 2^n$ identity matrix (n dropped where no ambiguity arises)

$\mathbf{J}^n$ 2^n column vector whose top entry takes the value 2^n, all remaining entries zero-valued

β $2^{n-m} - 1$, where m is $1 \leq m \leq n$, see sub-function definitions

$\bar{}$ Used above a binary variable or function to indicate the complement (negation)

General Transform of a Function $f(X)$

$$\mathbf{TY} = \mathbf{S}, \quad \mathbf{Y}\in\langle +1, -1\rangle, \qquad \text{or} \qquad \mathbf{TZ} = \mathbf{R}, \quad \mathbf{Z}\in\langle 0, 1\rangle$$

Sub-functions of a Binary Function $f(X)$

$f_u(x_1, \ldots, x_m) \equiv f(x_1, \ldots, x_m, u_1, \ldots, u_{n-m})$, where u is a constant and $(u_1, \ldots, u_{n-m})$ is the binary expansion of u, i.e., $u = \sum_{i=1}^{n-m} u_i 2^{i-1}$

$\mathbf{Y}_u$, $\mathbf{Z}_u$ minterm column vectors for f_u, $Y_u\in\langle +1, -1\rangle$, $Z_u\in\langle 0, 1\rangle$, respectively

$\mathbf{S}_u$, $\mathbf{R}_u$ spectrum for $\mathbf{Y}_u$, $\mathbf{Z}_u$, respectively

Decomposition g, h of a Function $f(X)$ (see Fig. 4.3)

$f(X) = h(X, g(X))$, where g, h may be any function of n input variables

$\mathbf{R}^g$, $\mathbf{S}^g$ the spectrum of function g in the above decomposition

$\hat{\mathbf{R}}$, $\hat{\mathbf{S}}$ the spectrum of function h in the above decomposition

Additional Superscript Notations

$\tilde{}$ In Chapters 4 and 5 used above an x_i input variable when it is a remapping from the original x_i input variables, for example, $\tilde{x}_2$

$\hat{}$ In Chapter 5, used above a function or spectral parameter where the given function $f(X)$ contains undefined ("don't-care") minterms, and when the don't-care minterms are all allocated the logic value 0, for example, $\hat{s}_3$

$\hat{\hat{}}$ In Chapter 5, as above but when all the don't-care minterms are allocated the logic value 1, for example, $\hat{\hat{s}}_3$

$*$ In Chapter 6, used above a function or spectral parameter to indicate that it is a parameter of a faulty function, for example, $\overset{*}{s}_1$

Mathematical Symbols

$+$ Arithmetic addition, or maximum of, = Boolean addition for $m = 2$ case, = logical OR (the context of use should identify which meaning is present)

$\times$ Arithmetic multiplication

Minimum of, = Boolean multiplication for the $p = 2$ case, = logical AND (symbol dropped where no ambiguity occurs)

$*$	Convolution
$\circledast$	Kronecker matrix product
$\oplus$	Addition mod$_p$, = Exclusive-OR for $p = 2$ case
$\otimes$	Multiplication or product mod$_p$
$-$	Cyclic negation, = NOT for the $p = 2$ case
j	$\sqrt{-1}$, $= 1.0\angle 90°$
a	$\exp(2\pi j/3) = -0.866 + j0.5 = 1.0\angle 120°$; used principally in Section 2.7
b	$\exp(2\pi j/p) = 1.0 \angle 360°/p$; used principally in Section 2.7
$H_i^q, H_i^{q,r}$	Haar functions

1

General Introduction

1.1 BOOLEAN AND SPECTRAL DOMAINS

In this book we are concerned with digital logic and with the design and analysis of switching circuits. The majority of existing methods are concerned with the properties of Boolean functions since it was proved by Shannon[1] that the Boolean domain provides a precise model for the analysis of switching circuits. We are only concerned with a two-valued Boolean algebra for practical applications, with these two values normally being represented by 0, 1, irrespective of their actual implementation. The behaviour of a device is represented by a function $f(x_1, x_2, \ldots, x_n)$ of its input variables $x_1, \ldots, x_n$. This function can most conveniently be defined by a table. For example, a function $f(x_1, x_2, x_3)$ of the three variables x_1, x_2, x_3 is illustrated in Table 1.1.

It is common to use the product and sum operators of the Boolean algebra together with negation to define such functions—for example, $f(x_1, x_2, x_3) = x_1\bar{x}_2\bar{x}_3 + \bar{x}_1x_2\bar{x}_3 + x_1x_2x_3$. The use of Boolean algebra for the manipulation and analysis of switching circuits is well known and is not part of our purpose in this book. We shall be looking at a different domain in which to express the information required to define and analyse a Boolean function.

Part of the problem with the definition in the Boolean domain is that each of the entries in the column for $f(x_1, x_2, x_3)$ in Table 1.1 tells us precisely the behaviour of the function at a single point but nothing of its behaviour for any other points. It is possible to give an alternate representation of a function where the information about the function is much more global in nature. This alternate representation is in the spectral domain, and it will be demonstrated in the later chapters that a number of properties are much more easily deduced in the spectral domain than in the Boolean one.

The basic idea of the spectral domain, and how to get there, is illustrated in Table 1.2. If we are to avoid losing information, we shall have to ensure that the transform can be reversed, that is, that we can move to and from the spectral domain without any loss of information.

Table 1.1. The function
$f(x_1, x_2, x_3) = x_1\bar{x}_2\bar{x}_3 + \bar{x}_1 x_2 \bar{x}_3 + x_1 x_2 x_3$.

x_3	x_2	x_1	$f(x_1, x_2, x_3)$
0	0	0	0
0	0	1	1
0	1	0	1
0	1	1	0
1	0	0	0
1	0	1	0
1	1	0	0
1	1	1	1

Table 1.2. The transform.

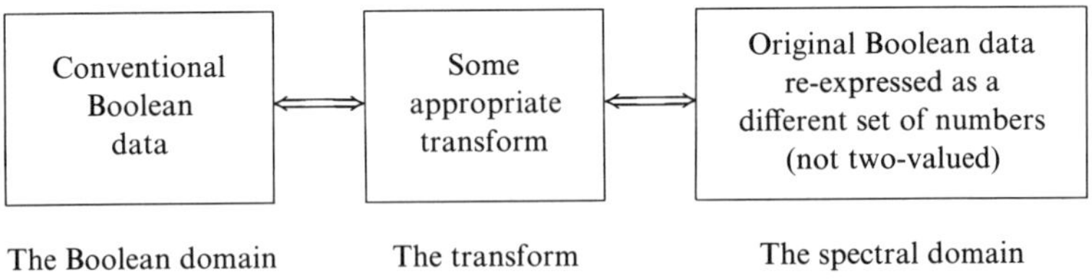

To transform the Boolean representation (such as that for $f(x_1, x_2, x_3)$ given in Table 1.1) into the spectral domain, we take the right-hand column from the defining table for a column vector **Z**. Using a particular square matrix **T** of the correct size as our transform gives the spectrum $\mathbf{R} = \mathbf{TZ}$ for the function. This vector **R** (of the same size as **Z**) is an alternate representation for the function as long as **T** has an inverse. For the particular transforms that we shall be using, the inverse is very straightforward. A detailed discussion of these is given in Chapter 2.

1.2 HISTORY

Here we shall only give a brief outline of the major developments that have led to the results that are described in the later chapters. The basis for the transforms goes back to Rademacher[2] and Walsh[3] and the transforms themselves are particular examples of Hadamard[4] matrices. Their early work was followed by others in the area of studying orthogonal functions.[5–8]

The question of the evaluation of the transform has been extensively studied, demonstrating that the evaluation can be limited to just additions and subtractions and moreover only involves $n \cdot 2^n$ operations for a $2^n \times 2^n$ transform.[9–12] These fast transform procedures are explained in Section 2.4.

The applications of orthogonal functions in the digital area came first in the areas of signal processing and the transmission of information. For a detailed discussion of this work, the reader is referred to Ahmed and Rao[13] and Harmuth.[14] In applications to digital logic and circuit design, analysis and synthesis, the first suggestion that orthogonal functions might be useful was made by Coleman.[15] He suggested that orthogonal functions might be useful for the design of circuits, and this theme recurs through much of Chapters 4 and 5. The appropriate references will be found in the chapters concerned. A number of books have addressed the use of orthogonal functions and Walsh functions in the digital area, and the reader is referred to them for related work in this area.[16–18]

Following the work of Dertouzos[19] and Lechner[20] it has been shown that spectral methods yield powerful classification techniques for functions (see Chapter 3). These classification methods can be used as the basis for synthesis algorithms and realization techniques.[21,22] They have also been used as a basis to design universal logic modules.[23] In the area of sequential machine design and state optimization some results have been reported.[24,25] Besslich[26,27] and Lloyd[28] have considered the idea of prime implicant extraction in the spectral domain and covering problems.

In 1955 Chrestenson[29] generalized the work of Walsh[3] to the many-valued case, that is, we are no longer transforming from the Boolean two-valued domain, but from a multiple-valued one (p-valued, $p \geq 2$). The procedure of using multiple-valued transforms is much more complex than two-valued and is considered briefly in Section 2.7, but is otherwise outside the scope of this book. Karpovsky[17] gives good coverage to multiple-valued results and the reader is also referred to the recent work of Moraga.[30–32]

This very brief summary is not intended to provide any detailed coverage for the various topics. This will be done as they are discussed in the later chapters. It does however illustrate some of the diversity of the areas of application of the spectral techniques.

1.3 MOTIVATION

The question arises as to the reason for considering the spectral domain and if there are any real purposes for its use. To understand the first difference between the Boolean and spectral domain, let us consider a Boolean function $f(X)$ of n variables. One row of the table defining this function provides complete and precise information about the behaviour of the function for one combination of the input variables. Of course, it does not tell us anything about the value of the function anywhere else. The combination of the knowledge of the behaviour of $f(X)$ for the 2^n rows of the table gives a complete definition of the function. Similarly, the spectrum for a function of n variables also contains 2^n values, which together completely define the function and can be used to recover its Boolean specification. Each of the 2^n values in the spectrum (the spectral coefficients) contains some information about the behaviour of the function at all 2^n points, but does not contain complete information about any of them. The combination of all the values in the spectrum does lead to complete information about the function, but each individual coefficient gives us some global information about the whole function. In this sense the spectral coefficients are giving us global information about the function, while the Boolean domain consists of local information. For some applications this global information is more directly useful than the Boolean representation of the function.

The easiest way to demonstrate the value of the spectral methods is to give a brief description of the areas that are covered in the rest of the book. Chapter 2 gives a complete explanation of the spectra of discrete functions and their calculation, together with some consideration to the use of other transformations using incomplete and non-orthogonal matrices.

The classification of Boolean functions is explored in Chapter 3, and an explanation is given of the way in which the spectral classification is connected to threshold functions and other properties of certain classes of functions. The synthesis and design of circuits is discussed in Chapters 4 and 5. A number of powerful techniques are described in Chapter 4, showing how certain properties of a function are easily detected in the spectral domain but may be very difficult in the Boolean domain. This enables a number of different approaches to design to be considered. Chapter 5 is entirely concerned with the detection of symmetries and partial symmetries in functions, and demonstrates their value in synthesis techniques.

The most recent application for spectral techniques has been in fault diagnosis. This exciting area is described in Chapter 6. One of the serious problems with digital circuits concerns their testing. There are two aspects to this—first to verify whether or not a circuit is performing correctly and, second, if there is a fault, to find its location (fault isolation). Of course, from the user's point of view there is not usually much purpose in locating a fault beyond identifying a chip that needs replacing. There is no need to identify the exact location of the fault on the chip. We shall be showing that a check of the correctness of certain spectral coefficients for a digital network ensures that the network is free of certain types of faults. The testing technique involved is straightforward and can be easily applied. The required coefficients for the circuit are called a signature, and the ease with which they can be derived using spectral techniques and the resulting high level of fault coverage give a number of promising new ideas for fault testing in digital circuits.

Therefore we hope to cover all aspects of the present state of the art of this subject area in the following chapters. Readers who may wish to be reminded of the properties of vectors and matrices, we refer to Appendix C in this text. Whilst this Appendix covers the general case, where the matrix entries may be complex numbers, in all our binary work we shall only need to be concerned with the particular case of real entries in the matrix operations. Complex entries become necessary only when higher-valued logic than binary is being considered, which is the subject of Section 2.7 only.

References

1. Shannon, C. E., A symbolic analysis of relay and switching circuits, *Trans. AIEE* **57,** 713–723 (1938).
2. Rademacher, H., Einige Sätze über Reihen von allgemeinen orthogonal funktionen, *Math. Ann.* **87,** 112–138 (1922).
3. Walsh, J. L., A closed set of orthogonal functions, *Amer. J. Math.* **45,** 5–24 (1923).
4. Wallis, J. S., "Hadamard Matrices" (Lecture Notes No. 292). Springer-Verlag, New York, 1972.
5. Paley, R. E. A. C., A remarkable series of orthogonal functions, *Proc. London Math. Soc.* **34**(2), 241–279 (1932).
6. Paley, R. E. A. C., On orthogonal matrices, *J. Math. Phys.* **12,** 311–320 (1933).
7. Fine, N. J., On the Walsh functions, *Trans. Amer. Math. Soc.* **65,** 372–414 (1949).
8. Selfridge, R. E., Generalized Walsh transforms, *Pacific J. Math.* **5,** 451–480 (1955).
9. Brown, R. D., A recursive algorithm for sequency-ordered fast Walsh transforms, *IEEE Trans. Comput.* **C-26,** 819–822. (1977).
10. Fino, B. J., and Algazi, V. R., Unified matrix treatment of the fast Walsh–Hadamard transform, *IEEE Trans. Comput.* **C-25,** 1142–1145.
11. Shanks, J., Computation of the fast Walsh–Fourier transform, *IEEE Trans.* **EC-18,** 457 459 (1969).
12. Ulman, L. J., Computation of the Hadamard transform and the R transform in ordered form, *IEEE Trans. Comput.* **EC-19,** 359–360 (1970).
13. Ahmed, N., and Rao, K. R., "Orthogonal Transforms for Digital Signal Processing." Springer-Verlag, New York, 1975.
14. Harmuth, H. F., "Transmission of Information by Orthogonal Functions." Springer-Verlag, New York, 1972.
15. Coleman, R. P., Orthogonal functions for the logical design of switching circuits, *IEEE Trans. Comput.* **EC-10,** 379–383 (1961).
16. Beauchamp, K. G., "Walsh Functions and Their Application." Academic Press, London, 1975.
17. Karpovsky, M. G., "Finite Orthogonal Series in the Design of Digital Devices." Wiley, New York, 1976.
18. Hurst, S. L., "The Logical Processing of Digital Signals." Crane-Russak, New York and Edward Arnold, London, 1978.

19. Dertouzos, M. L., "Threshold Logic: A Synthesis Approach." MIT Press, Cambridge, Massachusetts, 1965.
20. Lechner, R. J., Harmonic analysis of switching functions, *in* "Recent Developments in Switching Theory" (A. Mukhopadhay, ed.). Academic Press, New York, 1971.
21. Edwards, C. R., The application of the Rademacher–Walsh transform to Boolean function classification and threshold logic synthesis, *IEEE Trans. Comput.* **C-24,** 48–52 (1975).
22. Edwards, C. R., Characterization of threshold functions under the Walsh transform and linear translation, *Electron. Lett.* **11,** 563–565 (1975).
23. Edwards, C. R., A special class of universal logic gates and their evaluation under the Walsh transform, *Internat. J. Electron.* **44,** 49–59 (1978).
24. Edwards, C. R., and Hurst, S. L., Preliminary considerations of combinatorial and sequential digital systems under symmetry methods, *Internat. J. Electron.* **40,** 499–507 (1976).
25. Edwards, C. R., and Hurst, S. L., A digital synthesis procedure under function symmetries and mapping methods, *Trans. IEEE Comput.* **C-27,** 985–997 (1978).
26. Besslich, P. W., On the Walsh–Hadamard transform and prime implicant extraction, *IEEE Trans.* **EMC-20,** 516–519 (1978).
27. Besslich, P. W., Determination of the irredundant forms of a Boolean function using Walsh–Hadamard analysis and dyadic groups, *IEE Comp. Dig. Tech.* **1,** 143–150 (1978).
28. Lloyd, A. M., Spectral addition techniques for the synthesis of multivariable logic networks, *IEE Comp. Dig. Tech.* **1,** 152–164 (1978).
29. Chrestenson, H. E., A class of generalized Walsh functions, *Pacific J. Math.* **5,** 17–31 (1955).
30. Moraga, C., Introducing disjoint spectral translation in multiple-valued logic design, *Electron. Lett.* **14,** 241–243 (1978).
31. Moraga, C., Spectral characterisation of ternary threshold functions, *Electron. Lett.* **15,** 712–713 (1979).
32. Moraga, C., Characterisation of ternary threshold functions using a partial spectrum, *Electron. Lett.* **15,** 803–805 (1979).

2

The Spectra of Discrete Functions

2.1 INTRODUCTION

In the previous chapter we considered the shortcomings of the conventional methods of expressing and manipulating digital data, such shortcomings arising largely because of the discrete nature of the data format, preventing any global picture of the network or function under consideration being given.

Here we shall begin to consider formally the procedures for the transformation of our conventional digital data from the functional (Boolean) domain into the alternative spectral domain, and vice versa. The information content in the functional and spectral domains will be identical, the data in either domain being uniquely recreatable from the data in the other, but the meaning of the parameters in the two domains will be dissimilar. In particular, the discrete nature of the data in the functional domain will generally be replaced by data in the spectral domain which is global in nature, being influenced by the complete functional performance of the circuit or network under consideration.

The developments that we shall consider in the following sections will mainly be confined to two-valued (binary) functions. However, the mathematical principles of orthogonal transformation of discrete data apply to any-valued logic, and hence what we shall here consider in depth is but a particular case, the simplest case, of more general procedures applicable to multiple-valued logic. In Section 2.7 we shall return to comment upon the wider general case, and indicate how our considerations here and in the following chapters may be related to higher-valued logic situations.

2.2 THE TRANSFORMATION OF BINARY DATA

Any given combinatorial function $f(X)$ may be explicitly defined by its truthtable, which lists all 2^n input combinations, against each combination being given the local function output value $f(X)$. Conventionally, all x_i input variables and the output value of $f(X)$ are expressed in $\{0, 1\}$ notation. Note that each input combination and the corresponding output value of $f(X)$ is discrete data, no information about the output value of $f(X)$ at any input combination being obtainable from any other input condition.

With the order of tabulation of the input variables standardised, then the 2^n local output values of $f(X)$ may be treated as a column vector **Z** defining $f(X)$; the order of tabulation we shall here employ will be as indicated in Table 2.1(a). See also the introductory List of Symbols. Should we recode the binary logic values from $\{0, 1\}$ to $\{+1, -1\}$, respectively, then an alternative column vector **Y** defining $f(X)$ is given, as shown in Table 2.1(b). The reason for such recoding will be clear later.

Table 2.1. Truthtable formats for example three-variable function $f(X) = x_1\bar{x}_2\bar{x}_3 + \bar{x}_1 x_2 \bar{x}_3 + x_1 x_2 x_3$.[a]

(a) In $\{0, 1\}$ notation, function vector **Z**

	Truthtable				Output vector	
Minterm designation	x_3	x_2	x_1	$f(X)$		**Z**
m_0	0	0	0	0	z_0	0
m_1	0	0	1	1	z_1	1
m_2	0	1	0	1	z_2	1
m_3	0	1	1	0	z_3	0
m_4	1	0	0	0	z_4	0
m_5	1	0	1	0	z_5	0
m_6	1	1	0	0	z_6	0
m_7	1	1	1	1	z_7	1

(b) In $\{+1, -1\}$ notation, function vector **Y**

	Truthtable (recoded)				Output vector	
Minterm designation	x_3	x_2	x_1	$f(X)$		**Y**
m_0	1	1	1	1	y_0	1
m_1	1	1	−1	−1	y_1	−1
m_2	1	−1	1	−1	y_2	−1
m_3	1	−1	−1	1	y_3	1
m_4	−1	1	1	1	y_4	1
m_5	−1	1	−1	1	y_5	1
m_6	−1	−1	1	1	y_6	1
m_7	−1	−1	−1	−1	y_7	−1

[a] Note that x_1 = least-significant digit in the decimal notation for minterm m_j, $j = 0$ to $2^n - 1$.

In both **Y** and **Z** we have 2^n entries necessary to fully define an n-variable function $f(X)$. As an alternative to these 2^n entries, we may also specify 2^n coefficients that define $f(X)$, these coefficients being the "spectral coefficients" of $f(X)$; these coefficients may also be listed in a defined order to give a column vector representation, which we shall subsequently identify by **R** or **S**. As will be seen, the entries in **R** and **S** will no longer be binary values as in the local value truthtable vectors **Y** and **Z**.

The definition of the individual spectral coefficients in **R** or **S** for any $f(X)$ may be approached from several mathematical viewpoints. Here we shall first consider the Hadamard transform $\mathbf{T}^n$, which will directly generate **S** from **Y** or **R** from **Z**, following which we consider alternative but equivalent mathematical definitions and aspects.

2.2.1 The Hadamard Orthogonal Transform Matrix

The Hadamard transform is a complete orthogonal square matrix, with row and column entries $\in\{+1, -1\}$, and with a recursive structure as follows:

$$\mathbf{T}^n \triangleq \begin{bmatrix} \mathbf{T}^{n-1} & \mathbf{T}^{n-1} \\ \mathbf{T}^{n-1} & -\mathbf{T}^{n-1} \end{bmatrix} \tag{2.1}$$

Note that the dimensions of the transform are $2^n \times 2^n$ for any n. For increasing n we have the real integer values

$$\begin{aligned} \mathbf{T}^0 &\triangleq +1 \\ \mathbf{T}^1 &= \begin{bmatrix} 1 & 1 \\ 1 & -1 \end{bmatrix} \\ \mathbf{T}^2 &= \begin{bmatrix} 1 & 1 & 1 & 1 \\ 1 & -1 & 1 & -1 \\ 1 & 1 & -1 & -1 \\ 1 & -1 & -1 & 1 \end{bmatrix}, \quad \text{etc.} \end{aligned} \tag{2.2}$$

We may alternatively express this recursive structure by

$$\begin{aligned} \mathbf{T}^n &= \begin{bmatrix} 1 & 1 \\ 1 & -1 \end{bmatrix} \circledast \mathbf{T}^{n-1} \\ &= \begin{bmatrix} 1 & 1 \\ 1 & -1 \end{bmatrix} \circledast^n \end{aligned} \tag{2.3}$$

where $\circledast$ here denotes the Kronecker product operator.[1-3]

The value of each element t_{jk} in the Hadamard matrix may be individually defined by mathematically summing the element-by-element multiplication of the binary (mod 2) expansions of j and k (call this v) and taking the vth power of -1. For example, taking the third row ($j = 10$), fourth column ($k = 11$) entry t_{23} of $\mathbf{T}^n$, shown circled below, we have

$$\begin{array}{cc} & k(\text{binary}) \;\; 00 \quad 01 \quad 10 \quad 11 \\ j(\text{binary}) & \begin{array}{c} 00 \\ 01 \\ 10 \\ 11 \end{array} \begin{bmatrix} 1 & 1 & 1 & 1 \\ 1 & -1 & 1 & -1 \\ 1 & 1 & -1 & \textcircled{-1} \\ 1 & -1 & -1 & 1 \end{bmatrix} \end{array}$$

$$j = 10, \quad k = 11; \quad \therefore v = \{1.1 + 0.1\} = 1$$

$$\therefore t_{23} = (-1)^1 = -1$$

This may formally be expressed by

$$t_{jk} = (-1)^{\sum_{\eta=0}^{n-1} j_\eta k_\eta} \tag{2.4}$$

$$\equiv \exp\left\{\frac{2\pi}{2}\sqrt{-1}\sum_{\eta=0}^{n-1} j_\eta k_\eta\right\}^{\dagger} \tag{2.4a}$$

where j_η, k_η are determined by the binary expansions of j, k, respectively, $j, k \in 0$ to $2^n - 1$, and where

$$j = \{j_{n-1}2^{n-1} + j_{n-2}2^{n-2} + \cdots + j_0 2^0\}$$
$$k = \{k_{n-1}2^{n-1} + k_{n-2}2^{n-2} + \cdots + k_0 2^0\}$$

Note that the value for any particular t_{jk} is independent of n. However, in general the recursive definition of (2.1), rather than this individual-element definition, will be most relevant for our purposes herewith.

The individual row vectors $\mathbf{T}_{j*}$ of $\mathbf{T}^n$ define a complete orthogonal series of functions as shown in Fig. 2.1, which is closed under row multiplication. The inner product of any two matrix rows j_α, j_β has the orthogonal property

$$\sum_{k=0}^{2^n-1} t_{j_\alpha k} t_{j_\beta k} \begin{array}{ll} = 2^n & \text{if} \quad \alpha = \beta \\ = 0 & \text{otherwise} \end{array} \tag{2.5}$$

This inner product relationship holds also for any two-column vectors $\mathbf{T}_{*k}$.

† Recall $\exp(j\xi) = \cos\xi + j\sin\xi$, where here $j = \sqrt{-1}$; hence, when $\xi = (2\pi/2)\delta$, where δ is any real positive integer, we have either $\cos 0° + j\sin 0°$ or $\cos 180° + j\sin 180° = \pm 1 + j0$.

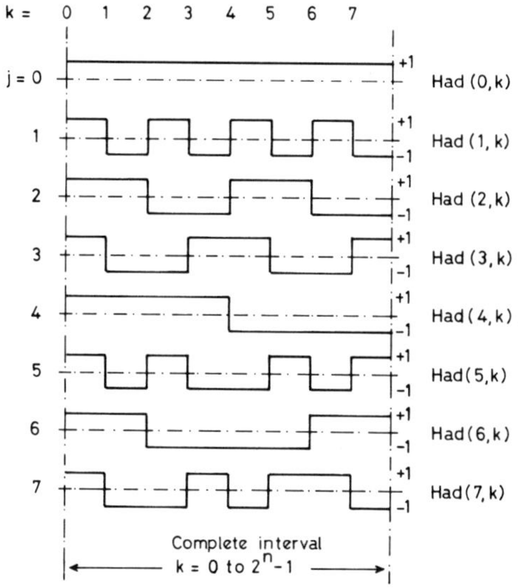

Fig. 2.1. The orthogonal Hadamard functions for $n = 3$.

The inverse of $\mathbf{T}^n$ is given by†

$$[\mathbf{T}^n]^{-1} = \frac{1}{K}[\mathbf{T}^n]^h$$

where $[\mathbf{T}^n]^h$ is the hermitian conjugate of $\mathbf{T}^n$, but as all entries in $\mathbf{T}^n$ are real positive or real negative numbers $[\mathbf{T}^n]^h \equiv [\mathbf{T}^n]^t$, giving

$$[\mathbf{T}^n]^{-1} = \frac{1}{K}[\mathbf{T}^n]^t \tag{2.6}$$

Finally it will be noted that the Hadamard transform has the additional property that its transpose is identical to itself, that is, $\mathbf{T}^n \equiv [\mathbf{T}^n]^t$, thus giving the simple inverse transform relationship for any n of

$$[\mathbf{T}^n]^{-1} = \frac{1}{K}[\mathbf{T}^n] \tag{2.7}$$

It is a trivial exercise to confirm that $K = 2^n$, finally giving us

$$[\mathbf{T}^n]^{-1} = \frac{1}{2^n}[\mathbf{T}^n] \tag{2.8}$$

† Recall that the inverse of any transform (if available) must be such that $[\mathbf{T}^n][\mathbf{T}^n]^{-1} = \mathbf{I}$, where $\mathbf{I}$ is the identity matrix.[2,3]

For further supporting details of orthogonal vectors and matrices, see cited references,[1–3,10,15] and other standard algebraic texts.

2.2.2 The Hadamard Transformation of Binary Data

The Hadamard matrix may be used to transform any 2^n column vector of numbers into an alternative data column vector. No information is lost in this transformation, the original data being fully recoverable from the resulting vector by application of the inverse transform $[\mathbf{T}^n]^{-1}$. However our main concern will be its use in the transformation of a column vector $\mathbf{Y}$ or $\mathbf{Z}$ representing the output truthtable of a given function $f(X)$.

Consider the three-variable example function given in Table 2.1. Taking the (conventional) 0, 1 coding we may transform $\mathbf{Z}$ with an $n = 3$ Hadamard transform as follows. For clarity we identify all entries in this transformation:

$$
\begin{array}{c}
 \\ \mathbf{T}_{0*} \\ \mathbf{T}_{1*} \\ \mathbf{T}_{2*} \\ \mathbf{T}_{3*} \\ \mathbf{T}_{4*} \\ \mathbf{T}_{5*} \\ \mathbf{T}_{6*} \\ \mathbf{T}_{7*}
\end{array}
\begin{array}{c}
\begin{array}{cccccccc}
\mathbf{T}_{*0} & \mathbf{T}_{*1} & \mathbf{T}_{*2} & \mathbf{T}_{*3} & \mathbf{T}_{*4} & \mathbf{T}_{*5} & \mathbf{T}_{*6} & \mathbf{T}_{*7}
\end{array} \\
\begin{bmatrix}
1 & 1 & 1 & 1 & 1 & 1 & 1 & 1 \\
1 & -1 & 1 & -1 & 1 & -1 & 1 & -1 \\
1 & 1 & -1 & -1 & 1 & 1 & -1 & -1 \\
1 & -1 & -1 & 1 & 1 & -1 & -1 & 1 \\
1 & 1 & 1 & 1 & -1 & -1 & -1 & -1 \\
1 & -1 & 1 & -1 & -1 & 1 & -1 & 1 \\
1 & 1 & -1 & -1 & -1 & -1 & 1 & 1 \\
1 & -1 & -1 & 1 & -1 & 1 & 1 & -1
\end{bmatrix} \\
\mathbf{T}^n
\end{array}
\begin{array}{c}
\begin{bmatrix} 0 \\ 1 \\ 1 \\ 0 \\ 0 \\ 0 \\ 0 \\ 1 \end{bmatrix}
\begin{array}{c} z_0 \\ z_1 \\ z_2 \\ z_3 \\ z_4 \\ z_5 \\ z_6 \\ z_7 \end{array} \\
\mathbf{Z}
\end{array}
\begin{array}{c} = \\ \\ = \end{array}
\begin{array}{c}
\begin{bmatrix} 3 \\ -1 \\ -1 \\ -1 \\ 1 \\ 1 \\ 1 \\ -3 \end{bmatrix}
\begin{array}{c} r_0 \\ r_1 \\ r_2 \\ r_{12} \\ r_3 \\ r_{13} \\ r_{23} \\ r_{123} \end{array} \\
\mathbf{R}
\end{array}
\qquad (2.9)
$$

If the output vector of $f(X)$ is recoded from $\{0, 1\}$ to $\{+1, 1\}$, see Table 2.1, then we have the alternative transformation:

$$
\begin{array}{c}
 \\ \mathbf{T}_{0*} \\ \mathbf{T}_{1*} \\ \mathbf{T}_{2*} \\ \mathbf{T}_{3*} \\ \mathbf{T}_{4*} \\ \mathbf{T}_{5*} \\ \mathbf{T}_{6*} \\ \mathbf{T}_{7*}
\end{array}
\begin{array}{c}
\begin{array}{cccccccc}
\mathbf{T}_{*0} & \mathbf{T}_{*1} & \mathbf{T}_{*2} & \mathbf{T}_{*3} & \mathbf{T}_{*4} & \mathbf{T}_{*5} & \mathbf{T}_{*6} & \mathbf{T}_{*7}
\end{array} \\
\begin{bmatrix}
1 & 1 & 1 & 1 & 1 & 1 & 1 & 1 \\
1 & -1 & 1 & -1 & 1 & -1 & 1 & -1 \\
1 & 1 & -1 & -1 & 1 & 1 & -1 & -1 \\
1 & -1 & -1 & 1 & 1 & -1 & -1 & 1 \\
1 & 1 & 1 & 1 & -1 & -1 & -1 & -1 \\
1 & -1 & 1 & -1 & -1 & 1 & -1 & 1 \\
1 & 1 & -1 & -1 & -1 & -1 & 1 & 1 \\
1 & -1 & -1 & 1 & -1 & 1 & 1 & -1
\end{bmatrix} \\
\mathbf{T}^n
\end{array}
\begin{array}{c}
\begin{bmatrix} 1 \\ -1 \\ -1 \\ 1 \\ 1 \\ 1 \\ 1 \\ -1 \end{bmatrix}
\begin{array}{c} y_0 \\ y_1 \\ y_2 \\ y_3 \\ y_4 \\ y_5 \\ y_6 \\ y_7 \end{array} \\
\mathbf{Y}
\end{array}
\begin{array}{c} = \\ \\ = \end{array}
\begin{array}{c}
\begin{bmatrix} 2 \\ 2 \\ 2 \\ 2 \\ -2 \\ -2 \\ -2 \\ 6 \end{bmatrix}
\begin{array}{c} s_0 \\ s_1 \\ s_2 \\ s_{12} \\ s_3 \\ s_{13} \\ s_{23} \\ s_{123} \end{array} \\
\mathbf{S}
\end{array}
\qquad (2.10)
$$

The resulting vectors (spectra) $\mathbf{R}$ and $\mathbf{S}$ each uniquely represent the given function $f(X)$. The individual coefficient values in **R** and **S** are linearly related due to the linear recoding between **Z** and **Y**, the relationships being:[†]

$$r_0 = \tfrac{1}{2}(2^n - s_0), \qquad r_\alpha, \alpha \neq 0, = -\tfrac{1}{2}s_\alpha \tag{2.11}$$

The inverse transformation back from the spectral domain to the Boolean function domain is directly given by

$$\frac{1}{2^n}[\mathbf{T}^n]^{-1}\mathbf{R} = \mathbf{Z} \quad \text{or} \quad \frac{1}{2^n}[\mathbf{T}^n]^{-1}\mathbf{S} = \mathbf{Y} \tag{2.12}$$

which in the case of the Hadamard transform is merely

$$\frac{1}{2^n}\mathbf{T}^n\mathbf{R} = \mathbf{Z} \quad \text{or} \quad \frac{1}{2^n}\mathbf{T}^n\mathbf{S} = \mathbf{Y} \tag{2.13}$$

Whilst the coefficients **R** or **S** of (2.9) or (2.10) uniquely represent a single combinatorial function $f(X)$, a joint spectrum of two or more functions may be computed. For example, consider the two functions $f_1(X)$ and $f_2(X)$ tabulated below. Let us encode and combine the two individual outputs as shown:

x_3	x_2	x_1	$f_1(X)$	$f_2(X)$	Combined $f_1(X)$ and $f_2(X)$
0	0	0	0	0	0
0	0	1	1	0	2
0	1	0	1	0	2
0	1	1	0	1	1
1	0	0	1	0	2
1	0	1	0	1	1
1	1	0	0	1	1
1	1	1	1	1	3

The Hadamard transform now gives

$$\underbrace{\begin{bmatrix} 1 & 1 & 1 & 1 & 1 & 1 & 1 & 1 \\ 1 & -1 & 1 & -1 & 1 & -1 & 1 & -1 \\ 1 & 1 & -1 & -1 & 1 & 1 & -1 & -1 \\ 1 & -1 & -1 & 1 & 1 & -1 & -1 & 1 \\ 1 & 1 & 1 & 1 & -1 & -1 & -1 & -1 \\ 1 & -1 & 1 & -1 & -1 & 1 & -1 & 1 \\ 1 & 1 & -1 & -1 & -1 & -1 & 1 & 1 \\ 1 & -1 & -1 & 1 & -1 & 1 & 1 & -1 \end{bmatrix}}_{\mathbf{T}^n} \underbrace{\begin{bmatrix} 0 \\ 2 \\ 2 \\ 1 \\ 2 \\ 1 \\ 1 \\ 3 \end{bmatrix}}_{\mathbf{Z}_{1+2}} = \underbrace{\begin{bmatrix} 12 \\ -2 \\ -2 \\ 0 \\ -2 \\ 0 \\ 0 \\ -6 \end{bmatrix}}_{\mathbf{R}_{1+2}} \tag{2.14}$$

[†] Note that the recoding between **Z** and **Y** is formally expressed by $y_j = 1 - 2z_j$ for all j; any recoding of the output truthtable of $f(X)$ could be used, but with no advantages over **Y** and **Z**.

In general we shall not pursue the use of the spectra of combined functions in this text, but shall largely concentrate on individual function spectra. However, the principle of the orthogonal transformation of *any* column data, be it a combinatorial logic function, a sequential output sequence, or any other discrete data, is straightforward.

2.2.3 The Meaning and Order of the Spectral Coefficients

The meaning of the resultant spectral coefficient values of a function $f(X)$ is now considered. Consider first the individual row vectors of $\mathbf{T}^n$. Reading $+1$ as equivalent to logic 0 and -1 as equivalent to logic 1, then the Hadamard row vectors may be seen to be related to the x_i input variables over the interval 0 to $2^n - 1$ as follows:

Dimension	Transform	Row vector meaning
$n = 1$	$\begin{bmatrix} 1 & 1 \\ 1 & -1 \end{bmatrix}$	Constant, say x_0 Input variable x_1
$n = 2$	$\begin{bmatrix} 1 & 1 & 1 & 1 \\ 1 & -1 & 1 & -1 \\ 1 & 1 & -1 & -1 \\ 1 & -1 & -1 & 1 \end{bmatrix}$	Constant x_0 Input variable x_1 Input variable x_2 Function $x_1 \oplus x_2$
$n = 3$	$\begin{bmatrix} 1 & 1 & 1 & 1 & 1 & 1 & 1 & 1 \\ 1 & -1 & 1 & -1 & 1 & -1 & 1 & -1 \\ 1 & 1 & -1 & -1 & 1 & 1 & -1 & -1 \\ 1 & -1 & -1 & 1 & 1 & -1 & -1 & 1 \\ 1 & 1 & 1 & 1 & -1 & -1 & -1 & -1 \\ 1 & -1 & 1 & -1 & -1 & 1 & -1 & 1 \\ 1 & 1 & -1 & -1 & -1 & -1 & 1 & 1 \\ 1 & -1 & -1 & 1 & -1 & 1 & 1 & -1 \end{bmatrix}$	Constant x_0 (2.15) Input variable x_1 Input variable x_2 Function $x_1 \oplus x_2$ Input variable x_3 Function $x_1 \oplus x_3$ Function $x_2 \oplus x_3$ Function $x_1 \oplus x_2 \oplus x_3$

The pattern for $n > 3$ should also be clear, each additional x_i input uniquely appearing at the appropriate row, with the subsequent rows then forming the Exclusive-OR of this additional input and all preceding functions.

Referring back to the example transform (2.10), it will also be apparent that the value of each resulting coefficient in $\mathbf{S}$ is equal to the number of occasions the minterm values in $\mathbf{Y}$ agree with the corresponding Hadamard row values minus the number of times the two values disagree. Each spectral coefficient in **S** therefore represents a correlation between the given function vector **Y** and the corresponding Hadamard row function.

As an example, consider the $n = 4$ function $f(X) = x_1x_2x_3 + x_1x_3\bar{x}_4 + \bar{x}_1x_2x_4 + \bar{x}_2\bar{x}_3x_4$. The spectrum **S** is as follows:

$$\begin{bmatrix}
1 & 1 & 1 & 1 & 1 & 1 & 1 & 1 & 1 & 1 & 1 & 1 & 1 & 1 & 1 & 1 \\
1 & -1 & 1 & -1 & 1 & -1 & 1 & -1 & 1 & -1 & 1 & -1 & 1 & -1 & 1 & -1 \\
1 & 1 & -1 & -1 & 1 & 1 & -1 & -1 & 1 & 1 & -1 & -1 & 1 & 1 & -1 & -1 \\
1 & -1 & -1 & 1 & 1 & -1 & -1 & 1 & 1 & -1 & -1 & 1 & 1 & -1 & -1 & 1 \\
1 & 1 & 1 & 1 & -1 & -1 & -1 & -1 & 1 & 1 & 1 & 1 & -1 & -1 & -1 & -1 \\
1 & -1 & 1 & -1 & -1 & 1 & -1 & 1 & 1 & -1 & 1 & -1 & -1 & 1 & -1 & 1 \\
1 & 1 & -1 & -1 & -1 & -1 & 1 & 1 & 1 & 1 & -1 & -1 & -1 & -1 & 1 & 1 \\
1 & -1 & -1 & 1 & -1 & 1 & 1 & -1 & 1 & -1 & -1 & 1 & -1 & 1 & 1 & -1 \\
1 & 1 & 1 & 1 & 1 & 1 & 1 & 1 & -1 & -1 & -1 & -1 & -1 & -1 & -1 & -1 \\
1 & -1 & 1 & -1 & 1 & -1 & 1 & -1 & -1 & 1 & -1 & 1 & -1 & 1 & -1 & 1 \\
1 & 1 & -1 & -1 & 1 & 1 & -1 & -1 & -1 & -1 & 1 & 1 & -1 & -1 & 1 & 1 \\
1 & -1 & -1 & 1 & 1 & -1 & -1 & 1 & -1 & 1 & 1 & -1 & -1 & 1 & 1 & -1 \\
1 & 1 & 1 & 1 & -1 & -1 & -1 & -1 & -1 & -1 & -1 & -1 & 1 & 1 & 1 & 1 \\
1 & -1 & 1 & -1 & -1 & 1 & -1 & 1 & -1 & 1 & -1 & 1 & 1 & -1 & 1 & -1 \\
1 & 1 & -1 & -1 & -1 & -1 & 1 & 1 & -1 & -1 & 1 & 1 & 1 & 1 & -1 & -1 \\
1 & -1 & -1 & 1 & -1 & 1 & 1 & -1 & -1 & 1 & 1 & -1 & 1 & -1 & -1 & 1
\end{bmatrix}
\begin{bmatrix} 1 \\ 1 \\ 1 \\ 1 \\ 1 \\ -1 \\ 1 \\ -1 \\ -1 \\ -1 \\ -1 \\ 1 \\ 1 \\ 1 \\ -1 \\ -1 \end{bmatrix}
=
\begin{bmatrix} 2 \\ 2 \\ 2 \\ 2 \\ 2 \\ -6 \\ -6 \\ 2 \\ 6 \\ 6 \\ -2 \\ -2 \\ 6 \\ -2 \\ 6 \\ -2 \end{bmatrix}
\begin{matrix} s_0 \\ s_1 \\ s_2 \\ s_{12} \\ s_3 \\ s_{13} \\ s_{23} \\ s_{123} \\ s_4 \\ s_{14} \\ s_{24} \\ s_{124} \\ s_{34} \\ s_{134} \\ s_{234} \\ s_{1234} \end{matrix}$$

The functional subscript identification used for the coefficients of **S** will now be clear, namely:

s_0 is the coefficient that indicates the correlation (likeness) of $f(X)$ to the constant function x_0, which is numerically equal to {the number of false minterms in $f(X)$ − the number of true minterms in $f(X)$};

s_i, $i = 1$ to n, is the correlation (likeness) between $f(X)$ and the input variable x_i;

s_{ij}, $ij = 1$ to n, $i \neq j$, is the correlation between $f(X)$ and the Exclusive-OR input function $x_i \oplus x_j$;

s_{ijk}, $i, j, k \in i$, $i = 1$ to n, $i \neq j \neq k$, is the correlation between $f(X)$ and the Exclusive-OR input function $x_i \oplus x_j \oplus x_k$;

and so on for higher coefficients. All coefficients are thus numerically equal to {$\sum$ agreements between output $f(X)$ and the appropriate input function − $\sum$ disagreements between $f(X)$ and the input function}.

The properties of the spectral coefficients of **S** include the following. We state them here without formal proof, all following from the orthogonality of the transform matrix. Formal developments and proofs may be found elsewhere.[1,4–10]

(i) The sum of all spectral coefficients s of **S** for any fully defined function $f(X)$ is $\pm 2^n$.[†]

(ii) The maximum value of any individual s is $\pm 2^n$; this occurs when $f(X)$ is identically equal to any row function in $\mathbf{T}^n$ or its complement. The range for each s is $\{-2^n, -2^n + 2, \ldots, 0, \ldots, 2^n - 2, 2^n\}$.

(iii) When any individual s is maximum-valued, all remaining $2^n - 1$ coefficients of **S** will be zero-valued.

(iv) When any input variable x_i is redundant in a given function $f(X)$, the 2^{n-1} spectral coefficients that contain i in their subscript identification will all be zero-valued.

Finally, because of the functional meaning of the individual coefficients of **S**, we frequently classify and regroup them in accordance with their logical significance, as follows:

s_0:	the zero-order coefficient (a "dc" term)
s_i, $i = 1$ to n:	the primary or first-order coefficients
s_{ij}, $ij = 12, 13, \ldots$:	second-order coefficients
s_{ijk}, $ijk = 123, 124, 134, \ldots$:	third-order coefficients
⋮	

Collectively, all second- and higher-order coefficients may jointly be termed the secondary coefficients.

For $n = 4$, we may therefore write the coefficients in the following order, as distinct from the column vector order by which they are generated by the Hadamard transform:

$$s_0;\quad s_1\, s_2\, s_3\, s_4;\quad s_{12}\, s_{13}\, s_{14}\, s_{23}\, s_{24}\, s_{34};\quad s_{123}\, s_{124}\, s_{134}\, s_{234};\quad s_{1234}$$

However, it is important to stress that the coefficients *must be reassembled in correct column vector order* should it be required to perform the inverse transformation $(1/2^n)[\mathbf{T}^n]^{-1}\mathbf{S}$ back from the spectral domain to the function domain, or any decomposition of the spectral data that depends upon the original order of generation.

The spectral coefficients of **R** possess exactly the same information content as the coefficients of **S**, but will not have the same magnitudes, particularly under the specific conditions defined in (i) to (iv). It is left as a simple exercise for the reader to apply the relationships of Eq. (2.11) to the specific conditions

[†] Should a given function $f(X)$ have any don't-care or undefined local minterm values, then it is possible to encode such minterms zero in **S**. In this case, $\sum^{2^n} s$ will not be $\pm 2^n$.

defined in (i)–(iv). It will be noted that unlike s_0 the value of r_0 cannot be negative, but must be between 0 and $+2^n$; the remaining **R** coefficient values are integer values between $\pm 2^{n-1}$. Again, the **R** coefficients are often written in functional order, e.g.,

$$r_0;\quad r_1 r_2 r_3 r_4;\quad r_{12} r_{13} r_{14} r_{23} r_{24} r_{34};\quad r_{123} r_{124} r_{134} r_{234};\quad r_{1234}$$

See also Appendix A for these **R** and **S** orderings.

2.3 FURTHER BINARY TRANSFORM MATRICES

The Hadamard transform matrix discussed in the preceding section will remain our principal mathematical transform in subsequent chapters of this volume. This is because its recursive structure, defined by Eq. (2.1), provides appropriate features for function decomposition and other logic operations which we may wish to pursue.

However, the Hadamard matrix is but one of many complete, square orthogonal transforms that may be proposed. Let us briefly look at certain further possibilities.

2.3.1 The Walsh and Rademacher–Walsh Transforms

The Walsh functions [1,5,6,8,11] are, by definition, a closed set of two-valued orthogonal functions, given by

$$\mathrm{Wal}(j, k) = \prod_{\eta=0}^{n-1} \{(-1)^{\{k_{n-\eta} + k_{n-1-\eta}\} j_\eta}\} \tag{2.16}$$

where j_η, k_η are determined by the binary expansions of j, k, respectively, $j, k \in 0$ to $2^n - 1$, where

$$j = \{j_{n-1}2^{n-1} + j_{n-2}2^{n-2} + \cdots + j_0 2^0\}$$
$$k = \{k_{n-1}2^{n-1} + k_{n-2}2^{n-2} + \cdots + k_0 2^0\}$$

and where $j_n, k_n = 0$ (do not exist). For $n = 3$, we therefore have

$$\mathrm{Wal}(j, k),\ j_2 j_1 j_0 = 000 \text{ through to } 111$$
$$k_2 k_1 k_0 = 000 \text{ through to } 111$$
$$= \{(-1)^{k_2 j_0} . (-1)^{\{k_2 + k_1\} j_1} . (-1)^{\{k_1 + k_0\} j_2}\} \tag{2.17}$$

which may equally be rearranged as

$$\mathrm{Wal}(j, k) = \{(-1)^{j_2 k_0} . (-1)^{\{j_2 + j_1\} k_1} . (-1)^{\{j_1 + j_0\} k_2}\} \tag{2.18}$$

Either expression gives the functions illustrated in Fig. 2.2, which when discretely sampled gives the following Walsh transform matrix. Note that due to the symmetry between Eqs. (2.17) and (2.18), the resulting Walsh transform is symmetric.

$$
\begin{array}{rcccccccccl}
k(\text{binary}) = & & 000 & 001 & 010 & 011 & 100 & 101 & 110 & 111 & \\
j(\text{binary}) = 000 & \left[\begin{array}{c} \\ \\ \\ \\ \\ \\ \\ \\ \end{array}\right. & 1 & 1 & 1 & 1 & 1 & 1 & 1 & 1 & \mathrm{Wal}(0,k) \\
001 & & 1 & 1 & 1 & 1 & -1 & -1 & -1 & -1 & \mathrm{Wal}(1,k) \\
010 & & 1 & 1 & -1 & -1 & -1 & -1 & 1 & 1 & \mathrm{Wal}(2,k) \\
011 & & 1 & 1 & -1 & -1 & 1 & 1 & -1 & -1 & \mathrm{Wal}(3,k) \\
100 & & 1 & -1 & -1 & 1 & 1 & -1 & -1 & 1 & \mathrm{Wal}(4,k) \\
101 & & 1 & -1 & -1 & 1 & -1 & 1 & 1 & -1 & \mathrm{Wal}(5,k) \\
110 & & 1 & -1 & 1 & -1 & -1 & 1 & -1 & 1 & \mathrm{Wal}(6,k) \\
111 & & 1 & -1 & 1 & -1 & 1 & -1 & 1 & -1 & \mathrm{Wal}(7,k)
\end{array}
$$

If we consider the number of waveform transitions σ from $+1$ to -1 and vice versa within the extremes of each Walsh function interval $[0, 2^n)$ (see Fig. 2.2), then it will be observed that σ progressively increases, being equal to the row identification number j. This property may be considered in a more rigorous mathematical sense, involving the definition of sequency, ψ, defined as one-half the number of sign changes per complete waveform interval,[†] together with sal (sine-Walsh) and cal (cosine-Walsh) symmetry relationships about the midpoint of the waveforms.[1,12–14] However, we shall not pursue these developments herewith, except to emphasize that sequency ψ is not the same as σ; for any n as j increases, we have the following tabulation:

Walsh row j	Internal transitions σ	Sequency ψ	
0	0	0	
1	1	1	(sal, compare sine)
2	2	1	(cal, compare cosine)
3	3	2	(sal, compare sine)
4	4	2	(cal, compare cosine)
5	5	3	(sal, compare sine)
⋮	⋮	⋮	

For our present purposes, the number of internal transitions σ as defined is a more useful identifier.

† Compare the frequency ω of a sinusoidal waveform, which by definition is one-half the number of zero crossings per unit time.

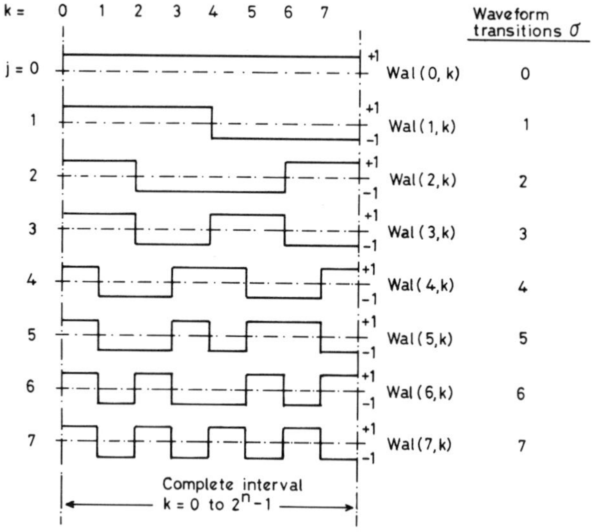

Fig. 2.2. The orthogonal Walsh functions for $n = 3$ in the defined Walsh order.

The 2^n Walsh functions for any n constitute a closed set of orthogonal functions, the multiplication of any two functions always generating a function within the set. However, prior to the publication by Walsh, Rademacher considered an incomplete set of $n + 1$ orthogonal functions,[4] which will be seen to be a subset of Walsh functions, and from which all 2^n Walsh functions can be generated by multiplication. The Rademacher functions may be variously defined; here we define them as follows:

$$\text{Rad}(j, k) = \text{sign}\{\sin 2^j \pi k\} 1.0 \tag{2.19}$$

where k is here taken over the continuous interval 0 to 1 rather than discrete

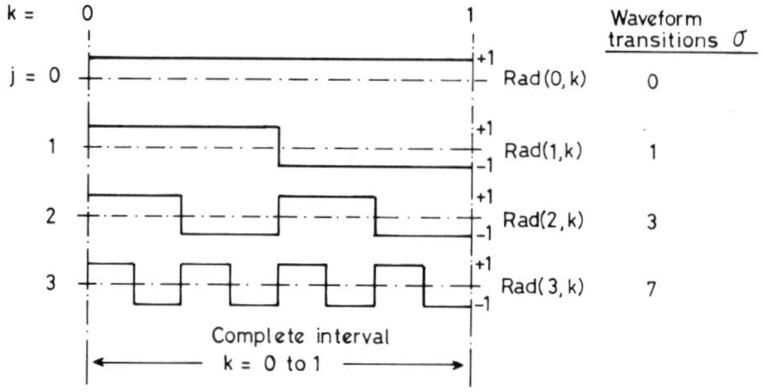

Fig. 2.3. The orthogonal Rademacher functions for $n = 3$, defined over the interval 0 to 1.

intervals 0 to $2^n - 1$. This equation, or its equivalents,[1,4,5] yields a series of equal mark–space ratio square-wave functions as illustrated in Fig. 2.3.

The Rademacher functions and the Walsh functions are related as follows:†

$$\begin{aligned}
\mathrm{Rad}(0, k) &= \mathrm{Wal}(0, k)\\
\mathrm{Rad}(1, k) &= \mathrm{Wal}(1, k)\\
\mathrm{Rad}(2, k) &= \mathrm{Wal}(3, k)\\
&\ \vdots\\
\mathrm{Rad}(j, k) &= \mathrm{Wal}(2^j - 1, k)
\end{aligned}$$

Taking the Rademacher functions as a basis set, the complete set of Walsh functions is the multiplicative closure of the set of Rademacher functions, as identified in Fig. 2.4, which when discretely sampled gives the following orthogonal transform, where × here indicates element-by-element multiplication:

$$\begin{bmatrix}
1 & 1 & 1 & 1 & 1 & 1 & 1 & 1\\
1 & 1 & 1 & 1 & -1 & -1 & -1 & -1\\
1 & 1 & -1 & -1 & 1 & 1 & -1 & -1\\
1 & -1 & 1 & -1 & 1 & -1 & 1 & -1\\
1 & 1 & -1 & -1 & -1 & -1 & 1 & 1\\
1 & -1 & 1 & -1 & -1 & 1 & -1 & 1\\
1 & -1 & -1 & 1 & 1 & -1 & -1 & 1\\
1 & -1 & -1 & 1 & -1 & 1 & 1 & -1
\end{bmatrix}
\begin{array}{l}
\mathrm{Rad}(0, k) = \mathrm{Wal}(0, k)\\
\mathrm{Rad}(1, k) = \mathrm{Wal}(1, k)\\
\mathrm{Rad}(2, k) = \mathrm{Wal}(3, k)\\
\mathrm{Rad}(3, k) = \mathrm{Wal}(7, k)\\
\mathrm{Rad}(1, k) \times (2, k) = \mathrm{Wal}(2, k)\\
\mathrm{Rad}(1, k) \times (3, k) = \mathrm{Wal}(6, k)\\
\mathrm{Rad}(2, k) \times (3, k) = \mathrm{Wal}(4, k)\\
\mathrm{Rad}(1, k) \times (2, k) \times (3, k) = \mathrm{Wal}(5, k)
\end{array}$$

It is evident that using the Rademacher functions as a basis set, the complete set of orthogonal Walsh functions is generated in an alternative order from the original Walsh order, being no longer in σ or sequency order. We shall term such revised-order matrices Rademacher–Walsh to distinguish them from the former sequency-ordered Walsh matrices.

The use of the Walsh or Rademacher–Walsh transform as an alternative to the Hadamard transform may be considered. If we adopt either in a function transformation $\mathbf{T}^n\mathbf{Y} = \mathbf{S}$ or $\mathbf{T}^n\mathbf{Z} = \mathbf{R}$, then we obtain the same spectral coefficients as in Section 2.2, but assembled in revised order. Consider again the example of Eq. (2.10), but using the $n = 3$ Rademacher–

† This general relationship between Rad(j) and Wal(j) will be found to be the basis for certain alternative definitions of Rad (j, k).

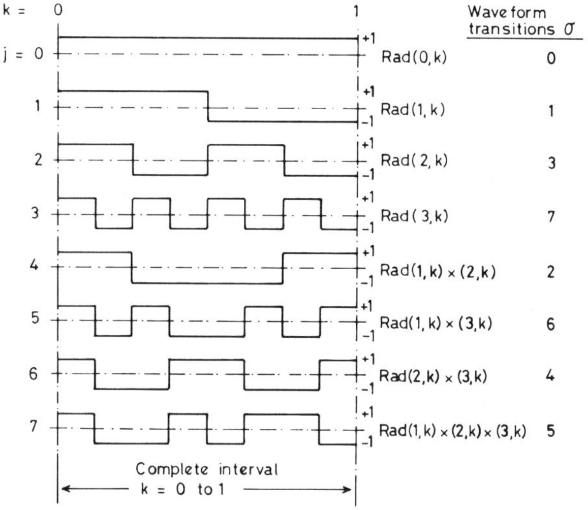

Fig. 2.4. The complete orthogonal set of Rademacher–Walsh functions for $n = 3$, generated from multiplication of the Rademacher basis set.

Walsh transform. We now have

$$\begin{bmatrix} 1 & 1 & 1 & 1 & 1 & 1 & 1 & 1 \\ 1 & 1 & 1 & 1 & -1 & -1 & -1 & -1 \\ 1 & 1 & -1 & -1 & 1 & 1 & -1 & -1 \\ 1 & -1 & 1 & -1 & 1 & -1 & 1 & -1 \\ 1 & 1 & -1 & -1 & -1 & -1 & 1 & 1 \\ 1 & -1 & 1 & -1 & -1 & 1 & -1 & 1 \\ 1 & -1 & -1 & 1 & 1 & -1 & -1 & 1 \\ 1 & -1 & -1 & 1 & -1 & 1 & 1 & -1 \end{bmatrix} \begin{bmatrix} 1 \\ -1 \\ -1 \\ 1 \\ 1 \\ 1 \\ 1 \\ -1 \end{bmatrix} \begin{matrix} y_0 \\ y_1 \\ y_2 \\ y_3 \\ y_4 \\ y_5 \\ y_6 \\ y_7 \end{matrix} = \begin{bmatrix} 2 \\ -2 \\ 2 \\ 2 \\ -2 \\ -2 \\ 2 \\ 6 \end{bmatrix} \begin{matrix} s_0 \\ s_3 \\ s_2 \\ s_1 \\ s_{23} \\ s_{13} \\ s_{12} \\ s_{123} \end{matrix}$$

$$\mathbf{T}^n \quad \mathbf{Y} = \mathbf{S}$$

It will be noted that the spectral coefficients are now generated in **S** in logical significance order, that is,

zero-order coefficient s_0
first-order coefficients s_1 to s_n (most significant coefficient s_n first) } primary coefficients
second-order coefficients
third-order coefficients
⋮
*n*th order coefficient $s_{12\ldots n}$ } secondary coefficients

The functional meaning of each spectral coefficient (see Section 2.2.3) remains unchanged, only the order of generation having been modified.

2.3.2 Further Hadamard–Walsh Variants

The Hadamard, Walsh and Rademacher–Walsh matrices represent three of many possible row orderings of the 2^n functions which collectively constitute this closed set. However, as well as row permutation, it is also possible to

(i) permutate the column ordering of any complete $2^n \times 2^n$ matrix, and
(ii) negate all entries of any one or more rows, or any one or more columns

without impairing either completeness or orthogonality. Hence in total there are many related variants within this one family of two-valued transforms.[10,15–21]

For comparison, the four most commonly cited orderings are given for $n = 3$ in Table 2.2. They include the three that we have already considered, plus one further variant (Walsh–Paley).[1,21] A brief resume of their principal features is as follows:

(a) Hadamard (sometimes termed Walsh–Hadamard)

$$\text{Symmetric, } [\mathbf{T}^n]^t = \mathbf{T}^n, \qquad \therefore \text{ inverse} = \frac{1}{2^n}[\mathbf{T}^n]$$

(b) Walsh (sometimes termed Walsh–Kaczmarz)

$$\text{Symmetric, } [\mathbf{T}^n]^t = \mathbf{T}^n, \qquad \therefore \text{ inverse} = \frac{1}{2^n}[\mathbf{T}^n]$$

Functions are in strict sequency order

(c) Rademacher–Walsh

$$\text{Not symmetric, } [\mathbf{T}^n]^t \neq \mathbf{T}^n, \qquad \therefore \text{ inverse} = \frac{1}{2^n}[\mathbf{T}^n]^t, \neq \frac{1}{2^n}[\mathbf{T}^n]$$

(d) Walsh–Paley

$$\text{Symmetric, } [\mathbf{T}^n]^t = \mathbf{T}^n, \qquad \therefore \text{ inverse} = \frac{1}{2^n}\mathrm{T}^n$$

Note that since all these binary matrices have non-complex entries, the transposed conjugate operation required in the general case of orthogonal transforms and their inverse[2,3] is not apparent in these cases.

All four examples will therefore generate identical spectral coefficients when used for function transformation, but with dissimilar coefficient order in the resulting vector **S** or **R**. However, note that

(i) only the Hadamard matrix has the recursive Kronecker product structure

$$\mathbf{T}^n = \begin{bmatrix} \mathbf{T}^{n-1} & \mathbf{T}^{n-1} \\ \mathbf{T}^{n-1} & -\mathbf{T}^{n-1} \end{bmatrix}$$

and for this reason will remain our preferred transform for developments in this text, and

(ii) should column permutations or row or column negation variants be considered, then the spectral coefficients will *not* remain the same; indeed

Table 2.2. Four possible $2^n \times 2^n$, $n = 3$ complete orthogonal transforms $\mathbf{T}^n$ related by row permutations, together with the transition sequence σ and resulting order of spectral coefficients in **S**.

(a) Hadamard

								σ	**S**
1	1	1	1	1	1	1	1	0	s_0
1	−1	1	−1	1	−1	1	−1	7	s_1
1	1	−1	−1	1	1	−1	−1	3	s_2
1	−1	−1	1	1	−1	−1	1	4	s_{12}
1	1	1	1	−1	−1	−1	−1	1	s_3
1	−1	1	−1	−1	1	−1	1	6	s_{13}
1	1	−1	−1	−1	−1	1	1	2	s_{23}
1	−1	−1	1	−1	1	1	−1	5	s_{123}

(b) Walsh

								σ	**S**
1	1	1	1	1	1	1	1	0	s_0
1	1	1	1	−1	−1	−1	−1	1	s_3
1	1	−1	−1	−1	−1	1	1	2	s_{23}
1	1	−1	−1	1	1	−1	−1	3	s_2
1	−1	−1	1	1	−1	−1	1	4	s_{12}
1	−1	−1	1	−1	1	1	−1	5	s_{123}
1	−1	1	−1	−1	1	−1	1	6	s_{13}
1	−1	1	−1	1	−1	1	−1	7	s_1

(c) Rademacher–Walsh

								σ	**S**
1	1	1	1	1	1	1	1	0	s_0
1	1	1	1	−1	−1	−1	−1	1	s_3
1	1	−1	−1	1	1	−1	−1	3	s_2
1	−1	1	−1	1	−1	1	−1	7	s_1
1	1	−1	−1	−1	−1	1	1	2	s_{23}
1	−1	1	−1	−1	1	−1	1	6	s_{13}
1	−1	−1	1	1	−1	−1	1	4	s_{12}
1	−1	−1	1	−1	1	1	−1	5	s_{123}

(d) Walsh–Paley

								σ	**S**
1	1	1	1	1	1	1	1	0	s_0
1	1	1	1	−1	−1	−1	−1	1	s_3
1	1	−1	−1	1	1	−1	−1	3	s_2
1	1	−1	−1	−1	−1	1	1	2	s_{23}
1	−1	1	−1	1	−1	1	−1	7	s_1
1	−1	1	−1	−1	1	−1	1	6	s_{13}
1	−1	−1	1	1	−1	−1	1	4	s_{12}
1	−1	−1	1	−1	1	1	−1	5	s_{123}

certain variants will no longer remain predominantly Exclusive-OR related, but may have NAND or other function correlation. For further discussion on such possibilities, see Lloyd.[17]

Finally it may be mentioned that there are still further possible two-valued complete orthogonal transforms, which are not obtainable by negation and/or row and column permutation from the above. These may be constructed by considering maximum-length, pseudo-random binary sequences that are then cyclically shifted, to which are added an additional all $+1$ row and column to complete the $2^n \times 2^n$ transform dimensions.[18] An example for $n = 3$ is

$$\begin{bmatrix} 1 & 1 & 1 & 1 & 1 & 1 & 1 & 1 \\ 1 & 1 & 1 & -1 & 1 & -1 & -1 & -1 \\ 1 & 1 & -1 & 1 & -1 & -1 & -1 & 1 \\ 1 & -1 & 1 & -1 & -1 & -1 & 1 & 1 \\ 1 & 1 & -1 & -1 & -1 & 1 & 1 & -1 \\ 1 & -1 & -1 & -1 & 1 & 1 & -1 & 1 \\ 1 & -1 & -1 & 1 & 1 & -1 & 1 & -1 \\ 1 & -1 & 1 & 1 & -1 & 1 & -1 & -1 \end{bmatrix} \tag{2.20}$$

2.3.3 The Haar Transform

The orthogonal Haar functions may be defined as follows,[1,5,22–26] where k is taken over the continuous interval 0 to 1:

$$\begin{aligned} H_0^0(k) &\triangleq +1.0 \\ H_i^q(k) &= |\sqrt{2}|^{i-1}(+1.0) \quad \text{for} \quad \frac{q}{2^{i-1}} \le k < \frac{q+\frac{1}{2}}{2^{i-1}} \\ &= |\sqrt{2}|^{i-1}(-1.0) \quad \text{for} \quad \frac{q+\frac{1}{2}}{2^{i-1}} \le k < \frac{q+1}{2^{i-1}} \\ &= 0 \text{ at all other points} \end{aligned} \tag{2.21}$$

where $i = 1, 2, \ldots, n$ and $q = 0, 1, \ldots, 2^{i-1} - 1$. This gives the functions illustrated in Fig. 2.5 for the $n = 3$ case. Note that for any n there will be 2^n Haar functions, with 2^n divisions in the complete interval.

Discrete sampling of the set of Haar functions gives the $2^n \times 2^n$ orthogonal

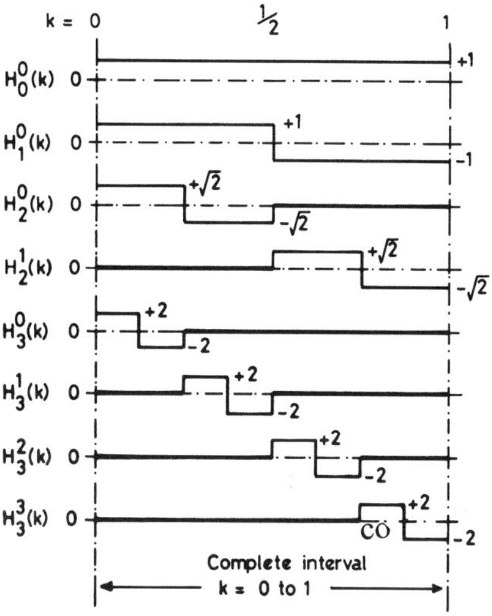

Fig. 2.5. The orthogonal Haar functions for $n = 3$, defined over the interval 0 to 1.

matrix as follows for $n = 3$:

$$\text{Haar } \mathbf{T}^n = \begin{bmatrix} 1 & 1 & 1 & 1 & 1 & 1 & 1 & 1 \\ 1 & 1 & 1 & 1 & -1 & -1 & -1 & -1 \\ \sqrt{2} & \sqrt{2} & -\sqrt{2} & -\sqrt{2} & 0 & 0 & 0 & 0 \\ 0 & 0 & 0 & 0 & \sqrt{2} & \sqrt{2} & -\sqrt{2} & -\sqrt{2} \\ 2 & -2 & 0 & 0 & 0 & 0 & 0 & 0 \\ 0 & 0 & 2 & -2 & 0 & 0 & 0 & 0 \\ 0 & 0 & 0 & 0 & 2 & -2 & 0 & 0 \\ 0 & 0 & 0 & 0 & 0 & 0 & 2 & -2 \end{bmatrix}$$

This is a complete, orthogonal matrix with the orthogonal properties

$$\sum_{k=0}^{2^n-1} t_{j_\alpha k} t_{j_\beta k} = 2^n \quad \text{if} \quad \alpha = \beta,$$
$$= 0 \qquad \text{otherwise} \tag{2.5}$$

$$[\mathbf{T}^n]^{-1} = \frac{1}{K}[\mathbf{T}^n]^h$$
$$= \frac{1}{K}[\mathbf{T}^n]^t \tag{2.6}$$

where $K = 1/2^n$. However, note that the matrix is not closed under row multiplication as are the previous matrices, and also is not symmetric, thus not allowing

$$[\mathbf{T}^n]^{-1} = \frac{1}{K}[\mathbf{T}^n]$$

Table 2.3. The normalised forward and inverse Haar orthogonal transform matrices $\mathbf{T}^n$.

(a) Forward transform, $n = 3$

$$\begin{bmatrix} 1 & 1 & 1 & 1 & 1 & 1 & 1 & 1 \\ 1 & 1 & 1 & 1 & -1 & -1 & -1 & -1 \\ 1 & 1 & -1 & -1 & 0 & 0 & 0 & 0 \\ 0 & 0 & 0 & 0 & 1 & 1 & -1 & -1 \\ 1 & -1 & 0 & 0 & 0 & 0 & 0 & 0 \\ 0 & 0 & 1 & -1 & 0 & 0 & 0 & 0 \\ 0 & 0 & 0 & 0 & 1 & -1 & 0 & 0 \\ 0 & 0 & 0 & 0 & 0 & 0 & 1 & -1 \end{bmatrix}$$

(b) Inverse transform, $n = 3$

$$\frac{1}{2^n}\begin{bmatrix} 1 & 1 & 2 & 0 & 4 & 0 & 0 & 0 \\ 1 & 1 & 2 & 0 & -4 & 0 & 0 & 0 \\ 1 & 1 & -2 & 0 & 0 & 4 & 0 & 0 \\ 1 & 1 & -2 & 0 & 0 & -4 & 0 & 0 \\ 1 & -1 & 0 & 2 & 0 & 0 & 4 & 0 \\ 1 & -1 & 0 & 2 & 0 & 0 & -4 & 0 \\ 1 & -1 & 0 & -2 & 0 & 0 & 0 & 4 \\ 1 & -1 & 0 & -2 & 0 & 0 & 0 & -4 \end{bmatrix}$$

(c) Forward transform, $n = 4$

$$\begin{bmatrix} 1 & 1 & 1 & 1 & 1 & 1 & 1 & 1 & 1 & 1 & 1 & 1 & 1 & 1 & 1 & 1 \\ 1 & 1 & 1 & 1 & 1 & 1 & 1 & 1 & -1 & -1 & -1 & -1 & -1 & -1 & -1 & -1 \\ 1 & 1 & 1 & 1 & -1 & -1 & -1 & -1 & 0 & 0 & 0 & 0 & 0 & 0 & 0 & 0 \\ 0 & 0 & 0 & 0 & 0 & 0 & 0 & 0 & 1 & 1 & 1 & 1 & -1 & -1 & -1 & -1 \\ 1 & 1 & -1 & -1 & 0 & 0 & 0 & 0 & 0 & 0 & 0 & 0 & 0 & 0 & 0 & 0 \\ 0 & 0 & 0 & 0 & 1 & 1 & -1 & -1 & 0 & 0 & 0 & 0 & 0 & 0 & 0 & 0 \\ 0 & 0 & 0 & 0 & 0 & 0 & 0 & 0 & 1 & 1 & -1 & -1 & 0 & 0 & 0 & 0 \\ 0 & 0 & 0 & 0 & 0 & 0 & 0 & 0 & 0 & 0 & 0 & 0 & 1 & 1 & -1 & -1 \\ 1 & -1 & 0 & 0 & 0 & 0 & 0 & 0 & 0 & 0 & 0 & 0 & 0 & 0 & 0 & 0 \\ 0 & 0 & 1 & -1 & 0 & 0 & 0 & 0 & 0 & 0 & 0 & 0 & 0 & 0 & 0 & 0 \\ 0 & 0 & 0 & 0 & 1 & -1 & 0 & 0 & 0 & 0 & 0 & 0 & 0 & 0 & 0 & 0 \\ 0 & 0 & 0 & 0 & 0 & 0 & 1 & -1 & 0 & 0 & 0 & 0 & 0 & 0 & 0 & 0 \\ 0 & 0 & 0 & 0 & 0 & 0 & 0 & 0 & 1 & -1 & 0 & 0 & 0 & 0 & 0 & 0 \\ 0 & 0 & 0 & 0 & 0 & 0 & 0 & 0 & 0 & 0 & 1 & -1 & 0 & 0 & 0 & 0 \\ 0 & 0 & 0 & 0 & 0 & 0 & 0 & 0 & 0 & 0 & 0 & 0 & 1 & -1 & 0 & 0 \\ 0 & 0 & 0 & 0 & 0 & 0 & 0 & 0 & 0 & 0 & 0 & 0 & 0 & 0 & 1 & -1 \end{bmatrix}$$

If we normalise the Haar functions to take the values $+1$, -1 in place of the proper integer or fractional positive or negative values, respectively, we obtain the modified Haar transform matrices for $n = 3$ and $n = 4$ as given in Table 2.3. The construction for any n should be clear. The rows of these modified Haar matrices maintain pairwise orthogonality, $\sum_{k=0}^{2n-1} t_{j_\alpha k} t_{j_\beta k} = 0$, $\alpha = \beta$, but since $\sum_{k=0}^{2n-1} t_{j_\alpha k} t_{j_\beta k}$ is no longer constant for all α, β $\alpha = \beta$, the inverse of $\mathbf{T}^n$ is no longer given by the transpose relationship of Eq. (2.6).† Column scaling factors now must be incorporated within the transpose, as illustrated in Table 2.3, in order to satisfy $[\mathbf{T}^n][\mathbf{T}^n]^{-1} = \mathbf{I}$.

Employing the Haar matrices of Table 2.3 for any function transformation $\mathbf{T}^n\mathbf{Y} = \mathbf{S}$ or $\mathbf{T}^n\mathbf{Z} = \mathbf{R}$ clearly will not result in spectral coefficients having the same meaning as previously detailed. In particular, apart from the first two coefficients, the remaining $2^n - 2$ coefficients will not be global parameters of the function $f(X)$, but instead will give local correlations corresponding to Shannon decompositions of the function. Examination of Table 2.3 will show that the coefficients in a resulting vector $\mathbf{S}$ have the following meaning:

s_0: exactly as before;

s_1: correlation with input x_n, maximum value $\pm 2^n$;

s_2: correlation with input function x_{n-1} within n-space $\bar{x}_n$, maximum value $\pm 2^{n-1}$;

s_3: correlation with input function x_{n-1} within n-space x_n, maximum value $\pm 2^{n-1}$;

s_4: correlation with input function x_{n-2} within n-space $\bar{x}_{n-1}\bar{x}_n$, maximum value $\pm 2^{n-2}$;

s_5: correlation with input function x_{n-2} within n-space $x_{n-1}\bar{x}_n$, maximum value $\pm 2^{n-2}$;

⋮

s_{2^n-1}: correlation with input function x_1 within n-space $x_2 \ldots x_n$, maximum value ± 2.

However, although these properties have considerable interest and attraction, the majority of publications to date have not employed the Haar transform. It is possible that further developments in this area to parallel our considerations in later chapters may be very profitable.[23]

2.3.4 Closing Considerations

The concentration in the preceding sections has been upon defining complete transform matrices $\mathbf{T}^n$, of dimensions $2^n \times 2^n$, used to obtain the 2^n

† Normalisation of the Haar transform entries may also be considered by making $\sum_{k=0}^{2n-1} (t_{jk})^2 = 1.0$ for all j. This now results in $[\mathbf{T}^n][\mathbf{T}^n]^t = \mathbf{I}$, giving $[\mathbf{T}^n]^{-1} = [\mathbf{T}^n]^t$.

spectral domain coefficients **R** or **S** for any given function $f(X)$. However, we have variously defined the matrices considered either by:

(a) defining the structure of the complete matrix, given that its individual element values are ± 1, e.g., Eqs. (2.1)–(2.3),

(b) defining the value of each matrix element t_{jk}, $j = 0$ to $2^n - 1$, $k = 0$ to $2^n - 1$, such definitions for the binary case involving powers of -1, or (equivalently) $\exp\{(2\pi/2)\sqrt{-1}\ldots\}$, see Eq. (2.4), or

(c) defining a series of orthogonal functions, which when sampled at 2^n discrete intervals give the $2^n \times 2^n$ matrix values t_{jk}.

These various homogeneous approaches may be found followed by different authors for possibly dissimilar reasons or applications.[1–5,8,10–14,27,28] Here we have generally adopted the most direct for our immediate purpose, that of defining a particular transform matrix $\mathbf{T}^n$.

It may also be appreciated that rather than define the individual t_{jk} entries of $\mathbf{T}^n$, it follows that it is equally possible to define directly the resultant spectral coefficient values which are given by the transformation of any given function vector.[5,8,28] For the spectral coefficients generated from the Hadamard transform, we have

$$r_j = \sum_{k=0}^{2^n-1} t_{jk}\mathbf{Z}]^t \tag{2.22}$$

where

$$t_{jk} = (-1)^{\sum_{\eta=0}^{n-1} j_\eta k_\eta} \qquad = \exp\left\{\frac{2\pi\sqrt{-1}}{2}\sum_{\eta=0}^{n-1} j_\eta k_\eta\right\} \tag{2.4}$$

and where $\mathbf{Z}]^t$ is the $\{0, 1\}$ output truthtable of $f(X)$ transposed so as to give a row vector $k = 0$ to $2^n - 1$. Similarly,

$$s_j = \sum_{k=0}^{2^n-1} t_{jk}\mathbf{Y}]^t, \tag{2.23}$$

where $\mathbf{Y}]^t$ is the transpose of the $\{+1, -1\}$ recoded truthtable vector of $f(X)$.

The spectral coefficients generated by other transforms with alternative t_{jk} values may likewise be defined by incorporating the appropriate equation for the t_{jk} elements in place of the Hadamard equation (2.4). Care should be taken in comparing published definitions, since $1/2^n$ scaling of the individual spectral coefficient values may be incorporated, such that

$$\mathbf{R} = \frac{1}{2^n}\{\mathbf{T}^n\mathbf{Z}\} \qquad \text{or} \qquad \mathbf{S} = \frac{1}{2^n}\{\mathbf{T}^n\mathbf{Y}\}$$

thus giving the inverse transform

$$[\mathbf{T}^n]^{-1}\mathbf{R} = \mathbf{Z} \qquad \text{and} \qquad [\mathbf{T}^n]^{-1}\mathbf{S} = \mathbf{Y}$$

respectively. Here we shall maintain the $1/2^n$ factor in the inverse transformation (see Eq. (2.12)), so as to avoid fractional values for the spectral coefficients.

A further aspect that we shall not specifically employ herewith is to consider any given Boolean function $f(X)$ (or combined functions $f_1(X), f_2(X), \ldots$; see Eq. (2.14)) as a continuous integer-valued step function $\Phi(k)$ of real arguments, defined over the interval $[0, 2^n)$, whose waveform may be defined by considering the weighted summation of its orthogonal components. This Fourier series approach corresponds to the sinusoidal Fourier series analysis of periodic ac waveforms, but here involving a finite series of orthogonal square-wave functions rather than an infinite series of orthogonal harmonically related sinusoids.[5,8,14,29]

As an example, consider again the function originally given in Table 2.1, namely, $f(X) = x_1\bar{x}_2\bar{x}_3 + \bar{x}_1x_2\bar{x}_3 + x_1x_2x_3$, whose spectrum is

$$\begin{array}{cccccccc} r_0; & r_1 & r_2 & r_3; & r_{12} & r_{13} & r_{23}; & r_{123} \\ 3 & -1 & -1 & 1 & -1 & 1 & 1 & -3 \end{array}$$

or

$$\begin{array}{cccccccc} s_0; & s_1 & s_2 & s_3; & s_{12} & s_{13} & s_{23}; & s_{123} \\ 2 & 2 & 2 & -2 & 2 & -2 & -2 & 6 \end{array}$$

Figures 2.6a and b illustrate the waveform $\Phi(k)$ of this function, while Figs. 2.6c and d show syntheses by the summation of the appropriately weighted Hadamard and Walsh orthogonal functions, respectively. The required $1/2^n$ scaling factor may be included in the individual spectral coefficient values if desired.[1,5,30]. The similarity of this synthesis with the possibly more familiar Fourier ac waveform synthesis, namely,

$$x(t) = \left\{ a_0 + \sum_{n=1}^{\infty} a_n \cos n\omega t + \sum_{n=1}^{\infty} b_n \sin n\omega t \right\}$$

(where $x(t)$ is the given real-valued signal periodic over a given time $T = 2\pi/\omega$ seconds; a_0, a_n, b_n are real-number coefficients (spectra of $x(t)$); ω is the fundamental angular frequency in radians per second; and $n\omega$, $n > 1$ are the harmonics of ω) will be apparent, including the frequency, sin, cos and sequency, sal, cal parameters.†

† Note that in many ac cases the phase relationship between the Fourier components of a periodic waveform is not required. Under such conditions, the amplitude spectrum only of $x(t)$ is employed, with spectral coefficients c_n, $n = 0$ to ∞, $c_0 = a_0$, $c_n(n \neq 0)$ = non-negative square root of $a_n^2 + b_n^2$.

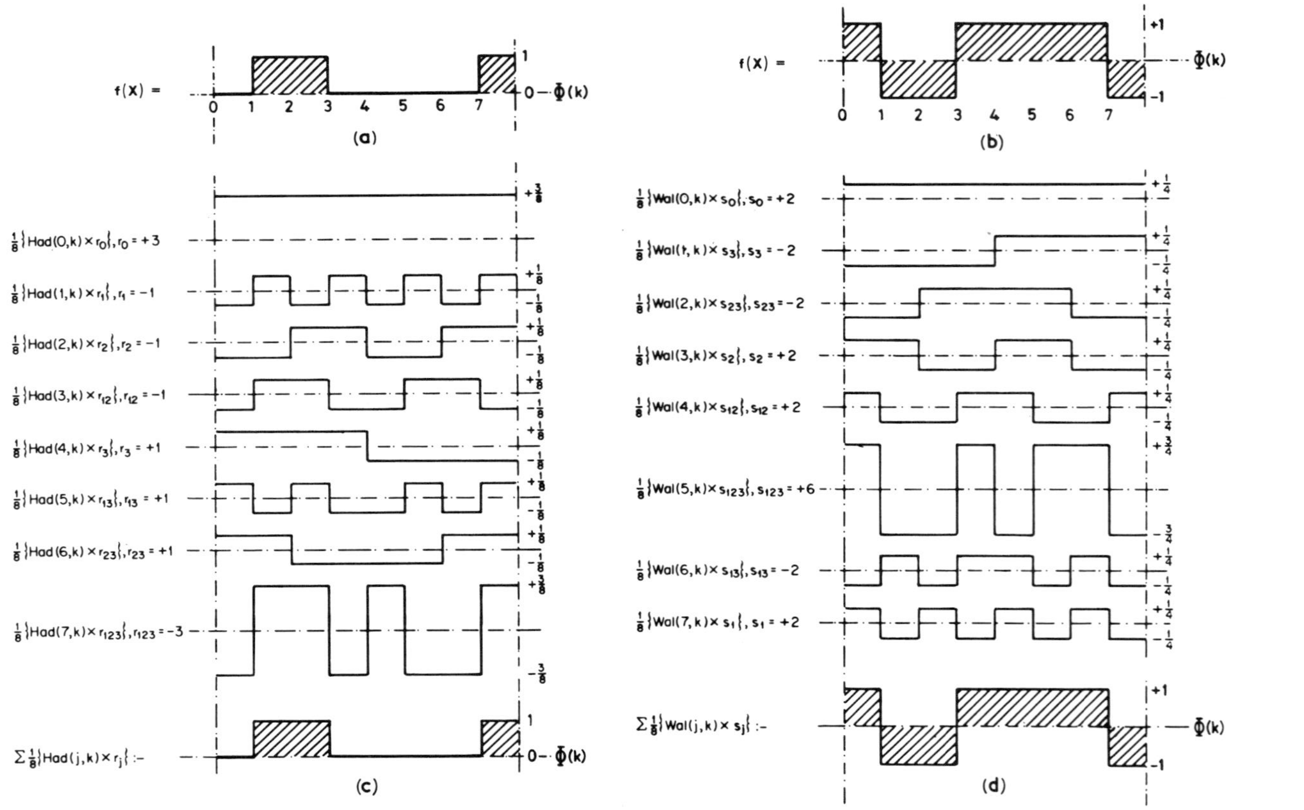

Fig. 2.6. The Fourier synthesis of $f(x)$ from the weighted summation of continuous orthogonal functions; (a) given function waveform coded $\{0, 1\}$; (b) given function waveform, recoded $\{+1, -1\}$; (c) example synthesis of waveform (a) using the Hadamard functions and the spectral coefficients of **R**; (d) example synthesis of waveform (b) using the Walsh functions and the spectral coefficients of **S**.

It is clear that the inverse transformation $(1/2^n)[\mathbf{T}^n]^{-1}\mathbf{R} = \mathbf{Z}$ or $(1/2^n)[\mathbf{T}^n]^{-1}\mathbf{S} = \mathbf{Y}$ (see Eq. (2.12)) is identical to discrete sampling at 2^n points of this Fourier synthesis procedure; the functions employed in this Fourier procedure must correspond to the row vectors of the appropriate inverse transform $[\mathbf{T}^n]^{-1}$, and not to the row vectors of the forward transform $\mathbf{T}^n$ which may be used to compute the spectral coefficient values. The Hadamard and sequency-ordered Walsh functions used in Fig. 2.6 both possess symmetrical properties (see Table 2.2), and hence this distinction does not arise.

Finally, we have concerned ourselves entirely in the foregoing developments with binary ($p = 2$) functions. This will remain our primary concern in subsequent chapters of this text. However, in Section 2.7, we shall very briefly consider ternary and other higher-valued logic ($p > 2$), at which time it will be made clear that the spectral techniques for binary functions are but particular cases of generalised multiple-valued procedures. In particular, we shall observe the following:

(a) The dimensions of complete orthogonal matrices for p-valued transforms will be $p^n \times p^n$.

(b) The real-positive, real-negative elements of our binary vectors and matrices will be superseded by complex numbers, generally taking the pth root of unity. Specifically, Eq. (2.4a) may be generalised to $\exp\{(2\pi/p)\sqrt{-1}\ldots\}$, which will give complex values on the unit circle, not confined to the two real axis values $+1.\ -1.$

(c) The recursive structure of the two-valued Hadamard matrix will be superseded by a generalised recursive structure involving complex number entries and relationships.

(d) The Walsh functions will be superseded by generalised Chrestenson functions, of which the Walsh are a particular case.

(e) The structure of the discontinuous Haar transform will be superseded by a generalised structure, also discontinuous, involving complex number entries.

2.4 FAST TRANSFORM PROCEDURES

2.4.1 The Full Transform

The step-by-step execution of any Hadamard or Walsh-variant forward or inverse transform involves the $\pm$ summation of a total of $2^n \times 2^n$ individual product terms. However, an examination of any such transform will show that subsets of product terms are common to several coefficient summations.

For example, in (2.9) the first four product terms of $\mathbf{T}_{0*}\mathbf{Z}$ are identical to the first four product terms of $\mathbf{T}_{4*}\mathbf{Z}$, and therefore it is a duplication of effort to perform all the individual $2^n \times 2^n$ product summations.

Methods of performing $2^n \times 2^n$ orthogonal transforms with a reduced number of operations, the "fast" transform procedures, may be found in many publications.[1,16,19,27,28,30–33] Instead of involving the total number of individual product terms, the total can be reduced to $2^n \times n$, which is frequently expressed as $N \log_2 N$, where $N = 2^n$.

A graphical representation of the fast procedure for the $n = 3$ Hadamard transform is shown in Fig. 2.7. Such diagrams are often referred to as "butterfly" diagrams. Two points of interest with the Hadamard transform order may be noticed, namely,

(a) the additions and subtractions on each line of this topology follows a reversed binary sequence, e.g., top line $+ + + \cdots$, next line $- + + \cdots$, down to $- - - \cdots$ on the lowest line;

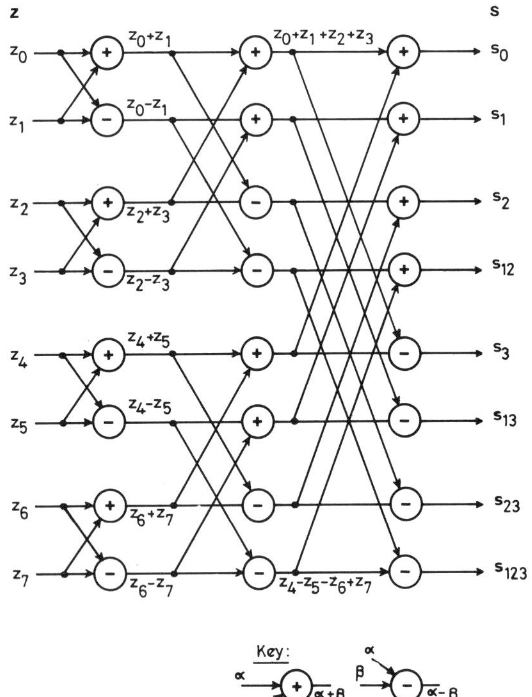

Fig. 2.7. The flow chart of the fast Hadamard transform for $n = 3$, compare Eq. (2.10); generalised flow chart dimensions = 2^n additions/subtractions high × n addition/subtractions wide.

(b) for $n = 2$, the bottom half of this schematic is deleted; for $n = 4$, we vertically duplicate this schematic and complete the additional RH column of $\pm$ summations.

Transforms in alternative row order to the Hadamard will produce similar butterfly diagrams, but in some alternative $+,-$ order along each row. The complete $2^n \times n$ matrix of additions and subtractions also forms the principle of all "fast" hardware circuits for generating the Hadamard and Walsh waveforms and resulting coefficients.[14,27,28]

The zero values in the row functions of the Haar transform lead to a dissimilar flow-chart topology to the preceding, which may be implemented in a total of $2N - 2$ additions/subtractions, where $N = 2^n$ as before. Figure 2.8 illustrates the fast Haar transform flow chart for $n = 3$. Further details

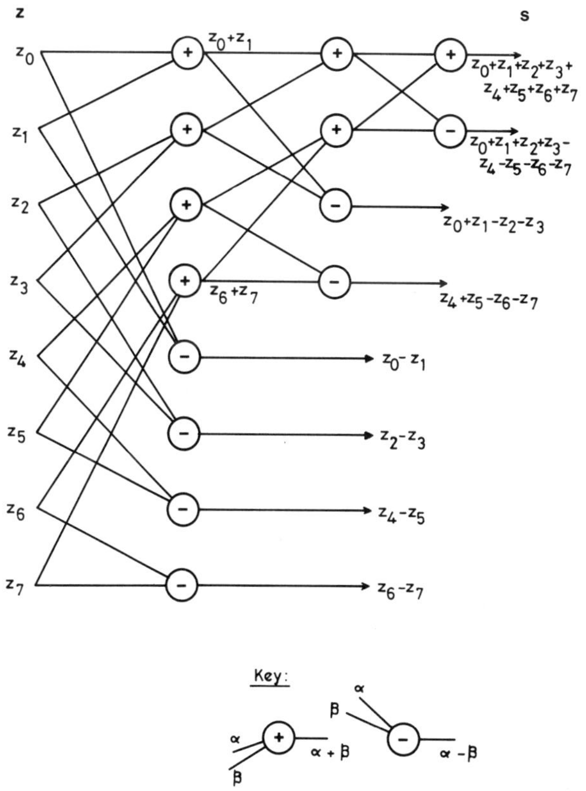

Fig. 2.8. The flow chart of the fast Haar transform for $n = 3$, see Table 2.3(a). Note that scaling (multiplication) of the additions/subtractions is required with the unnormalised transform defined by Eq. (2.21).

of the Cooley–Tukey and other algorithms that generate fast transform topologies may be found in Ahmed and Rao and other publications.[1,26,34–38]

2.4.2 Determination of Subsets of Coefficients

Should the full set of 2^n spectral coefficients not be wanted for a particular application, then to execute the full complement of operations shown in Fig. 2.7 or Fig. 2.8 is unnecessary. For example, in certain logic design applications it may be adequate to consider the zero- and first-order coefficients $s_0, s_1, \ldots, s_n$ only of a Hadamard or equivalent transformation and ignore all higher-order coefficients. Under these circumstances a reduced minimum set of operations would be advantageous in order to save computation time.

If one coefficient only is required, then it is evident that $2^n - 1$ additions/subtractions are necessary with any Hadamard–Walsh-variant transform. If any two are required, then the minimum number becomes 2^n, although this requires an appropriate reorganisation of the flow chart from that explicitly given in Fig. 2.7 except for the case of coefficients s_0 and s_3, s_1 and s_{13}, s_2 and s_{23}, and s_{12} and s_{123}, respectively. However, when complete ordered subsets are required, then the minimum necessary number of additions/subtractions becomes more complex to evaluate.

This problem has been investigated and details have been published.[33,39] The number of addition/subtraction operations necessary to calculate all the spectral coefficients up to order ζ, $= AS_n^\zeta$, $\zeta \in \{1 \text{ to } n\}$, is given by

$$AS_n^\zeta = n2^n - \sum_{r=1}^{n-\zeta} \{r2^{r-1}({}^{n-r}C_\zeta)\} \tag{2.24}$$

This equation may be rearranged to incorporate the $n2^n$ term within the summation, but we shall leave it here in the form of {(total no. of additions/subtractions for all 2^n coefficients) − (the unnecessary additions/subtractions for spectral coefficients of higher order than ζ)}. We may also relate the minimum number of necessary additions/subtractions for any given ζ, n with the number for the corresponding $\zeta, n - 1$ case.[39] For example, considering the two coefficient subsets of (a) zero, first and second order ($\zeta = 2$), and (b) zero, first, second, and third order ($\zeta = 3$), we have

$$\begin{aligned} &\text{(a)}\ AS_n^2 = 2(AS_{n-1}^2) + \tfrac{1}{2}n(n+1) + 1 \\ &\text{(b)}\ AS_n^3 = 2(AS_{n-1}^3) + \tfrac{1}{6}(n+1) \end{aligned} \tag{2.25}$$

Table 2.4 details the values obtained for various cases by applying the preceding results.

Table 2.4. Tabulation of the reduced number of operations necessary to compute the spectral coefficients of **S** when only the first ζ orders of coefficients are required.

n[a]	$n2^n$[b]	AS_n^ζ[c]				
		1	2	3	4	5
3	24	18	23	24	—	—
4	64	41	57	63	64	—
5	160	88	130	152	159	160
6	384	183	282	346	375	383
7	896	374	593	756	849	886
8	2048	793	1223	1605	1861	1991
9	4608	1560	2492	3340	3978	4364
10	10240	3059	5040	6856	8342	9366
11	22528	6130	10147	13944	17246	19756
12	49152	12273	20373	28187	35286	41098

[a] n = number of variables.
[b] $n2^n$ = total number of additions/subtractions in the complete transform.
[c] AS_n^ζ = minimum number of necessary additions/subtractions for increasing order of subsets of **S**.

Savings with the Haar transform when a reduced number of spectral coefficients are required are more inconvenient to specify mathematically, since the number of additions/subtractions fundamentally differs depending upon which spectral coefficients are required. However, having specified a particular subset of interest, then the minimum number of operations can generally be determined from an examination of the computation procedure on the flow chart of Fig. 2.8.

2.4.3 Determination of Largest-Magnitude Coefficients

Because large-magnitude spectral coefficients of **R** or **S**, excluding $r_0(s_0)$, indicate the "most important" input conditions that control a function output, it may be advantageous to know the largest-magnitude spectral coefficients that are present in the complete spectrum. The zero-order coefficient $r_0(s_0)$ may always be separately calculated. Such a need may be important for fault diagnosis and test signature purposes, particularly as n increases.

Ideally we should like to determine the largest-magnitude coefficient(s) without having to compute the complete vector **R** or **S**. However, to a large extent, the optimum procedure is dependent upon the form in which the function $f(X)$ is given, which may be either

(a) already in a minimised sum-of-products Boolean expression, or
(b) as a list of true ($f(X) = 1$) minterms, or alternatively a list of false ($f(X) = 0$) minterms.†

We shall assume that we do not wish to perform any Boolean procedures to convert (a) to (b) or vice versa, but use the given data in the form provided.

Suppose we are given a list of η true minterms that fully define $f(X)$. Then the spectrum of each true minterm is given by

$$\begin{aligned} s_0 &= 2^n - 2 \\ \text{all remaining } 2^n - 1 \text{ coefficients} &= \pm 2, \end{aligned} \tag{2.26}$$

as may be confirmed by example from Eq. (2.10). The signs of each s_α, $\alpha \neq 0$, are given by the following:

(a) If in the given true minterm the x_i variable is uncomplemented, then the sign of the corresponding first-order coefficient s_i is positive; if complemented, then the sign of the corresponding first-order coefficient s_i is negative;
(b) The signs of the second- and higher-even-order coefficients (i.e., 2nd, 4th, ..., order) will be given by $-\{$the multiplication of the signs of the associated first-order coefficients$\}$,
(c) The signs of the third- and higher-odd-order coefficients (i.e., 3rd, 5th, ..., order) will be given by $+\{$the multiplication of the signs of the associated first-order coefficients$\}$.‡

Because the minterms of $f(X)$ are by definition disjoint functions, then the spectrum **S** of the complete function $f(X)$, excluding s_0, is given by the simple arithmetic summation of the corresponding spectral coefficients of the η true minterms,[33] that is,

$$\begin{aligned} s_1 &= \sum\{s_1^1 + s_1^2 + \cdots + s_1^\eta\} \\ s_2 &= \sum\{s_2^1 + s_2^2 + \cdots + s_2^\eta\} \\ &\vdots \\ s_{12\ldots n} &= \sum\{s_{12\ldots n}^1 + s_{12\ldots n}^2 + \cdots + s_{12\ldots n}^\eta\} \end{aligned}$$

The value of s_0 is most readily given by $2^n - 2\eta$, since the presence of each

† We shall assume in this discussion fully defined functions, in which there are no unspecified "don't-care" minterms.

‡ The sign relationships between first- and higher-order coefficients given by (b) and (c) are only valid when minterm coefficients are being considered. Also if $f(X)$ is defined by a list of η false minterms, then *all* the sign conventions (a), (b) and (c) require negating.

true minterm reduces the value of this d.c. spectral coefficient by 2 from its maximum value of $+2^n$. This may equally be given by $\sum\{s_0^1 + s_0^2 + \cdots + s_0^\eta\} - (\eta - 1)2^n$, the $(\eta - 1)2^n$ term correcting for the summation over the false minterms of $f(X)$ η times in the $\sum s_0$ summation.[33]

This procedure is clearly advantageous with very sparsely defined functions. As η increases above n, then the full fast transform procedure becomes more economic in total computational effort, although the minterm summation technique frequently remains convenient for hand computation.

Where $f(X)$ is given in a minimised sum-of-products form, we may adopt a strategy based upon the preceding discussion, which without final correction can give approximate spectral coefficient values. In outline, we treat each product term as a minterm within its particular reduced n-space of $f(X)$, generate its weighted "minterm" spectra based upon Eq. (2.26), and finally sum to give an (approximate) overall spectrum for $f(X)$. The procedure is best illustrated by example.

Consider the function of six variables fully defined by the eight prime implicants of Table 2.5. Each prime implicant contains v defining variables, either true or complemented, as denoted by 1 or 0, respectively. For convenience we have assembled them in increasing v order.

Each prime implicant in the array can contribute only to the spectral coefficients associated with the variables present in the implicant. For example, the first entry can contribute to all spectral coefficients of $f(X)$ which contain 1, 2 and/or 3 in their subscript identification, but contributes zero to all coefficients involving subscripts 4, 5 and 6. Further, if we consider each prime implicant as a minterm within its own restricted v-space, then the

Table 2.5. The on array ($f(X) = 1$) of the six-variable function.[a]

Prime implicant	x_1	x_2	x_3	x_4	x_5	x_6	No. of defining variables v	Weighting factor 2^{n-v}
1	0	0	0	—	—	—	3	×8
2	1	1	1	—	—	—	3	×8
3	—	—	—	1	1	1	3	×8
4	—	1	0	—	0	0	4	×4
5	1	0	0	—	0	0	5	×2
6	0	0	1	—	0	0	5	×2
7	1	0	0	0	0	1	6	×1
8	1	0	0	1	1	0	6	×1

[a] $f(X) = \bar{x}_1\bar{x}_2\bar{x}_3 + x_1x_2x_3 + x_4x_5x_6 + x_2\bar{x}_3\bar{x}_5\bar{x}_6 + x_1\bar{x}_2\bar{x}_3\bar{x}_5\bar{x}_6 + \bar{x}_1\bar{x}_2x_3\bar{x}_5\bar{x}_6 + x_1\bar{x}_2\bar{x}_3\bar{x}_4\bar{x}_5x_6 + x_1\bar{x}_2\bar{x}_3x_4x_5\bar{x}_6$.

contribution of this v-variable term to the full n-space spectrum of $f(X)$ is related as follows:

$$s_0 \text{ in full } n\text{-space} = \{2^n - 2(2^{n-v})\}$$
$$s_i \text{ in full } n\text{-space} \quad (2.27)$$
$$i \neq 0 = \{s_i \text{ in } v\text{-space} \times (2^{n-v})\}$$

The factor 2^{n-v} is the weighting factor to expand the restricted v-space to the full n-space of $f(X)$. For example, the spectral coefficients of the first prime implicant in Table 2.5 are as follows:

(a) within its own v-space, treated as a single minterm:

$$\begin{matrix} s_0; & s_1 & s_2 & s_3; & s_{12} & s_{13} & s_{23}; & s_{123} \\ 6 & -2 & -2 & -2 & -2 & -2 & -2 & -2 \end{matrix}$$

(b) within the full n-space of $f(X)$:

$$\begin{matrix} s_0; & s_1 & s_2 & s_3 & s_4 & s_5 & s_6; & s_{12} & s_{13} & s_{14} \cdots; & s_{123} & s_{124} \cdots \\ 48 & -16 & -16 & -16 & 0 & 0 & 0 & -16 & -16 & 0 \cdots & -16 & 0 \ldots \end{matrix}$$

Applying this procedure to all prime implicants of the given on array, we obtain the arithmetic summation shown in Table 2.6. Note that since the spectral coefficients associated with the first three prime implicants will dominate, the values of the cross-spectral coefficients such as s_{14}, s_{15}, s_{24}, etc. will be small; for this reason we have omitted tabulating them, since our present objective is to determine the dominating spectral coefficients.

The initial arithmetic summation given in Table 2.6 is not necessarily the true spectrum of $f(X)$, since no correction has yet been made for any minterms that are common between two or more implicants, and that have therefore been included more than once in this initial total. However, provided the given prime implicants are largely disjoint, as may be estimated by examination of the on array, then this initial summation may be taken as an indicator of the dominant spectral coefficients of $f(X)$.

Thus we may see from this initial summation that s_{12}, s_{23} and s_{56} are the dominating spectral coefficients of $f(X)$, indicating that $f(X)$ is strongly determined by the input functions $\overline{x_1 \oplus x_2}$, $\overline{x_2 \oplus x_3}$ and $\overline{x_5 \oplus x_6}$, respectively. This is not obvious from the given on-array data. Hence, these input functions may be good diagnostic signals with which to exercise a circuit realisation of $f(X)$.

If we wish to correct this initial summation to give true spectral values, then examination of the on array will show two minterms of $f(X)$ that have been included twice in this working example, namely, (i) minterm 000111,

Table 2.6. Summation of the dominant spectral coefficients of the prime implicants given in the on array of Table 2.5. The value of s_0 is not required in these considerations.

	s_1	s_2	s_3;	s_{12}	s_{13}	s_{23};	s_{123}	s_4	s_5	s_6;	s_{45}	s_{46}	s_{56};	s_{456}
Prime implicant 1	−16	−16	−16	−16	−16	−16	−16							
2	16	16	16	−16	−16	−16	16							
3								16	16	16	−16	−16	−16	16
4		8	−8			8			−8	−8			−8	
5	4	−4	−4	4	4	−4	4		−4	−4			−4	
6	−4	−4	4	−4	4	4	4		−4	−4			−4	
7	2	−2	−2	2	2	−2	2	−2	−2	2	−2	2	2	2
8	2	−2	−2	2	2	−2	2	2	2	−2	−2	2	2	−2
Initial arithmetic summation	4	−4	−12	−28	−20	−28	12	16	0	0	−20	−12	−28	16
Spectra of the cojoint minterms (i)	−2	−2	−2	−2	−2	−2	−2	2	2	2	−2	−2	−2	2
(ii)	2	2	2	−2	−2	−2	2	2	2	2	−2	−2	−2	2
S of $f(X)$	4	−4	−12	−24	−16	−24	12	12	−4	−4	−16	−8	−24	12

included in implicants (a) and (c), and (ii) minterm 111111, included in implicants (b) and (c). Correction for these two minterms is shown in Table 2.6. However, provided the given function $f(X)$ is defined by a scatter of prime implicants, then this correction may be omitted if only identification or approximate values of the dominant spectral coefficients is sought.

Further elaboration of these principles, including the consideration of on arrays and off arrays for functions that are incompletely specified, is available.[33,40]

2.5 THE COMPUTATION OF COMPOSITE SPECTRA

In the preceding sections we have considered the computation of spectral coefficients from a truthtable vector of the given function $f(X)$ and a possible technique for a function given in a sum-of-products form, where the product terms are (ideally) disjoint. A function expressed in a sum-of-minterms form is a specific case. However, we have not considered the general case where $f(X)$ is given in algebraic form, or as a gate-level description of a network that realises the function. Given such a presentation, it is clearly possible to compile the truthtable vector for $f(X)$ and apply the normal transform $\mathbf{T}^n$, but we may wish to proceed so that all computations from the decomposition level are performed in the spectral domain, as considered for the particular cases in the preceding section. This approach may be of particular interest when the spectra of the subfunctions are required in addition to the total function spectrum. This will be the case in Chapter 6.

Eris[41] and Muzio[42,43] have investigated methods for determining the spectrum of the Boolean sum (OR) and product (AND) of two functions directly from their spectra. The former's approach is somewhat involved and effectively introduces the inverse transform, a step which in essence translates the problem back into its Boolean equivalent. The later publication of Muzio[43] has presented a generalised approach that includes negation and Exclusive-OR of functions in addition to the sum and product. A completely general coding scheme is adopted in this work; here, however, we shall concentrate our attention on the conventional $\{0, 1\}$ and $\{+1, -1\}$ codings.

2.5.1 Composite Spectra in $\{0, 1\}$ Coding

For the moment, let f and g be any pair of real-valued functions on V_n, the vector space of all binary n-tuples. Lechner[8] has defined the *convolution* *sum* $f * g$ of f and g, where for each point $j \in V_n$

$$(f * g)_j = \sum_{k=0}^{2^n-1} f_{j \oplus k} g_k \tag{2.28}$$

where $j \oplus k$ is the decimal integer whose binary expansion is the bit-wise mod.2 sum of the binary expansions of j and k. Here we use an exactly similar convolution sum of two N-place vectors, where $N = 2^n$. The latter has also been used by several authors, specifically Karpovsky[5] and Pichler.[13,44]

Thus, given any two Boolean functions $f^1(X)$ and $f^2(X)$ with spectra $\mathbf{R}^1$ and $\mathbf{R}^2$, respectively, we define the vector $\mathbf{R}^1 * \mathbf{R}^2$ by

$$\mathbf{R}^1 * \mathbf{R}^2]_j = \sum_{k=0}^{2^n-1} r^1_{j \oplus k} r^2_k \tag{2.29}$$

Note that in Eq. (2.29) and the rest of this section, all given spectra are in Hadamard order, and, unless otherwise specified, all spectral coefficients are identified by a decimal number subscript $j = 0$ to $N - 1$ or $k = 0$ to $N - 1$, reading down the coefficient vector, for example, $r_0, r_1, r_2, r_3, \ldots, r_{N-2}, r_{N-1}$.

Let $\mathbf{Z}^1$ and $\mathbf{Z}^2$ be the $\{0, 1\}$ truthtable column vectors of $f^1(X)$ and $f^2(X)$, respectively. Lechner[8] has shown that

$$\mathbf{Z}^1 * \mathbf{Z}^2 = \frac{1}{N}\{\mathbf{T}^n\mathbf{A}\} \tag{2.30}$$

where the jth element of $\mathbf{A}$ is $r^1_j r^2_j$. The dual result[43] gives

$$\mathbf{R}^1 * \mathbf{R}^2 = N\{\mathbf{T}^n\mathbf{Z}^{12}\} \tag{2.31}$$

where the jth element of $\mathbf{Z}^{12}$ is the arithmetic product $f^1_j f^2_j$. Therefore $\mathbf{Z}^{12}$ represents the Boolean product (AND) of the given functions $f^1(X)$ and $f^2(X)$. Consequently, letting $\mathbf{R}^{12}$ represent the AND spectrum, we have

$$\mathbf{R}^1 * \mathbf{R}^2 = N\mathbf{R}^{12},$$

whence

$$\mathbf{R}^{12} = \frac{1}{N}(\mathbf{R}^1 * \mathbf{R}^2) \tag{2.32}$$

This equation gives the relation between the spectrum of the product of any two functions and the spectra of those functions.

The $\{0, 1\}$ truthtable vector of the Boolean sum (OR) of two functions $f^1(X)$ and $f^2(X)$ is given by

$$\mathbf{Z}^{1+2} = \mathbf{Z}^1 + \mathbf{Z}^2 - \mathbf{Z}^{12}$$

whilst the $\{0, 1\}$ truthtable vector of the Exclusive-OR of $f^1(X)$ and $f^2(X)$ is given by

$$\mathbf{Z}^{1\oplus 2} = \mathbf{Z}^1 + \mathbf{Z}^2 - 2\mathbf{Z}^{12}$$

Applying the transform $\mathbf{T}^n$ to both sides and substituting Eq. (2.32) gives

$$\mathbf{R}^{1+2} = \mathbf{R}^1 + \mathbf{R}^2 - \frac{1}{N}(\mathbf{R}^1 * \mathbf{R}^2) \tag{2.33}$$

and

$$\mathbf{R}^{1\oplus 2} = \mathbf{R}^1 + \mathbf{R}^2 - \frac{2}{N}(\mathbf{R}^1 * \mathbf{R}^2) \tag{2.34}$$

as the spectra for the Boolean sum and Exclusive-OR of $f^1(X)$ and $f^2(X)$, respectively.

The final operation to be considered is complementation. Given $\mathbf{R}$, the spectrum of $f(X)$, then $\bar{\mathbf{R}}$ the spectrum of $\overline{f(X)}$ is such that

$$\begin{aligned} \bar{r}_0 &= 2^n - r_0 \\ \bar{r}_j, j \neq 0, &= -r_j \end{aligned} \tag{2.35}$$

which can be written as

$$\bar{\mathbf{R}} = \mathbf{J}^n - \mathbf{R} \tag{2.36}$$

where $\mathbf{J}^n$ is defined in the List of Symbols.

As an example of these relationships, consider the two simple functions

$$f^1(X) = \bar{x}_3(x_2 + x_1) + x_2x_1$$

and

$$f^2(X) = \bar{x}_3x_2 + x_3\bar{x}_2 + \bar{x}_2x_1$$

The corresponding spectra are

$$\mathbf{R}^1 = 4\ -2\ -2\quad 0\ 2\quad 0\quad 0\ -2]^t$$

$$\mathbf{R}^2 = 5\ -1\quad 1\ -1\ 1\ -1\ -3\ -1]^t$$

From Eq. (2.36) we find the spectra of the complements of $f^1(X)$ and $f^2(X)$ to be

$$\bar{\mathbf{R}}^1 = 4\ 2\quad 2\ 0\ -2\ 0\ 0\ 2]^t$$

and

$$\bar{\mathbf{R}}^2 = 3\ 1\ -1\ 1\ -1\ 1\ 3\ 1]^t$$

The convolution of $\mathbf{R}^1$ and $\mathbf{R}^2$ is given by

$$\mathbf{R}^1 * \mathbf{R}^2 = \begin{bmatrix} 4 & -2 & -2 & 0 & 2 & 0 & 0 & -2 \\ -2 & 4 & 0 & -2 & 0 & 2 & -2 & 0 \\ -2 & 0 & 4 & -2 & 0 & -2 & 2 & 0 \\ 0 & -2 & -2 & 4 & -2 & 0 & 0 & 2 \\ 2 & 0 & 0 & -2 & 4 & -2 & -2 & 0 \\ 0 & 2 & -2 & 0 & -2 & 4 & 0 & -2 \\ 0 & -2 & 2 & 0 & -2 & 0 & 4 & -2 \\ -2 & 0 & 0 & 2 & 0 & -2 & -2 & 4 \end{bmatrix} \begin{bmatrix} 5 \\ -1 \\ 1 \\ -1 \\ 1 \\ -1 \\ -3 \\ -1 \end{bmatrix} = \begin{bmatrix} 24 \\ -8 \\ -8 \\ -8 \\ 24 \\ -8 \\ -8 \\ -8 \end{bmatrix}$$

Note that the preceding matrix operating on vector $\mathbf{R}^2$ is constructed from $\mathbf{R}^1$ such that the entry in the jth row and kth column is $r^1_{j \oplus k}$. An identical final result for $\mathbf{R}^1 * \mathbf{R}^2$ would be obtained by constructing the matrix with j, k entries taken from $\mathbf{R}^2$ and operating upon vector $\mathbf{R}^1$.

It therefore follows that the spectra of the Boolean product, sum and Exclusive-OR of $f^1(X)$ and $f^2(X)$ are given by the following:

$$\begin{matrix}\text{product } \mathbf{R}^{12} \\ \text{(AND)}\end{matrix} = \frac{1}{8} \begin{bmatrix} 24 \\ -8 \\ -8 \\ -8 \\ 24 \\ -8 \\ -8 \\ -8 \end{bmatrix} = \begin{bmatrix} 3 \\ -1 \\ -1 \\ -1 \\ 3 \\ -1 \\ -1 \\ -1 \end{bmatrix}$$

(see Eq. (2.32));

$$\begin{matrix}\text{sum } \mathbf{R}^{1+2} \\ \text{(OR)}\end{matrix} = \begin{bmatrix} 4 \\ -2 \\ -2 \\ 0 \\ 2 \\ 0 \\ 0 \\ -2 \end{bmatrix} + \begin{bmatrix} 5 \\ -1 \\ 1 \\ -1 \\ 1 \\ -1 \\ -3 \\ -1 \end{bmatrix} - \begin{bmatrix} 3 \\ -1 \\ -1 \\ -1 \\ 3 \\ -1 \\ -1 \\ -1 \end{bmatrix} = \begin{bmatrix} 6 \\ -2 \\ 0 \\ 0 \\ 0 \\ 0 \\ -2 \\ -2 \end{bmatrix}$$

(see Eq. (2.33));

$$\text{Exclusive-OR } \mathbf{R}^{1\oplus 2} = \begin{bmatrix} 4 \\ -2 \\ -2 \\ 0 \\ 2 \\ 0 \\ 0 \\ -2 \end{bmatrix} + \begin{bmatrix} 5 \\ -1 \\ 1 \\ -1 \\ 1 \\ -1 \\ -3 \\ -1 \end{bmatrix} - 2 \times \begin{bmatrix} 3 \\ -1 \\ -1 \\ -1 \\ 3 \\ -1 \\ -1 \\ -1 \end{bmatrix} = \begin{bmatrix} 3 \\ -1 \\ 1 \\ 1 \\ -3 \\ 1 \\ -1 \\ -1 \end{bmatrix}$$

(see Eq. (2.34)).

It is left as a simple exercise for the reader to confirm these results by determining the resulting functions in the Boolean domain and executing the usual transform $\mathbf{T}^n\mathbf{Z} = \mathbf{R}$.

Finally, note that the spectrum of a single variable x_i has $r_0 = 2^{n-1}$ and the coefficient corresponding to $x_i = -2^{n-1}$, all remaining coefficients zero. The spectrum of $\bar{x}_i$ is identical except that the coefficient corresponding to x_i now equals $+2^{n-1}$. Given these basic spectra and the results of Eqs. (2.32)–(2.36), the computation of the total spectrum of any arbitrary algebraic expression or gate equivalent network is straightforward. The amount of data to be manipulated by hand may be somewhat tedious, but is straightforward by computer-aided means.

2.5.2 Composite Spectra in $\{+1, -1\}$ Coding

Any function truthtable vector can be recoded from $\{0, 1\}$ to $\{+1, -1\}$ as follows:

$$y_j = 1 - 2z_j, \qquad j = 0 \text{ to } 2^n - 1$$

or, equivalently,

$$\mathbf{Y} = 1]^N - 2\mathbf{Z}$$

where $1]^N$ represents a column vector of 2^n 1's. It therefore follows that the spectral coefficient vectors of $\mathbf{Y}$ and $\mathbf{Z}$ are related by

$$\mathbf{S} = \mathbf{J}^n - 2\mathbf{R} \tag{2.37}$$

and, conversely,

$$\mathbf{R} = \tfrac{1}{2}(\mathbf{J}^n - \mathbf{S}) \tag{2.38}$$

Applying these results to the simple case of complementation, from Eqs. (2.36) and (2.38), we have

$$\bar{\mathbf{R}} = \mathbf{J}^n - \mathbf{R}$$

$$\tfrac{1}{2}(\mathbf{J}^n - \bar{\mathbf{S}}) = \mathbf{J}^n - \tfrac{1}{2}(\mathbf{J}^n - \mathbf{S})$$

$$= \tfrac{1}{2}(\mathbf{J}^n + \mathbf{S})$$

which reduces to the well-known result of

$$\bar{\mathbf{S}} = -\mathbf{S} \tag{2.39}$$

For the product, sum and Exclusive-OR relationships, we first apply Eqs. (2.37) and (2.38) to give the following relationships:

AND $\quad \mathbf{S}^{12} = \mathbf{J}^n - 2\mathbf{R}^{12}$

OR $\quad \mathbf{S}^{1+2} = \mathbf{J}^n - 2\mathbf{R}^{1+2}$

Exclusive-OR $\quad \mathbf{S}^{1\oplus 2} = \mathbf{J}^n - 2\mathbf{R}^{1\oplus 2}$

Convolution $\quad \mathbf{R}^1 * \mathbf{R}^2 = \tfrac{1}{2}(\mathbf{J}^n - \mathbf{S}^1) * \tfrac{1}{2}(\mathbf{J}^n - \mathbf{S}^2)$

$$= \tfrac{1}{4}(\mathbf{J}^n * \mathbf{J}^n - \mathbf{J}^n * \mathbf{S}^1 - \mathbf{J}^n * \mathbf{S}^2 + \mathbf{S}^1 * \mathbf{S}^2)$$

$$= \tfrac{1}{4}(N\mathbf{J}^n - N\mathbf{S}^1 - N\mathbf{S}^2 + \mathbf{S}^1 * \mathbf{S}^2)$$

Thence for the product (AND) case we have

$$\begin{aligned}\mathbf{S}^{12} &= \mathbf{J}^n - 2\mathbf{R}^{12}\\ &= \mathbf{J}^n - 2\left\{\frac{1}{N}(\mathbf{R}^1 * \mathbf{R}^2)\right\}\\ &= \mathbf{J}^n - \frac{2}{N}\left\{\frac{1}{4}(N\mathbf{J}^n - N\mathbf{S}^1 - N\mathbf{S}^2 + \mathbf{S}^1 * \mathbf{S}^2)\right\}\\ &= \frac{1}{2}\left\{\mathbf{S}^1 + \mathbf{S}^2 - \frac{1}{N}(\mathbf{S}^1 * \mathbf{S}^2) + \mathbf{J}^n\right\}\end{aligned} \tag{2.40}$$

Similarly for the sum (OR) and Exclusive-OR cases, we find

$$\mathbf{S}^{1+2} = \frac{1}{2}\left\{\mathbf{S}^1 + \mathbf{S}^2 + \frac{1}{N}(\mathbf{S}^1 * \mathbf{S}^2) - \mathbf{J}^n\right\} \tag{2.41}$$

and

$$\mathbf{S}^{1\oplus 2} = \frac{1}{N}(\mathbf{S}^1 * \mathbf{S}^2) \tag{2.42}$$

These equations are the equivalent of their $\{0, 1\}$ coding counterparts previously given and are used in the same fashion,

For completeness we include the general formulae given by Muzio[43] without proof. Here logic 0 is coded as a_0, logic 1 is coded as a_1, with $b = a_1 - a_0$ and $c = a_1 + a_0$.

$$\begin{aligned}
\bar{\mathbf{R}}^1 &= c\mathbf{J}^n - \mathbf{R}^1 \\
b\mathbf{R}^{12} &= \frac{1}{N}(\mathbf{R}^1 * \mathbf{R}^2) - a_0(\mathbf{R}^1 + \mathbf{R}^2) + a_0 a_1 \mathbf{J}^n \\
b\mathbf{R}^{1+2} &= -\frac{1}{N}(\mathbf{R}^1 * \mathbf{R}^2) + a_1(\mathbf{R}^1 + \mathbf{R}^2) - a_0 a_1 \mathbf{J}^n \\
b\mathbf{R}^{1\oplus 2} &= -\frac{2}{N}(\mathbf{R}^1 * \mathbf{R}^2) + c(\mathbf{R}^1 + \mathbf{R}^2) - a_0 c \mathbf{J}^n
\end{aligned} \tag{2.43}$$

In these general equations, **R** denotes a generic spectrum whose coding depends upon the choice of a_0 and a_1. The results presented earlier are a particular case of the above. In the following chapters, we shall only have need to refer to the particular case, normally using the **R** spectra. Finally, note that in these developments we have been considering without restriction any two functions $f^1(X)$ and $f^2(X)$. However, where $f^1(X)$ and $f^2(X)$ are disjoint and contain no true minterms in common, then the considerable simplification of the OR case outlined in Section 2.4 is available. Clearly in this case the OR and Exclusive-OR are identical, with the AND always logic 0 over all minterms. It is also left as an instructive exercise for the reader to apply the spectral equations developed in this section when $f^1(X)$ and $f^2(X)$ are identical functions.

2.6 TRANSFORMATIONS USING INCOMPLETE AND NON-ORTHOGONAL MATRICES

2.6.1 Incomplete Orthogonal Matrices

If transformations both from the functional domain to the spectral domain and back from spectral data to functional data are required, then it is essential to employ complete $2^n \times 2^n$ orthogonal transforms. Failure to do so entails loss of information from one domain to the other, except under restricted conditions such as when one spectral coefficient is maximum-valued, which implies all remaining coefficients are zero.

If we merely wish to compute the value of a particular spectral coefficient, then the selection of the appropriate row from $\mathbf{T}^n$ and the row/column multiplication and summation of Eq. (2.22) or (2.23) is available. Similarly, if we wish to compute a particular subset of coefficients, we are always free to compose a particular transform of less than 2^n rows to perform this duty, but this will not in general possess any fast transform capability such as discussed in Section 2.4.1.

However, there is one reduced set of spectral coefficients of particular logical significance that we shall have occasion to refer to in detail in Chapter 3. This set is the zero- and first-order coefficients $s_0, s_1, \ldots, s_n$ (or $r_0, r_1, \ldots, r_n$), which give information concerning the number of true/false minterms in $f(X)$ and the importance of each input variable x_i in controlling the function output. This particular set of $n + 1$ coefficients will be referred to in Chapter 3 as the Chow parameters and have a particular significance for linearly separable (threshold-logic) functions.[6,7,45,46]

The transform for the determination of these coefficients is the Rademacher transform considered earlier and illustrated for $n = 3$ in Fig. 2.3. For $n = 4$, we therefore have

$$\begin{bmatrix} 1 & 1 & 1 & 1 & 1 & 1 & 1 & 1 & 1 & 1 & 1 & 1 & 1 & 1 & 1 & 1 \\ 1 & 1 & 1 & 1 & 1 & 1 & 1 & 1 & -1 & -1 & -1 & -1 & -1 & -1 & -1 & -1 \\ 1 & 1 & 1 & 1 & -1 & -1 & -1 & -1 & 1 & 1 & 1 & 1 & -1 & -1 & -1 & -1 \\ 1 & 1 & -1 & -1 & 1 & 1 & -1 & -1 & 1 & 1 & -1 & -1 & 1 & 1 & -1 & -1 \\ 1 & -1 & 1 & -1 & 1 & -1 & 1 & -1 & 1 & -1 & 1 & -1 & 1 & -1 & 1 & -1 \end{bmatrix} \begin{bmatrix} \\ \\ \mathbf{Y} \\ \\ \\ \end{bmatrix} = \begin{bmatrix} s_0 \\ s_1 \\ s_2 \\ s_3 \\ s_4 \end{bmatrix}$$

See also Eqs. (3.10) and (3.11) (Chapter 3). Note that it is not possible to transform from the resulting $n + 1$ spectral coefficients back to the given function vector **Y**, since except under special circumstances the information content of $s_0, s_1, \ldots, s_n$ does not uniquely represent **Y**. This may be readily demonstrated by determining the spectral coefficients for, say, $f(X) = x_1\bar{x}_3 + x_2x_3 + \bar{x}_1x_3$ and $x_1\bar{x}_2 + x_2x_3 + \bar{x}_1x_2$, or any Exclusive-OR function,[41] However, $s_0, s_1, \ldots, s_n$ maintain the functional meanings that we have previously considered.

For the particular purposes of threshold-logic classification and synthesis, this restricted set of $n + 1$ spectral coefficients may be relabelled. Instead of $s_0, s_1, \ldots, s_n$, we may use **B**, $\mathbf{B} = b_0, b_1, \ldots, b_n$, respectively. This gives the transform designation

$$\mathbf{TY} = \mathbf{B} \tag{2.44}$$

where **T** is of dimensions $n + 1 \times 2^n$. Further, since the coefficients b_1 to b_n represent the correlation of the function with its independent input variables x_1 to x_n, Eq. (2.23) may be found in simplified form as

$$b_i, i = 1 \text{ to } n = \sum_{m=0}^{2^n - 1} \{f(X) \cdot x_i\} \tag{2.45}$$

where $\sum_{m=0}^{2^n-1} \{f(X) \cdot x_i\}$ is here read as the number of agreements between $f(X)$ and x_i minus the number of disagreements, summed over all minterms m_0 to m_{2^n-1} of $f(X)$.

2.6.2 Non-Orthogonal Matrices

Provided no inverse transform availability is required, and only a global or a local correlation meaning from each spectral coefficient value is required, then we are free to compile a transform matrix whose rows correspond to any functions of $f(X)$ in which we may be interested. These may be orthogonal functions selected from any Hadamard, Haar or other complete orthogonal transform, but equally may correspond to other functions. For example, should we be interested in the correlation of a given function with, say, $\overline{x_1 + x_2x_3}$, then one row of the transform would be

$$\begin{bmatrix} -1 & 1 & -1 & 1 & -1 & 1 & 1 & 1 \\ & & & & & & & \end{bmatrix} \begin{bmatrix} \\ \mathbf{Y} \\ \\ \\ \end{bmatrix} = \begin{bmatrix} \\ \mathbf{S} \\ \\ \\ \end{bmatrix}$$

Additionally, the row vectors may be composed to give correlation information over a subcube of $f(X)$, as in the Haar transform, by appropriate t_{jk} entries $\in\{+1, 0, -1\}$.

It will be appreciated that we have here departed entirely from the properties of orthogonal functions and complete orthogonal matrices and are concerned only with the result of individual row and column vector multiplications, where the row and column entries are $+1$, -1 (or $+1$, 0, -1). The use of such incomplete-orthogonal or non-orthogonal transforms for logic design purposes has received consideration,[17,47] but with the exception of the Chow parameters for linearly separable functions[6,7] has yet to find a significant place in logic theory and design.

2.7 TERNARY AND HIGHER-VALUED SPECTRA

In Section 2.3.4 we commented that the developments that were then being considered for binary systems could be extended to the general p-valued case, $p \geq 2$. We further noted that the $2^n \times 2^n$ complete orthogonal transforms of the binary case would be superseded by $p^n \times p^n$ transforms for the p-valued case, the entries in which would (in general) be complex numbers taking the pth root of unity.

In this section we shall briefly cover these generalised p-valued developments. The continuing absence of any commercial circuits and systems for $p > 2$ largely makes these developments of theoretical interest only; indeed the following chapters will provide an indication of the concentration of application to date on the binary case. However, all the techniques considered later have their generalised extension, albeit with greater mathematical complexity.

2.7.1 p-Valued Complex Numbers

Consider the unit circle, with origin zero and radius unity, divided into p segments as illustrated in Fig. 2.9, with one division always lying along the real positive axis. We may now consider the points lying on the unit circle at the division between segments as the truth values for a p-valued logic system.

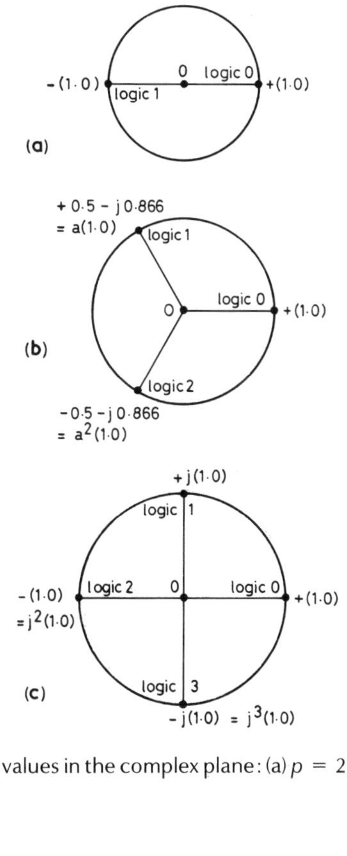

Fig. 2.9. p-Valued truth values in the complex plane: (a) $p = 2$ (binary); (b) $p = 3$ (ternary); (c) $p = 4$ (quaternary).

For increasing p, we therefore have the truth values variously defined as follows, where j is $\sqrt{-1} = 1.0 \angle +90°$, and a is $\sqrt[3]{1} = 1.0 \angle +120°$:

$$p = 2: \quad \text{roots of } \sqrt{1} = \{+1; -1\}$$

$$p = 3: \quad \text{roots of } \sqrt[3]{1} = \{+1; -0.5 + j0.866; -0.5 - j0.866\} = \{+1; a; a^2\}$$

$$p = 4: \quad \text{roots of } \sqrt[4]{1} = \{+1; j1.0; -1; -j1.0\} = \{+1, j; j^2; j^3\}$$

and similarly for higher p. These may be collectively summarised by

$$\exp\left\{\frac{2\pi}{p}\sqrt{-1}\,\delta\right\} \tag{2.46}$$

where δ takes the integer values $0, 1, \ldots, p-1$,

$$\equiv \left\{\cos\frac{2\pi}{p}\delta + j\sin\frac{2\pi}{p}\delta\right\} \tag{2.46a}$$

See also the footnote to Eq. (2.4a). Note also that the truth value recoding of $\{0, 1\}$ to $\{+1, -1\}$ for $\mathbf{Y}$, first introduced in Eq. (2.10) and subsequently used in all binary transforms $\mathbf{T}^n\mathbf{Y} = \mathbf{S}$, is formally given by Eq. (2.46).

Because of the increasing size of matrix dimensions with increasing p, we shall generally confine our following discussions to the $p = 3$ (ternary) case; extensions to $p > 3$ should be apparent.

2.7.2 Chrestenson Functions

Chrestenson functions are orthogonal p-valued functions defined over the interval $[0, p^n)$ by

$$\begin{aligned} \mathrm{Ch}^{(p)}(j,k) &= \exp\left\{\frac{2\pi}{p}\cdot\sqrt{-1}\,.\,C(j,k)\right\} \\ C(j,k) &= \sum_{\eta=0}^{n-1} j_\eta k_\eta \end{aligned} \tag{2.47}$$

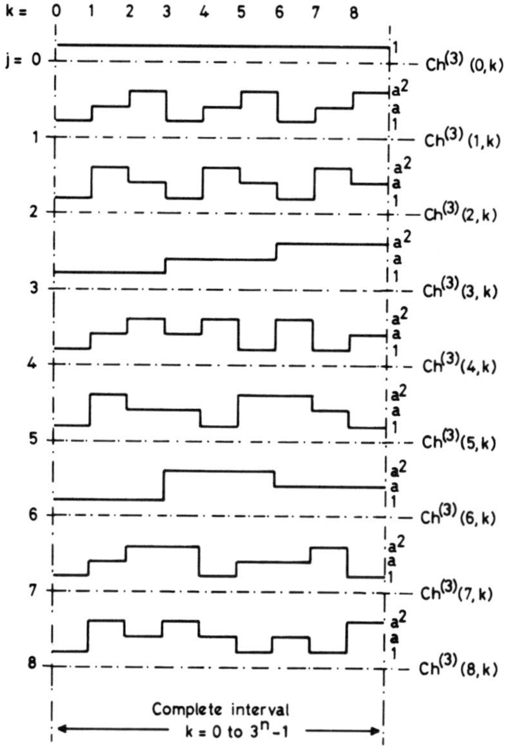

Fig. 2.10. The closed orthogonal set of Chrestenson functions for $p = 3$, $n = 2$, based upon Eq. (2.47), where $a = 1.0 \angle +120°$.

where j_η, k_η are the p-ary expansions of j and k, respectively, and where

$$j(\text{decimal}) = \{j_{n-1}p^{n-1} + j_{n-2}p^{n-2} + \cdots + j_0p^0\}$$

$$k(\text{decimal}) = \{k_{n-1}p^{n-1} + k_{n-2}p^{n-2} + \cdots + k_0p^0\}$$

For $p = 3, n = 2$, this gives the orthogonal functions shown in Fig. 2.10, which when sampled at discrete intervals gives the transform matrix of Table 2.7.

The salient features of this complete orthogonal matrix are

(i) its dimensions are $p^n \times p^n$;
(ii) it is symmetric, giving $[\mathbf{T}_p^n]^t \equiv [\mathbf{T}_p^n]$;†

† We shall use the symbol $\mathbf{T}_p^n$ for a complete orthogonal $p^n \times p^n$ p-valued matrix, rather than $\mathbf{T}^n$ as previously. For the $p = 2$ (binary) case, $\mathbf{T}_p^n \equiv \mathbf{T}^n$.

(iii) Its inverse is given by

$$[\mathbf{T}_p^n]^{-1} = \frac{1}{K}[\mathbf{T}_p^n]^*$$

where * indicates the transposed conjugate[1–3] (transjugate) of $\mathbf{T}_p^n$, and where $K = p^n$,

(iv) it has the recursive structure for the ternary case of

$$\begin{bmatrix} \mathbf{T}_p^{n-1} & \mathbf{T}_p^{n-1} & \mathbf{T}_p^{n-1} \\ \mathbf{T}_p^{n-1} & a\mathbf{T}_p^{n-1} & a^2\mathbf{T}_p^{n-1} \\ \mathbf{T}_p^{n-1} & a^2\mathbf{T}_p^{n-1} & a\mathbf{T}_p^{n-1} \end{bmatrix}$$

and similarly for all p, $\mathbf{T}_p^0 \triangleq +1.0$.

When $p = 2$, this will be recognised as the Hadamard matrix discussed earlier; in particular, see Eqs. (2.1)–(2.8) inclusive. Note also that for the binary case, the transposed conjugate of $\mathbf{T}_p^n$ is merely the transpose of $\mathbf{T}_p^n$, since only real positive or real negative numbers are involved. See also Appendix C.

An alternative definition as follows for the Chrestenson functions yields the same complete set, but in dissimilar order:

$$\mathrm{Ch}^{(p)}(j,k) = \exp\left\{\frac{2\pi}{p}\sqrt{-1}\,C(j,k)\right\}$$
$$C(j,k) = \sum_{\eta=0}^{n-1} j_\eta k_{n-1-\eta} \tag{2.48}$$

where j_η, k_η are the p-ary expansions of j, k as previously defined.

This definition gives the alternative transform matrix for $p = 3$, $n = 2$ shown in Table 2.8. Note that $k_{n-1-\eta}$ in Eq. (2.48) bit-reverses the p-ary expansion of k, since

$$k,\ k = k_{n-1}, k_{n-2}, \ldots, k_0$$

becomes

$$k',\ k' = k_0, k_1, \ldots, k_{n-1}$$

Hence the modified p-ary values across the matrix of Table 2.8 become

00 10 20 01 11 21 02 12 22

for the purpose of evaluating Eq. (2.48), rather than the p-ary values shown in parentheses in Table 2.7.

It will be noted that this reordered Chrestenson function matrix remains symmetric, but does not possess the recursive structure of the former

Table 2.7. The complete orthogonal transform for the ternary two-variable case.[a]

$k =$	0	1	2	3	4	5	6	7	8
	(00)	(01)	(02)	(10)	(11)	(12)	(20)	(21)	(22)
$j =$ 0(00)	1	1	1	1	1	1	1	1	1
1(01)	1	a	a^2	1	a	a^2	1	a	a^2
2(02)	1	a^2	a	1	a^2	a	1	a^2	a
3(10)	1	1	1	a	a	a	a^2	a^2	a^2
4(11)	1	a	a^2	a	a^2	1	a^2	1	a
5(12)	1	a^2	a	a	1	a^2	a^2	a	1
6(20)	1	1	1	a^2	a^2	a^2	a	a	a
7(21)	1	a	a^2	a^2	1	a	a	a^2	1
8(22)	a	a^2	a	a^2	a	1	a	1	a^2

[a] Based upon the Chrestenson functions of Eq. (2.47), Fig. 2.10.

ordering. Since it is symmetric, $C(j, k)$ in Eq. (2.48) may equally be written $\sum_{\eta=0}^{n-1} j_{n-1-\eta}k_{\eta}$. For the $p = 2$ (binary) case, it may readily be seen that Eq. (2.48) is identical to the Walsh functions defined in Section 2.3: it is left as an easy exercise for the reader to confirm that the waveforms shown in Fig. 2.2 are generated by $p = 2, n = 3$, in Eq. (2.48).

Table 2.8. The alternative row order for $p = 3$, $n = 2$ produced by Eq. (2.48).[a]

$k =$	0	1	2	3	4	5	6	7	8
$k' =$	(00)	(10)	(20)	(01)	(11)	(21)	(02)	(12)	(22)
$j =$ 0(00)	1	1	1	1	1	1	1	1	1
1(01)	1	1	1	a	a	a	a^2	a^2	a^2
2(02)	1	1	1	a^2	a^2	a^2	a	a	a
3(10)	1	a	a^2	1	a	a^2	1	a	a^2
4(11)	1	a	a^2	a	a^2	1	a^2	1	a
5(12)	1	a	a^2	a^2	1	a	a	a^2	1
6(20)	1	a^2	a	1	a^2	a	1	a^2	a
7(21)	1	a^2	a	a	1	a^2	a^2	a	1
8(22)	1	a^2	a	a^2	a	1	a	1	a^2

[a] Compare with Table 2.7.

The orthogonal Chrestenson functions therefore represent the general case of the Hadamard and Walsh binary functions considered earlier. Like the Hadamard and the Walsh functions, the Chrestenson functions may be defined in several dissimilar but functionally identical ways, which can yield different row orders in the complete orthogonal set.

Finally, we may also consider a subset of the Chrestenson functions for any p, n, which constitute the generalisation of the Rademacher functions in Section 2.3.1, and from which the complete set of orthogonal functions for the given p, n can be generated by element-by-element multiplication. The generalised Rademacher functions are defined by

$$\begin{aligned} \mathrm{Rad}^{(p)}(j,k) &= \exp\left\{\frac{2\pi}{p}\sqrt{-1}C(j',k)\right\} \\ C(j',k) &= \sum_{\eta=0}^{n-1} j'_\eta k_\eta \end{aligned} \tag{2.49}$$

compare Eq. (2.47), where k_η is as before, but where j'_η is here a subset of j whose decimal identification numbers are 0, 1 and all higher values of j that are divisible by a power of p. For $p = 3$, $n = 2$, we therefore have the functions illustrated in Fig. 2.11. The closed set of Chrestenson functions for $p = 3$, $n = 2$ given in Fig. 2.10 is generated from this reduced set by the element-by-element complex number multiplications:

$$\mathrm{Ch}^{(3)}(0,k) = \mathrm{Rad}^{(3)}(0,k)$$

$$\mathrm{Ch}^{(3)}(1,k) = \mathrm{Rad}^{(3)}(1,k)$$

$$\mathrm{Ch}^{(3)}(2,k) = \mathrm{Rad}^{(3)}(1,k) \times \mathrm{Rad}^{(3)}(1,k)$$

$$\mathrm{Ch}^{(3)}(3,k) = \mathrm{Rad}^{(3)}(3,k)$$

$$\mathrm{Ch}^{(3)}(4,k) = \mathrm{Rad}^{(3)}(1,k) \times \mathrm{Rad}^{(3)}(3,k)$$

$$\mathrm{Ch}^{(3)}(5,k) = \mathrm{Rad}^{(3)}(1,k) \times \mathrm{Rad}^{(3)}(1,k) \times \mathrm{Rad}^{(3)}(3,k)$$

$$\mathrm{Ch}^{(3)}(6,k) = \mathrm{Rad}^{(3)}(3,k) \times \mathrm{Rad}^{(3)}(3,k)$$

$$\mathrm{Ch}^{(3)}(7,k) = \mathrm{Rad}^{(3)}(1,k) \times \mathrm{Rad}^{(3)}(3,k) \times \mathrm{Rad}^{(3)}(3,k)$$

$$\mathrm{Ch}^{(3)}(8,k) = \mathrm{Rad}^{(3)}(1,k) \times \mathrm{Rad}^{(3)}(1,k) \times \mathrm{Rad}^{(3)}(3,k) \times \mathrm{Rad}^{(3)}(3,k)$$

Further details of Chrestenson functions, their generation and their properties may be found in available literature.[1,2,5,48–57]

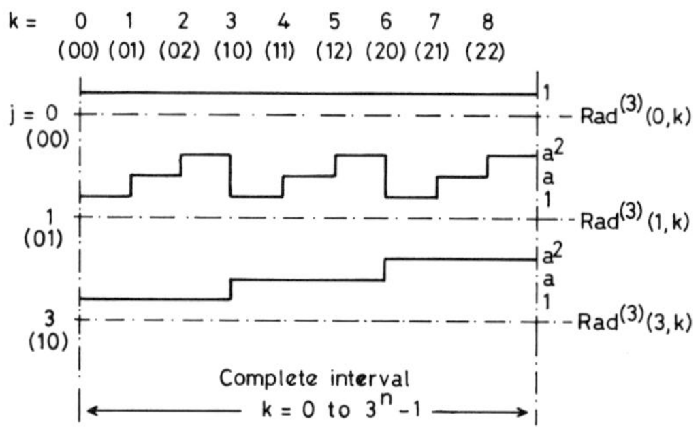

Fig. 2.11. The incomplete orthogonal set of generalised Rademacher functions for $p = 3$, $n = 2$, based upon Eq. (2.49).

2.7.3 The Generalised Haar Functions

The generalised Haar functions for any p, n are defined as follows, where k is taken over the continuous interval 0 to 1:

$$H_0^{0,0}(k) \triangleq +1.0$$

$$\begin{aligned} H_i^{q,r}(k) &= |\sqrt{p}|^{i-1} \exp\left\{\frac{2\pi}{p}.0.r\right\} && \text{for} \quad \frac{q}{p^{i-1}} \le k < \frac{q+(1/p)}{p^{i-1}} \\ &= |\sqrt{p}|^{i-1} \exp\left\{\frac{2\pi}{p}.1.r\right\} && \text{for} \quad \frac{q+(1/p)}{p^{i-1}} \le k < \frac{q+(2/p)}{p^{i-1}} \\ &\vdots \\ &= |\sqrt{p}|^{i-1} \exp\left\{\frac{2\pi}{p}(p-1)r\right\} && \text{for} \quad \frac{q+((p-1)/p)}{p^{i-1}} \le k < \frac{q+1}{p^{i-1}} \\ &= 0 \text{ at all other points} \end{aligned} \tag{2.50}$$

where $i = 1, 2, \ldots, n$; $q = 0, 1, \ldots, p^{i-1} - 1$; $r = 1, \ldots. p - 1$. For the $p = 2$ (binary) case, these equations collapse back to that given in Eq. (2.21).

The Haar functions for $p = 3, n = 2$ are illustrated in Fig. 2.12; compare the binary waveforms given in Fig. 2.5. Note that the denominator p^{i-1} in the divisions of each $H_i^{q,r}(k)$ subdivides k in the appropriate equal-length segments of k for any p, n, i, as shown in Figs. 2.5 and 2.12.

Discrete sampling of the waveforms of Fig. 2.12 gives the $2^3 \times 2^3$ complete orthogonal matrix $\mathbf{T}_p^n$ of Table 2.9. All the properties previously considered for the binary case carry over to the generalised case. In particular it will be

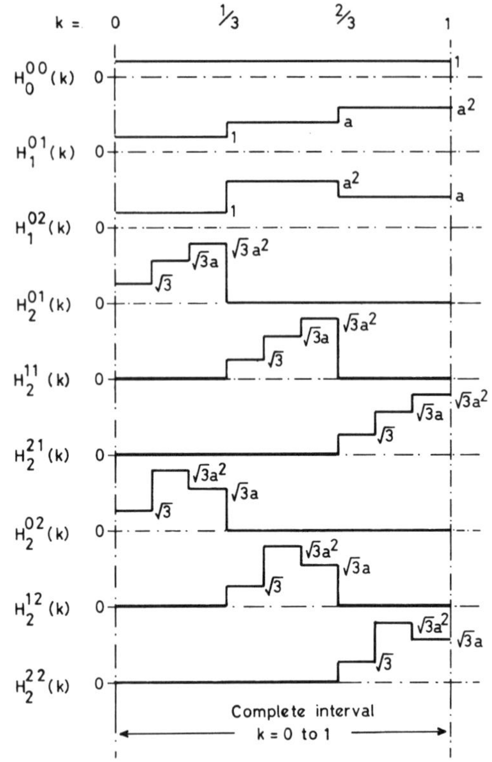

Fig. 2.12. The closed orthogonal set of Haar functions for $p = 3$, $n = 2$ based upon Eq. (2.50).

Table 2.9. The complete orthogonal Haar transform for the ternary two-variable case.[a]

1	1	1	1	1	1	1	1	1
1	1	1	a	a	a	a^2	a^2	a^2
1	1	1	a^2	a^2	a^2	a	a	a
$\sqrt{3}$	$\sqrt{3}a$	$\sqrt{3}a^2$	0	0	0	0	0	0
0	0	0	$\sqrt{3}$	$\sqrt{3}a$	$\sqrt{3}a^2$	0	0	0
0	0	0	0	0	0	$\sqrt{3}$	$\sqrt{3}a$	$\sqrt{3}a^2$
$\sqrt{3}$	$\sqrt{3}a^2$	$\sqrt{3}a$	0	0	0	0	0	0
0	0	0	$\sqrt{3}$	$\sqrt{3}a^2$	$\sqrt{3}a$	0	0	0
0	0	0	0	0	0	$\sqrt{3}$	$\sqrt{3}a^2$	$\sqrt{3}a$

[a] See Eq. (2.50), Fig. 2.12.

Table 2.10. The complete orthogonal Haar transform for the quaternary $n = 2$ case.

$$
\begin{bmatrix}
1 & 1 & 1 & 1 & 1 & 1 & 1 & 1 & 1 & 1 & 1 & 1 & 1 & 1 & 1 & 1 \\
1 & 1 & 1 & 1 & j & j & j & j & -1 & -1 & -1 & -1 & -j & -j & -j & -j \\
1 & 1 & 1 & 1 & -1 & -1 & -1 & -1 & 1 & 1 & 1 & 1 & -1 & -1 & -1 & -1 \\
1 & 1 & 1 & 1 & -j & -j & -j & -j & -1 & -1 & -1 & -1 & j & j & j & j \\
2 & 2j & -2 & -2j & 0 & 0 & 0 & 0 & 0 & 0 & 0 & 0 & 0 & 0 & 0 & 0 \\
0 & 0 & 0 & 0 & 2 & 2j & -2 & -2j & 0 & 0 & 0 & 0 & 0 & 0 & 0 & 0 \\
0 & 0 & 0 & 0 & 0 & 0 & 0 & 0 & 2 & 2j & -2 & -2j & 0 & 0 & 0 & 0 \\
0 & 0 & 0 & 0 & 0 & 0 & 0 & 0 & 0 & 0 & 0 & 0 & 2 & 2j & -2 & -2j \\
2 & -2 & 2 & -2 & 0 & 0 & 0 & 0 & 0 & 0 & 0 & 0 & 0 & 0 & 0 & 0 \\
0 & 0 & 0 & 0 & 2 & -2 & 2 & -2 & 0 & 0 & 0 & 0 & 0 & 0 & 0 & 0 \\
0 & 0 & 0 & 0 & 0 & 0 & 0 & 0 & 2 & -2 & 2 & -2 & 0 & 0 & 0 & 0 \\
0 & 0 & 0 & 0 & 0 & 0 & 0 & 0 & 0 & 0 & 0 & 0 & 2 & -2 & 2 & -2 \\
2 & -2j & -2 & 2j & 0 & 0 & 0 & 0 & 0 & 0 & 0 & 0 & 0 & 0 & 0 & 0 \\
0 & 0 & 0 & 0 & 2 & -2j & -2 & 2j & 0 & 0 & 0 & 0 & 0 & 0 & 0 & 0 \\
0 & 0 & 0 & 0 & 0 & 0 & 0 & 0 & 2 & -2j & -2 & 2j & 0 & 0 & 0 & 0 \\
0 & 0 & 0 & 0 & 0 & 0 & 0 & 0 & 0 & 0 & 0 & 0 & 2 & -2j & -2 & 2j
\end{bmatrix}
$$

observed that the Haar transform is globally sensitive for the first p of the p^n row vectors, but locally sensitive for all subsequent vectors.

The simple structure of the resulting Haar transform means that the transform for any p, n can be readily compiled. For example, for $p = 4$ (quaternary), $n = 2$, we have the $2^4 \times 2^4$ matrix of Table 2.10, where j *is* here equal to $\sqrt{-1}$.

2.7.4 The Spectrum of p-Valued Functions

The p-valued transforms may be used to transform any p-valued vector $\mathbf{Y}$ into its unique spectrum

$$\mathbf{T}_p^n \mathbf{Y} = \mathbf{S}$$

where the resulting values in $\mathbf{S}$ will be p-complex. The inverse

$$[\mathbf{T}_p^n]^{-1} \mathbf{S} = \mathbf{Y}$$

where $[\mathbf{T}_p^n]^{-1} = \dfrac{1}{p^n} [\mathbf{T}_p^n]^*$ (see Section 2.7.2) is immediately available.

However, in the binary considerations of earlier sections, we particularly related the row vectors of $\mathbf{T}_p^n$ ($\mathbf{T}^n$) to the x_i input variables, and also the resulting spectral coefficients to a correlation measurement between a given function and the row vectors of $\mathbf{T}_p^n$ ($\mathbf{T}^n$). Let us extend this logical interpretation to the general p-valued case.

The generalised Chrestenson functions of Section 2.7.1 have an immediate relationship to the x_i inputs of an n-variable, p-valued system. For example, consider the ternary transform of Table 2.7. Examination of the row entries

Table 2.11. The normalised forward and inverse Haar transforms for $p = 3$, $n = 2$.[a]

(a) Forward transform $\mathbf{T}_p^n$

$$
\begin{bmatrix}
1 & 1 & 1 & 1 & 1 & 1 & 1 & 1 & 1 \\
1 & 1 & 1 & a & a & a & a^2 & a^2 & a^2 \\
1 & 1 & 1 & a^2 & a^2 & a^2 & a & a & a \\
1 & a & a^2 & 0 & 0 & 0 & 0 & 0 & 0 \\
0 & 0 & 0 & 1 & a & a^2 & 0 & 0 & 0 \\
0 & 0 & 0 & 0 & 0 & 0 & 1 & a & a^2 \\
1 & a^2 & a & 0 & 0 & 0 & 0 & 0 & 0 \\
0 & 0 & 0 & 1 & a^2 & a & 0 & 0 & 0 \\
0 & 0 & 0 & 0 & 0 & 0 & 1 & a^2 & a
\end{bmatrix}
$$

(b) Inverse transform $[\mathbf{T}_p^n]^{-1}$

$$
\begin{bmatrix}
1 & 1 & 1 & 3 & 0 & 0 & 3 & 0 & 0 \\
1 & 1 & 1 & 3a^2 & 0 & 0 & 3a & 0 & 0 \\
1 & 1 & 1 & 3a & 0 & 0 & 3a^2 & 0 & 0 \\
1 & a^2 & a & 0 & 3 & 0 & 0 & 3 & 0 \\
1 & a^2 & a & 0 & 3a^2 & 0 & 0 & 3a & 0 \\
1 & a^2 & a & 0 & 3a & 0 & 0 & 3a^2 & 0 \\
1 & a & a^2 & 0 & 0 & 3 & 0 & 0 & 3 \\
1 & a & a^2 & 0 & 0 & 3a^2 & 0 & 0 & 3a \\
1 & a & a^2 & 0 & 0 & 3a & 0 & 0 & 3a^2
\end{bmatrix}
$$

[a] Compare Table 2.9.

will show the following relationships when the logic values $\{0, 1, 2\}$ are recoded $\{1, a, a^2\}$ as in Fig. 2.9:

Row vector	Logical equivalence
$j = 0$	constant x_0 (logic 0)
1	input variable x_1
2	function $x_1 \oplus x_1$
3	input variable x_2
4	function $x_1 \oplus x_2$
5	function $x_1 \oplus x_1 \oplus x_2$
6	function $x_2 \oplus x_2$
7	function $x_1 \oplus x_2 \oplus x_2$
8	function $x_1 \oplus x_1 \oplus x_2 \oplus x_2$

Note that the mod_p addition operator $\oplus$ in the logical $\{0, 1, 2\}$ domain is equivalent to mod_p multiplication in the recoded $\{1, a, a^2\}$ domain. This may be compared directly with the binary case, see Eq. (2.15), of Section 2.2.3.

The row vectors of Table 2.8 correspond to an appropriate reordering of the preceding tabulation of logical equivalence meanings.

The logical meaning of the Haar transform vectors, however, first involves rescaling, as has been seen for the restricted binary case in Section 2.3.3. Considering, for example, the rows of Table 2.9, there are six numerical values within this complete ternary matrix; in general the complete matrix for any p, n will contain n times as many numerical values as the system radix p. However, if we remove the scaling factor $|\sqrt{p}|^{i-1}$ of the defining equations (2.50), then for $p = 3, n = 2$, we obtain the normalised Haar transform given in Table 2.11(a). The logical equivalence of each row vector is now

Modified Haar row vector	Logical equivalence
$j = 0$	constant x_0 (logic 0)
1	input variable x_2
2	function $x_2 \oplus x_2$
3	input variable x_1 within $x_2 = 0$
4	input variable x_1 within $x_2 = 1$
5	input variable x_1 within $x_2 = 2$
6	function $x_1 \oplus x_1$ within $x_2 = 0$
7	function $x_1 \oplus x_1$ within $x_2 = 1$
8	function $x_1 \oplus x_1$ within $x_2 = 2$

However, to maintain $[\mathbf{T}_p^n][\mathbf{T}_p^n]^{-1} = \mathbf{I}$, the inverse transform must compensate for the removal of these scaling factors from the forward transform, as in Table 2.11(b). The corresponding binary case will be recalled in Table 2.3 (Section 2.3.3).

Turning to the meaning of the resulting spectral values in **S**, should we require a maximum value in a given spectral coefficient to indicate exact agreement of the given function $f(X)$ with one row of the transform $\mathbf{T}_p^n$, this being a key feature in all our binary developments, we have the following considerations.

Consider a given function $f(X) = $ logic 0. Recoding to the usual function values in vector **Y**, we have for any of the transforms previously detailed:

$$\underset{\mathbf{T}_p^n}{\begin{bmatrix} 1 & 1 & 1 & 1 & 1 & \cdots \\ & & & & & \\ & & & & & \\ & & & & & \\ & & & & & \\ & & & & & \end{bmatrix}} \underset{\mathbf{Y}}{\begin{bmatrix} 1 \\ 1 \\ 1 \\ 1 \\ 1 \\ \vdots \end{bmatrix}} = \underset{\mathbf{S}}{\begin{bmatrix} p^n \\ 0 \\ 0 \\ 0 \\ 0 \\ \vdots \end{bmatrix}}$$

The maximum value of p^n indicates exact agreement. However, considering a ternary case, should $f(X) = $ logic 1, we now have

$$\begin{bmatrix} 1 & 1 & 1 & 1 & 1 & \cdots \\ & & & & & \\ & & & & & \\ & & & & & \\ & & & & & \\ & & & & & \\ & & & & & \end{bmatrix} \begin{bmatrix} a \\ a \\ a \\ a \\ a \\ \vdots \\ \end{bmatrix} = \begin{bmatrix} ap^n \\ 0 \\ 0 \\ 0 \\ 0 \\ \vdots \\ 0 \end{bmatrix}$$

Similarly, $f(X) = $ logic 2 will give a^2p^n as the first spectral coefficient value, all remaining coefficients being zero-valued. Hence, the modulus value p^n is the indicator of agreement between a fixed function and the first row of the transform for any p, n.

Continuing, should $f(X) = x_1$, then we anticipate a similar maximum-valued spectral coefficient, all other coefficients zero, assuming $\mathbf{T}_p^n$ contains the function x_1 as one of its row vectors. Taking the ternary transform of Table 2.7, we have

$$\begin{array}{r} x_0 \\ x_1 \\ x_1 \oplus x_1 \\ x_2 \\ x_1 \oplus x_2 \\ x_1 \oplus x_1 \oplus x_2 \\ x_2 \oplus x_2 \\ x_1 \oplus x_2 \oplus x_2 \\ x_1 \oplus x_1 \oplus x_2 \oplus x_2 \end{array} \begin{bmatrix} 1 & 1 & 1 & 1 & 1 & 1 & 1 & 1 & 1 \\ 1 & a & a^2 & 1 & a & a^2 & 1 & a & a^2 \\ 1 & a^2 & a & 1 & a^2 & a & 1 & a^2 & a \\ 1 & 1 & 1 & a & a & a & a^2 & a^2 & a^2 \\ 1 & a & a^2 & a & a^2 & 1 & a^2 & 1 & a \\ 1 & a^2 & a & a & 1 & a^2 & a^2 & a & 1 \\ 1 & 1 & 1 & a^2 & a^2 & a^2 & a & a & a \\ 1 & a & a^2 & a^2 & 1 & a & a & a^2 & 1 \\ 1 & a^2 & a & a^2 & a & 1 & a & 1 & a^2 \end{bmatrix} \begin{bmatrix} 1 \\ a \\ a^2 \\ 1 \\ a \\ a^2 \\ 1 \\ a \\ a^2 \end{bmatrix} = \begin{bmatrix} 0 \\ 0 \\ 9 \\ 0 \\ 0 \\ 0 \\ 0 \\ 0 \\ 0 \end{bmatrix}$$

$$\mathbf{T}_p^n \qquad \mathbf{Y} \quad = \mathbf{S}$$

Hence, recoding $f(X)$ from $\{0, 1, 2\}$ to $\{1, a, a^2\}$ in vector $\mathbf{Y}$ does not give the required result.

To obtain the required maximum value of p^n in the second spectral coefficient, zero values elsewhere, it is necessary that $\mathbf{Y}$ be recoded $\{1, a^2, a\}$ rather than $\{1, a, a^2\}$, that is, be the *complex conjugate* of the encoding of the forward transform matrix. This is consistent with the meaning of orthogonality of complex functions.

This requirement may be treated in two opposite ways, namely,

(i) we may define our forward transforms $\mathbf{T}_p^n$ as has been done in Sections 2.7.2 and 2.7.3, and then encode our function vector $\mathbf{Y}$ to be the complex conjugate of the logic coding given by Fig. 2.9, Eq. (2.46); or

(ii) we may retain the encoding of the function vector $\mathbf{Y}$ as in Fig. 2.9, Eq. (2.46), but define all entries in our forward transform $\mathbf{T}_p^n$ to the complex conjugates of our previous definitions.

If we adopt the latter method, then the inverse transform $[\mathbf{T}_p^n]^{-1}$ becomes the simple transpose of our previous forward-transform matrices. This distinction does not arise in the binary systems considered in the main body of this text, but only arises when $p \geq 3$. Care must be taken to appreciate which convention is being used in various publications,[5,29,51,53,56,58,59] and also whether the scaling factor of $1/p^n$ is incorporated in the forward transform $\mathbf{T}_p^n$ ($\mathbf{T}^n$), or the inverse transform $[\mathbf{T}_p^n]^{-1}$ ($[\mathbf{T}^n]^{-1}$).

As an illustration of two conventions, consider the arithmetic ternary function taken from Tokmen,[29] whose truthtable is shown in the following tabulation.

x_2	x_1	$f(X)$	$\mathbf{Y}$, direct encoding $\{1, a, a^2\}$	$\mathbf{Y}$, conjugate encoding $\{1, a^2, a\}$
0	0	2	a^2	a
0	1	1	a	a^2
0	2	0	1	1
1	0	1	a	a^2
1	1	0	1	1
1	2	0	1	1
2	0	0	1	1
2	1	0	1	1
2	2	0	1	1

First, using the Chrestenson transform of Table 2.7 with the conjugate recoding $\{1, a^2, a\}$ for $\mathbf{Y}$, and relegating the $1/3^2$ scaling factor to the inverse rather than the forward transform, execution of $\mathbf{T}_p^n \mathbf{Y} = \mathbf{S}$ gives the following spectral coefficients in $\mathbf{S}$, in descending order:†

$$a^2 + 5; \quad 2a^2 + a; \quad 2a^2 + 4a; \quad 2a^2 + a; \quad a^2 + 2; \quad 2a + 1; \quad 2a^2 + 4a;$$
$$2a + 1; \quad 5a + 1$$

Alternatively, using the conjugate of this transform, which has the recursive structure of

$$\mathbf{T}_p^n = \begin{bmatrix} \mathbf{T}_p^{n-1} & \mathbf{T}_p^{n-1} & \mathbf{T}_p^{n-1} \\ \mathbf{T}_p^{n-1} & a^2\mathbf{T}_p^{n-1} & a\mathbf{T}_p^{n-1} \\ \mathbf{T}_p^{n-1} & a\mathbf{T}_p^{n-1} & a^2\mathbf{T}_p^{n-1} \end{bmatrix}$$

with the direct encoding $\{1, a, a^2\}$ for $\mathbf{Y}$, and incorporating the $1/3^2$ scaling factor in the forward transform, we have the alternative spectral coefficients:

$$\frac{a+5}{9}; \quad \frac{a^2+2a}{9}; \quad \frac{4a^2+2a}{9}; \quad \frac{a^2+2a}{9}; \quad \frac{a+2}{9};$$
$$\frac{2a^2+1}{9}; \quad \frac{4a^2+2a}{9}; \quad \frac{2a^2+1}{9}; \quad \frac{5a^2+1}{9}$$

It is left as an exercise for the reader to confirm these coefficients.

† Recall $a + a = 2a$; $a^2 + a^2 = 2a^2$; $a^2 + a + 1 = 0$.

It is apparent that the simplicity of the spectral domain data for the binary case cannot be maintained as p increases. Nevertheless, all the principles and techniques investigated for the binary case have their generalisation to higher-valued logic. Reference to available literature will show the following aspects have received consideration:

(a) the classification of ternary functions by spectral means, particularly for linearly separable (threshold) functions;[29,50,59,60,62,63]

(b) spectral translation techniques for ternary function synthesis;[29,52,57,58,62,63] and

(c) spectral decomposition techniques for ternary function synthesis.[29,57,61]

The parallel between these aspects and the corresponding binary developments considered in the subsequent chapters of this volume may be found.

Whilst the majority of publications to date for multiple-valued logic, $p > 2$, have concentrated upon ternary developments, it may possibly be that $p = 4$ (quaternary) is a more practical situation for the next radix higher than binary. Compatibility considerations with existing binary expertise and peripherals are involved in such considerations. As far as is known, no work on quaternary spectral application has been published to date, but the use of quaternary signals in memory and encoding/decoding circuits has received consideration.[64–66]

2.8 CHAPTER SUMMARY

This chapter has covered the fundamental background information on orthogonal transforms and their use in generating spectral data, which will form the basis for subsequent chapters of this volume.

We have seen how conventional binary data, exemplified by Boolean algebraic expressions, or truthtables, or equivalent representations, may be transformed into the spectral domain, the coefficients of which may represent some global information on the function or functions being considered. This is in contrast to conventional functional–domain data, where the information is presented in discrete form, no information concerning the overall function or functions being available from an isolated item of data.

The close parallel between the Fourier series analysis and synthesis of periodic ac waveforms, involving orthogonal harmonically related sinusoids, and the analysis and synthesis of digital functions using orthogonal Walsh or other square-wave functions has also been noted. However, in the following chapters we shall not have occasion to use this aspect of the spectral domain to any great extent, the reasons for which will be given in the introduction of Chapter 4. Instead we shall largely concentrate upon using the information content of the individual spectral coefficients, and subsets of coefficients, in order to recognise features of the function(s) being considered, and from such features construct appropriate syntheses using a choice of digital logic gates. The power of the spectral methods will frequently mean that the Boolean domain constraint of AND and OR connectives is no longer present.

Finally, we should note that we are here considering only one area of application of the whole arena of Walsh, Haar and related orthogonal functions. Areas such as image coding, pattern recognition, data compression, Walsh filtering and other digital signal processing applications lie entirely outside our present considerations. For details of these other areas, see the various Proceedings on the Applications of Walsh Functions and other references.[1,12–15,18,26,31,35] Extensive bibliographies may be found, particularly in the former.

References

1. Ahmed, N., and Rao, K. R., "Orthogonal Transforms for Digital Signal Processing." Springer-Verlag, New York, 1975.
2. Hohn, F. E., "Elementary Matrix Algebra." Macmillan, New York, 1964.
3. Ayres, F., "Theory and Problems of Matrices." Shaum's Outline Series, McGraw-Hill, New York, 1962.
4. Rademacher, H., Einige Sätze über Reihen von allgemeinen orthogonal funktionen, *Math. Ann.* **87,** 112–138 (1922).
5. Karpovsky, M. G., "Finite Orthogonal Series in the Design of Digital Devices." Wiley, New York, 1976.
6. Hurst, S. L., "The Logical Processing of Digital Signals." Crane-Russak, New York and Edward Arnold, London, 1978.
7. Dertouzos, M. L., "Threshold Logic: A Synthesis Approach." MIT Press, Cambridge, Massachusetts, 1965.
8. Lechner, R. J., Harmonic analysis of switching functions, *in* "Recent Developments in Switching Theory" (A. Mukhopadhay, ed.). Academic Press, New York, 1971.
9. Edwards, C. R., Matrix methods in combinational logic design, Ph.D. thesis, University of Bath, United Kingdom, 1973.
10. Wallis, J. S., "Hadamard matrices," (Lecture Notes No. 292), Springer-Verlag. New York, 1972.
11. Walsh, J. L., A closed set of orthogonal functions, *Amer. J. Math.* **45,** 5–24 (1923).
12. Harmuth, H. F., "Transmission of Information by Orthogonal Functions." Springer, New York, 1972.
13. Pichler, F., Walsh functions and linear system theory, *Proc. Symp. Applic. Walsh Functions, Washington, D.C.*, 175–182 (1970).

14. Schreiber, H. G., and Sandy, F. (eds.), "Application of Walsh Functions and Sequency Theory." IEEE, New York, 1974.
15. Beauchamp, K. G., "Applications of Walsh and Related Functions." Academic Press, London, 1984.
16. Ulman, J. L., Computation of the Hadamard transform and the *R* transform in ordered form, *Trans. IEEE* **EC-19,** 359–360 (1970).
17. Lloyd, A. M., A consideration of orthogonal matrices, other than the Rademacher–Walsh types, for the synthesis of digital networks, *Internat. J. Electron.* **47,** 205–212 (1979).
18. Clarke, C. K. P., Hadamard transformation: Walsh spectral analysis of television signals, British Broadcasting Corporation Research Department of Engineering, Division Report No. BBC/RD/1975/26, London, 1975.
19. Tzafestas, S., Frangakis, G., and Pimenidis, T., Global Walsh function generators, *Electron. Engin.* **48,** 45–49 (1976).
20. Whiteman, A. L., An infinite family of Hadamard matrices of Williamson type, *J. Comb. Theory Ser. A* **14,** 334–340 (1973).
21. Paley, R. E. A. C., A remarkable series of orthogonal functions, *Proc. London Math. Soc.* **34**(2), 241–279 (1932).
22. Haar, A., Zur Theorie der orthogonalen Funktionensysteme, *Math. Ann.* **69,** 331–371 (1914).
23. Hurst, S. L., The Haar transform in digital network synthesis, *Proc. IEEE 11th Internat. Symp. Multiple-Valued Logic*, 10–18 (1981).
24. Kremer, H., Algorithms for the Haar functions and the fast Haar transform, *Proc. Symp. Theory Applic. Walsh Functions, Hatfield, United Kingdom*, 1–13 (1971).
25. Rao, K. R., Narasimham, M. A., and Revuluri, K., A family of discrete Haar transforms, *Comput. Electr. Engrg.* **2,** 367–368 (1975).
26. Wendling, S., Gagneux, G., and Stamon, G., Use of the Haar transform and some of its properties in character recognition, *Proc. Internat. Joint Conf. Pattern Recognition, Colorado*, 844–848 (1976).
27. Durrani, T. S., and Nightingale, J. M., Sequential generation of binary orthogonal functions, *IEE Electron. Lett.* **13,** 377–380 (1971).
28. Davies, A. C., On the definition and generation of Walsh functions, *Trans. IEEE* **C-21,** 187–189 (1972).
29. Tokmen, V. H., An investigation into the properties of multi-valued spectral logic, Ph.D. thesis, University of Bath, United Kingdom, 1980.
30. Langheld, E., and Hurst, S. L., Die spektrale Darstellung binärer Logikfunctionen, *Elektronik* **No. 13,** 61–66; **No. 14,** 69–74 (1981).
31. Beauchamp, K. G., "Walsh Functions and Their Applications." Academic Press, London, 1975.
32. Shanks, J., Computation of the fast Walsh–Fourier transform, *Trans. IEEE* **EC-18,** 457–459 (1969).
33. Muzio, J. C., and Hurst, S. L., The computation of complete and reduced sets of orthogonal spectral coefficients for logic design and pattern recognition purposes, *Comput. and Electr. Engrg.* **5,** 231–249 (1978).
34. Cooley, J. W., and Tukey, J. W., An algorithm for the machine calculation of complex Fourier series, *Math. Comp.* **19,** 297–301 (1965).
35. Pratt, W. K., Andrews, H. C., and Kane, J., Hadamard transform image processing, *Proc. IEEE* **57,** 58–68 (1969).
36. Andrews, H. C., and Caspari, K. C., A generalized technique for spectral analysis, *Trans. IEEE* **C-19,** 16–25 (1970).

37. Geadah, Y. A., and Corinthios, M. J. G., Natural dyadic and sequency order algorithms and processors for the Walsh–Hadamard transform, *Trans. IEEE* **C-26,** 435–442 (1977).
38. Kremer, H., Algorithms for the Haar functions and the fast Haar transform, *Proc. Symp. Theory Applic. Walsh Functions, Hatfield, United Kingdom,* (1971).
39. Muzio, J. C., Concerning low-order spectral coefficients, *IEE Comp. Dig. Tech.* **2,** 179–202 (1979).
40. Muzio, J. C., The evaluation of large magnitude orthogonal spectral coefficients, University of Manitoba Scientific Report No. 87, Winnipeg, 1977.
41. Eris, E., Relationships between Rademacher–Walsh spectra of Boolean functions, *IEE Comp. Dig. Tech.* **1,** 45–48 (1978).
42. Muzio, J. C., Evaluation of the spectra of sum and product functions, *IEE Comp. Dig. Tech.* **1,** 113–118 (1978).
43. Muzio, J. C., Composite spectra and the analysis of switching circuits, *Trans. IEEE* **C-29,** 750–753 (1980).
44. Pichler, F. R., On discrete dyadic systems, *Proc. Symp. Theory Applic. Walsh Functions, Hatfield, United Kingdom,* 1–17 (1971).
45. Hurst, S. L., The application of Chow parameters and Rademacher–Walsh matrices in the synthesis of binary functions. *Comput. J.* **16,** 165–173 (1973).
46. Winder, R. O., Threshold functions through $n \leq 7$, Scientific Report No. 7, Air Force Cambridge Research Laboratory, Contract AF19 (604)-8423, Cambridge, Massachusetts, October 1969.
47. Muzio, J. C., Non-orthogonal transforms for logical design, Technical Report CS.77006-R, Virginia Polytechnic Institute and State University, Blacksburg, Virginia, February 1977.
48. Chrestenson, H. E., A class of generalized Walsh functions, *Pacific J. Math.* **5,** 17–31 (1955).
49. Moraga, C., Ternary spectral logic, *Proc. IEEE Seventh Internat. Multiple-Valued Logic Symp.* 7–12 (1977).
50. Moraga, C., Complex spectral logic, *Proc. IEEE Eighth Internat. Multiple-Valued Logic Symp.* 149–156 (1978).
51. Moraga, C., On a property of the Chrestenson spectrum, Report No. AIUD/MVL/8102, University of Dortmund, Dortmund, Federal Republic of Germany, 1981.
52. Tokmen, V. H., Some properties of the spectra of ternary logic functions, *Proc. IEEE Ninth Internat. Multiple-Valued Logic Symp.*, 88–93 (1979).
53. Muzio, J. C., Concerning transforms for three-valued systems, University of Manitoba Technical Report CS.77001/R, Winnipeg, 1977.
54. Hurst, S. L., An engineering consideration of spectral transforms for ternary logic synthesis, *Comput. J.* **22,** 173–183 (1979).
55. Liebler, M. E., and Roesser, R. P., Multiple real-valued Walsh functions, *Proc. IEEE Theory Applic. Multiple-Valued Logic Design,* 84–102 (1971).
56. Kitahasi, T., and Tanaka, A., Orthogonal expansion of many-valued logical functions and its application to their realization with single-threshold element. *Trans. IEEE* **C-21,** 211–218 (1972).
57. Tokmen, V. H., Disjoint decomposability of multi-valued functions by spectral means, *Proc. IEEE Tenth Internat. Multiple-Valued Logic Symp.*, 88–93 (1980).
58. Moraga, C., Introducing disjoint spectral translation in multiple-valued logic design, *Electron. Lett.* **14,** 241–243 (1978).
59. Moraga, C., Spectral characterisation of ternary threshold functions, *Electron. Lett.* **15,** 712–713 (1979).
60. Moraga, C., Characterisation of ternary threshold functions using a partial spectrum, *Electron. Lett.* **15,** 803–805 (1979).

61. Tokmen, V. H., Evaluation of the spectrum of multiple-valued logic networks, *Comput. Electr. Engrg.* **6,** 233–237 (1979).
62. Moraga, C., Ternary spectral logic, Report No. AIUD/23/76, University of Dortmund, Dortmund, Federal Republic of Germany, 1976.
63. Moraga, C., Spectral logic design, Report No. AIUD 57/78, University of Dortmund, Dortmund, Federal Republic of Germany, 1978.
64. Edwards, C. R., I^2L threshold circuits for binary/quaternary encoding and decoding, *Internat. J. Electron.* **44,** 445–448 (1978).
65. Tull, M. P., and Lee, S. C., A new method of realizing parallel processing machines using multiple-valued logic, *Proc. IEEE Tenth Internat. Symp. Multiple-Valued Logic*, 36–44 (1980).
66. Wheaton, L. B., and Wayne Current, K., A threshold logic modulo-four multiplier circuit for residue number system nonrecursive digital filters, *Proc. IEEE 11th Internat. Symp. Multiple-Value Logic*, 48–53 (1981).

3

The Classification of Boolean Functions

3.1 INTRODUCTION

The number of possible different functions of n independent input variables is 2^{2^n}, including degenerate functions.† Clearly this becomes a very large number for n greater than, say, three, and certainly too large to be individually listed in any algebraic or similar format.

The normal method of defining any particular function $f(X)$ is either by means of some Boolean expression or by a truthtable with 2^n entries to define the individual relationships between each input word (minterm) and the function output. There is, however, an alternative means for defining a given function, which is very compact, but requires expansion to define in detail the function specification. This is the octal numbering method of Hellerman,[1] which can be modified to a higher radix such as hexadecimal if desired.

As a simple example, consider the three-variable function

$$f(X) = \bar{x}_1\bar{x}_2 + x_1x_2 + x_2\bar{x}_3$$

whose truthtable is given in detail in Table 3.1. If we now transpose the output vector $f(X)$ to a string of numbers, with the value at minterm m_0 as the least-significant digit, we have

$$10 \quad 011 \quad 101$$

Reading this as an octal number, we may simply define the given function $f(X)$ as function No. 235.

It is clear that for $n = 3$, the 256 possible functions range from function

† Degenerate functions of n input variables are functions that do not depend upon all n inputs to determine the function output, that is, one or more are redundant. Extreme degenerate functions are merely $f(X) = 0$ or 1.

No. 0 to function No. 377, with certain numbers such as 188 being outside the meaningful list. It is equally clear that we may partition the output vector $f(X)$ into, say, a hexadecimal number rather than an octal, which in the simple case given in Table 3.1 will give us $f(X) =$ function No. 9D.

There are, however, a number of functions that have an identical realising network, the relationship between them being, for example, some permutation of the x_i input variables. Considering the example of Table 3.1, should inputs x_2 and x_3 be interchanged into the network that realises this function, then we obtain the alternative function

$$\begin{aligned} f(X) &= \bar{x}_1\bar{x}_3 + x_1x_3 + \bar{x}_2x_3, \\ &= \text{octal function No. 265} \end{aligned}$$

Thus it will be apparent that whilst the octal (or hexadecimal) functional numbering method provides a compact means of defining any given function, it does not give any direct indication of functions of similar structure or complexity.

Some means of classification of "similar" functions, however, is useful from several aspects. First, there is the increased understanding of functions that have essentially identical circuit realisations, leading to the classification of all 2^{2^n} functions of $\leq n$ variables in some compact manner. This may then have practical implications in that testing and fault diagnosis procedures may be standardised for each classification entry. Second, there is the possibility of establishing a small set of "standard functions" or "prototype functions," from which any particular function may be realised by the implementation of appropriate operations corresponding to the classification procedure. This in turn leads on to the consideration of universal logic elements, from which all the prototype functions and hence all possible

Table 3.1. Three-variable function $f(X) = \bar{x}_1\bar{x}_2 + x_1x_2 + x_2\bar{x}_3$.

Input			Minterm	Output
x_3	x_2	x_1	m_j	$f(X)$
0	0	0	m_0	1
0	0	1	m_1	0
0	1	0	m_2	1
0	1	1	m_3	1
1	0	0	m_4	1
1	0	1	m_5	0
1	1	0	m_6	0
1	1	1	m_7	1

2^{2^n} functions may be realised. We shall pursue the latter consideration in Section 3.7.

3.2 ALGEBRAIC CLASSIFICATION METHODS

The principal algebraic classification method involves the consideration of the following operations, taken individually or collectively:

(a) negation (N) of one or more of the x_i input variables of the function;
(b) permutation (P) of two or more of the x_i input variables of the function;
and
(c) negation (N) of the output of the function.

Functions that are equivalent under the three operations (a), (b) and (c) are said to be NPN-equivalent functions, or to belong to the same NPN-equivalence class. Functions that are equivalent under combinations of operations (a) and (b) only are said to be NP-equivalent functions, or in the same NP-equivalence class. Finally, N-equivalent functions (operation (a) only) and P-equivalent functions (operation (b) only) are similarly defined.

As a simple example, consider the following functions:

$$f_1(X) = x_1\bar{x}_2 + x_2x_3$$

$$f_2(X) = x_1x_2 + \bar{x}_2x_3$$

$$f_3(X) = x_1x_3 + x_2\bar{x}_3$$

$$f_4(X) = \bar{x}_1x_3 + \bar{x}_2\bar{x}_3$$

Functions $f_1(X)$ and $f_2(X)$ are in the same P-equivalence class, since they are related by the input permutation $x_1 \leftrightarrow x_3$; functions $f_1(X)$ and $f_3(X)$ are in the same NP-equivalence class, being related by the negation $x_2 \rightarrow \bar{x}_2$ and the permutation $x_2 \leftrightarrow x_3$, and, finally, $f_1(X)$ and $f_4(X)$ are in the same NPN-equivalence class, $f_4(X)$ being the overall negation of function $f_3(X)$. Thus all four functions belong to the same NPN-equivalence class.

It is clear that the NPN-equivalence class is the strongest and hence most compact of these three algebraic classifications. Equally, these simple examples show that it is not always easy to see from Boolean expressions the equivalence relationship that may exist between given functions, particularly if the functions are expressed in some alternative Boolean form. Hence some means of tabulation of all 2^{2^n} functions for any given n into their equivalence classes is desirable, but this becomes algebraically unmanageable for n greater than, say, 3. As will be shown later, spectral coefficients provide a more convenient, and also compact, form of classification in comparison with algebraic means.

Table 3.2. Classification statistics for functions of $\leq n$ variables.

	No. of input variables n					
	1	2	3	4	5	6
Total no. of functions of $\leq n$ variables	4	16	256	65536	$\simeq 4.3 \times 10^9$	$\simeq 1.8 \times 10^{19}$
Total no. of functions of exactly n variables	2	10	218	64594	$\simeq 4.3 \times 10^9$	$\simeq 1.8 \times 10^{19}$
No. of P-equivalent classes of functions of $\leq n$ variables	4	12	80	3984	$\simeq 3.7 \times 10^7$	—
No. of NP-equivalent classes of functions of $\leq n$ variables	3	6	22	402	1228158	$\simeq 4 \times 10^{14}$
No. of NPN-equivalent classes of functions of $\leq n$ variables	2	4	14	222	616126	$\simeq 2 \times 10^{14}$

The number of equivalence classes for functions of up to five variables has been extensively determined, although generally tabulated for only $n \leq 3$.[2–5] The basic statistics are given in Table 3.2; further details may be found in Muroga's work.[5] The rapid increase in the number of functions per classification class with n will be apparent.

The classification entry normally listed as the representative function of each class is the canonic, positive function for the class, that is, the input variables in the canonic order $x_1, \ldots, x_n$, all uncomplemented (positive) except in the P-classification case. For example, we have the following entries to cover all possible $n \leq 2$ functions.

P-classification representative functions:

$f(X) = 0$ (one function per entry)
$f(X) = 1$ (one function per entry)
$f(X) = x_1$ (two functions per entry)
$f(X) = \bar{x}_1$ (two functions per entry)
$f(X) = x_1 x_2$ (one function per entry)
$f(X) = x_1 \bar{x}_2$ (two functions per entry)
$f(X) = \bar{x}_1 \bar{x}_2$ (one function per entry)
$f(X) = x_1 + x_2$ (one function per entry)
$f(X) = x_1 + \bar{x}_2$ (two functions per entry)
$f(X) = \bar{x}_1 + \bar{x}_2$ (one function per entry)
$f(X) = x_1 \oplus x_2$ (one function per entry)
$f(X) = x_1 \oplus \bar{x}_2$ (one function per entry)

NP-classification representative functions:

$f(X) = 0$	(one function per entry)
$f(X) = 1$	(one function per entry)
$f(X) = x_1$	(four functions per entry)
$f(X) = x_1x_2$	(four functions per entry)
$f(X) = x_1 + x_2$	(four functions per entry)
$f(X) = x_1 \oplus x_2$	(two functions per entry)

NPN-classification representative functions:

$f(X) = 1$	(two functions per entry)
$f(X) = x_1$	(four functions per entry)
$f(X) = x_1 + x_2$	(eight functions per entry)
$f(X) = x_1 \oplus x_2$	(two functions per entry)

The corresponding NP and NPN algebraic listings for $n \leq 3$ may be found elsewhere,[4,5] but clearly the number of function entries to be listed is becoming prohibitive as n increases.

There is one further stronger algebraic classification which may be invoked, this being the SD (self-dualized) classification of Goto and Takahasi.[5,6] This classification was particularly relevant for the restricted class of linearly separable (threshold) functions, but is applicable to all functions whether linearly separable or not.

The definition of the SD classification first requires the introduction of dual and self-dual functions. The dual of any function $f(X)$ is given by

$$f^{\mathrm{D}}(X) \triangleq \overline{f(\bar{X})}$$

where the bar within the parentheses indicates that all the x_i variables of the function are individually complemented, whilst the bar over the whole function indicates the usual output function negation. For example, the dual of the function $f(X) = \bar{x}_1x_2 + \bar{x}_2x_3$ is $f^{\mathrm{D}}(X) = \overline{x_1\bar{x}_2 + x_2\bar{x}_3}$, which may be reorganized into some alternative form such as $(\bar{x}_1 + x_2)(\bar{x}_2 + x_3)$ if desired. Should $f^{\mathrm{D}}(X) \equiv f(X)$, then $f(X)$ is termed a self-dual function; $f(X) = x_1x_2 + x_2x_3 + x_1x_3$ is an example of such a function. In the more general case of $f^{\mathrm{D}}(X) \neq f(X)$, then $f(X)$ is a non-self-dual function.

However, if a self-dual function $f^{\mathrm{SD}}(X_{n+1})$ of $n + 1$ variables $x_1, \ldots, x_n, x_{n+1}$ is now considered, it may be decomposed into two n-variable functions about any x_i input variable, for example,

$$f_0(x_1, \ldots, x_n), \qquad x_{n+1} = 0$$

and

$$f_1(x_1, \ldots, x_n), \qquad x_{n+1} = 1$$

The property of a self-dual function, however, is that these two decomposition functions are duals of each other. Thus we have

$$f^{\mathrm{SD}}(X_{n+1}) \triangleq x_{n+1} f(X) + \overline{x_{n+1}} f^{\mathrm{D}}(X)$$

where $f(X) = f(x_1, \ldots, x_n)$. This relationship holds for the decomposition of the self-dual function about *any* x_i, $i = 1$ to $n + 1$.

Now a function $f(X)$ and its dual $f^{\mathrm{D}}(X)$ lie within the same NPN-equivalence classification, being related by input and output negation operations, but the self-dual function $f^{\mathrm{SD}}(X_{n+1})$, sometimes termed the hyperfunction $f^h(X_{n+1})$, lies above this NPN classification.

However, we may consider $f^{\mathrm{SD}}(X_{n+1})$ as being the representative function for all the NPN-equivalent functions obtained by the decomposition of $f^{\mathrm{SD}}(X_{n+1})$ about *each* of its x_i input variables, $i = 1$ to $n + 1$. Thus $f^{\mathrm{SD}}(X_{n+1})$ becomes a more compact algebraic classification than the preceding NPN-equivalence classification.

As an example, the two $n = 3$ canonic functions $f_1(X) = x_1 + x_2 + x_3$ and $f_2(X) = x_1(x_2 + x_3)$ do not belong to the same NPN classification. However, both belong to the same self-dualized classification, as may be shown by the development

$$\begin{aligned} f_1^{\mathrm{SD}}(X_{n+1}) &= (x_1 + x_2 + x_3) . x_4 + \overline{(\bar{x}_1 + \bar{x}_2 + \bar{x}_3)} . \bar{x}_4 \\ &= (x_1 + x_2 + x_3) x_4 + x_1 x_2 x_3, \end{aligned}$$

and

$$\begin{aligned} f_2^{\mathrm{SD}}(X_{n+1}) &= x_1(x_2 + x_3) . x_4 + \overline{\bar{x}_1(\bar{x}_2 + \bar{x}_3)} . \bar{x}_4 \\ &= x_1 x_2 x_4 + x_1 x_3 x_4 + (x_1 + x_2 x_3) . \bar{x}_4 \\ &= x_1 x_2 x_4 + x_1 x_3 x_4 + x_1 \bar{x}_4 + x_2 x_3 \bar{x}_4 \\ &= (x_2 + x_3 + \bar{x}_4) x_1 + x_2 x_3 \bar{x}_4 \end{aligned}$$

which under appropriate P and N operations becomes the same as $f_1^{\mathrm{SD}}(X_{n+1})$. Thus both functions may be classed under the same SD classification entry.†

† We shall briefly return to the SD classification in Section 3.6 when we have considered a numerical classification procedure covering all switching functions. We shall see that Exclusive-OR relationships are present in the SD classification, although this is not apparent from the normal algebraic development.

The canonic SD representative functions for $n \leq 3$ are

(i) x_1 (4 functions per entry)
(ii) $x_1x_2 + x_2x_3 + x_1x_3$ (32 functions per entry)
(iii) $x_1 \oplus x_2 \oplus x_3$ (8 functions per entry)
(iv) $(x_1 + x_2 + x_3)x_4 + x_1x_2x_3\bar{x}_4$ (128 functions per entry)
(v) $(x_1x_2x_3 + \bar{x}_1\bar{x}_2\bar{x}_3)x_4 + (x_1 + x_2 + x_3)(\bar{x}_1 + \bar{x}_2 + \bar{x}_3)\bar{x}_4$ (64 functions per entry)
(vi) $x_1(x_2x_3 + \bar{x}_2\bar{x}_3)x_4 + (x_1 + x_2\bar{x}_3 + \bar{x}_2x_3)\bar{x}_4$ (96 functions per entry)
(vii) $(\bar{x}_1x_2x_3 + x_1\bar{x}_2x_3 + x_1x_2\bar{x}_3)x_4 + (x_1x_2 + x_2x_3 + x_1x_3 + \bar{x}_1\bar{x}_2\bar{x}_3)\bar{x}_4$ (128 functions per entry)

Hence the 256 functions of $n \leq 3$ may be classified within these seven SD entries. Note that the total number of functions enumerated includes all self-dual functions of four variables, in addition to all $n \leq 3$ functions.

The practical significance of this final SD classification is that should we have a logic network that realises any one of these classification entries, then all functions covered by this representative function may be realised by the four operations of

(i) selectively applying logic 0 or 1 to one of the function inputs;
(ii) permutating and negating the given input variables $x_1, \ldots, x_n$ to the remaining function inputs, as necessary; and
(iii) negating the function output, if required.

This is illustrated in Fig. 3.1 together with the previous NPN-classification case. Note that there is still no one master or universal function from which all the functions of $\leq n$ input variables may be realised; consideration of a single universal function for any given n will be made in Section 3.7.

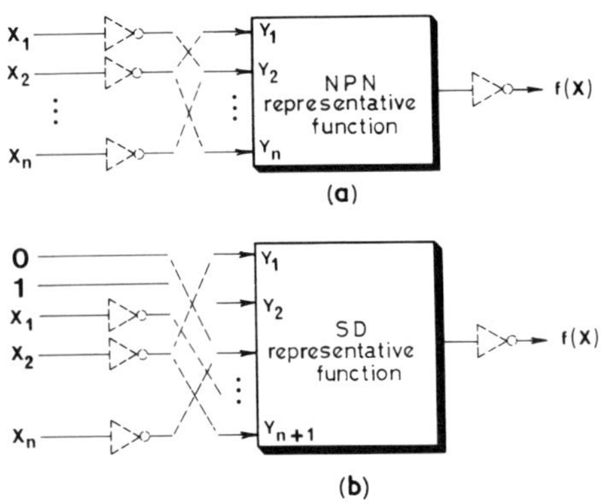

Fig. 3.1. The realisation of any function $f(X)$ from an appropriate representative function: (a) NPN-representative function; (b) SD-representative function.

Table 3.3. Equivalent P classification of all 256 functions of ≤3 variables in terms of the 80 canonic P-classified entries.[a]

	0	1	2	3	4	5	6	7
0	A000 *	A001	A002	A003 *	C002	C003	A006	A007
10	A010	A011	A012 *	A013	D012	D013	A016	A017 *
20	B002	B003	B006	B007	C006	C007	A026	A027
30	A030	A031	A032	A033	D032	D033	A036	A037
40	B010	B011	F012	F013	B030	B031	F032	F033
50	A050	A051	A052	A053	A054	A055	A056	A057
60	B012	B013	B016	B017	B032	B033	B036	B037
70	F054	F055	F056	F057	A074 *	A075	A076	A077 *
100	C010	C011	C030	C031	C012	C013	C032	C033
110	C050	C051	D054	D055	C052	C053	D056	D057
120	E012	E013	E032	E033	C016	C017	C036	C037
130	C054	C055	C074	C075	C056	C057	C076	C077
140	B050	B051	B054	B055	E054	E055	B074	B075
150	A150	A151	A152	A153	C152	C153	A156	A157
160	B052	B053	B056	B057	E056	E057	B076	B077
170	B152	B153	B156	B157	C156	C157	A176	A177
200	A200	A201	A202	A203	C202	C203	A206	A207
210	A210 *	A211	A212	A213	D212	D213	A216	A217
220	B202	B203	B206	B207	C206	C207	A226	A227
230	A230	A231 *	A232	A233	D232	D233	A236	A237
240	B210	B211	F212	F213	B230	B231	F232	F233
250	A250	A251	A252 *	A253	A254	A255	A256	A257 *
260	B212	B213	B216	B217	B232	B233	B236	B237
270	F254	F255	F256	F257	A274	A275	A276	A277
300	C210	C211	C230	C231	C212	C213	C232	C233
310	C250	C251	D254	D255	C252	C253	D256	D257
320	E212	E213	E232	E233	C216	C217	C236	C237
330	C254	C255	C274	C275	C256	C257	C276	C277
340	B250	B251	B254	B255	E254	E255	B274	B275
350	A350	A351	A352	A353	C352	C353	A356 *	A357
360	B252	B253	B256	B257	E256	E257	B276	B277
370	B352	B353	B356	B357	C356	C357	A376	A377 *

[a] A = canonic P-classified function, with inputs *a*, *b*, *c*; B = input permutation *b*, *c*, *a*; C = input permutation *c*, *a*, *b*; D = input permutation *a*, *c*, *b*; E = input permutation *c*, *b*, *a*; F = input permutation *b*, *a*, *c*.

Octal function 321, for example, is given at the intersection of row 320 and column 1, = P-classification E213. This means that function 321 is a P-variant on function 213, with input permutation E.

For clarity the P-classified functions themselves, i.e., A000 and A001, are underlined.

An asterisk marks all functions of less than three inputs.

Further details and statistics of these various levels of algebraic classification may be found.[4–7] The octal reference number for all 256 functions of $n \leq 3$ may be tabulated within the P classification, as shown in Table 3.3. Finally, the inconvenience of handling algebraic classification procedures beyond the relatively simple situations of $n \leq 3$ may be apparent from these discussions, and hence alternative means of classification, such as will be considered in the following sections, are particularly advantageous.

3.3 LINEARLY SEPARABLE (THRESHOLD) FUNCTIONS

Linearly separable or threshold functions form a particular subgroup of all functions for which a numerical as distinct from an algebraic classification procedure is particularly straightforward. However, threshold functions of n variables form a diminishingly small proportion of all functions of n variables as n increases (see Table 3.5), and hence have a somewhat restricted practical significance,† but in order to lead into the general case considered in the following sections, it is useful to first consider this special case.

Linearly separable functions are functions that, when their minterms are considered as 2^n equispaced nodes in n-dimensional space, possess a plane surface that unambiguously divides all true ($f(X) = 1$) nodes from all false ($f(X) = 0$) nodes. This is illustrated in Fig. 3.2a.

The equation of this plane in n-dimensional Euclidean space with axes $x_1, x_2, \ldots, x_n$ is

$$a_1x_1 + a_2x_2 + \cdots + a_nx_n = d \tag{3.1}$$

where $a_1, a_2, \ldots,$ and d are constants. Hence for any point lying on or to the origin side of this plane, we have

$$a_1x_1 + a_2x_2 + \cdots + a_nx_n \leq d \tag{3.2}$$

whilst for any point on or beyond the non-origin side, we have

$$a_1x_1 + a_2x_2 + \cdots + a_nx_n \geq d \tag{3.3}$$

From this basic consideration it readily follows that the mathematical equation for a linearly separable (threshold) function may be written

$$f(X) = 0 \text{ if } a_1x_1 + a_2x_2 + \cdots + a_nx_n < d \tag{3.4}$$

and

$$f(X) = 1 \text{ if } a_1x_1 + a_2x_2 + \cdots + a_nx_n \geq d \tag{3.5}$$

assuming the separating plane just passes through the lowest summation

† This statement must be treated with caution, since a large number of practical functions have threshold relationships, for example, arithmetic requirements and decision mechanisms.

true minterm(s). This functional relationship may be paraphrased by the notation

$$f(X) = \langle a_1x_1 + a_2x_2 + \cdots + a_nx_n \rangle_d \tag{3.6}$$

Note that normal arithmetic relationships hold in these considerations and not modulo 2.

The threshold-logic gate is illustrated in Fig. 3.2b. The parameters $a_1, a_2, \ldots$ are termed the "weights" of the respective x_i input variables, giving their relative importance in determining the gate output state, whilst d is termed the gate threshold discrimination, or merely threshold. The value $\sum_{i=1}^{n} a_ix_i$ at each minterm is termed the gate input summation, which must equal or exceed d to realise $f(X) = 1$.

Further details of threshold logic theory, including tolerance considerations of the input weights and gate threshold discrimination, may be found, particularly in Muroga.[4,5,8–12] It may also be found that if the difference (gap) between the highest input summation $\sum_{i=1}^{n} a_1x_1$ for $f(X) = 0$ and the lowest input summation $\sum_{i=1}^{n} a_ix_i$ for $f(X) = 1$ is made unity (the

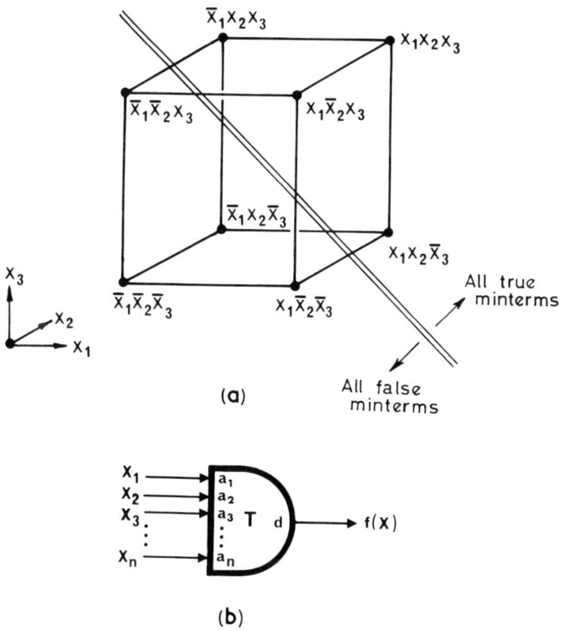

Fig. 3.2. Concepts of linear-separability and threshold functions: (a) the hypercube construction with a separating plane; function

$$f(X) = x_1x_2 + x_2x_3 + x_1x_3 = \langle 1.x_1 + 1.x_2 + 1.x_3 \rangle_2$$

illustrated; (b) the general symbol for a threshold-logic gate.

normalised integer criteria), then for all threshold functions of $n \leq 8$, the a_i and d gate parameters will themselves be integer values. For $n \geq 9$, however, fractional a_i weighting values occur under this normalised integer criteria. Further, all functions of $n \leq 8$ have unique a_i values, but above $n = 8$ alternative integer a_i values have been shown to be possible.

The values of the $n + 1$ parameters $a_1, \ldots, a_n; d$ uniquely define any linearly separable function,[4,5,8] and therefore may be investigated as a possible numerical basis for the classification of such functions. For example, consider the four functions illustrated in Fig. 3.3, the threshold realisations of which may be determined from the Boolean equations by inspection or by other more academic means. The logical relationship between $f_1(X)$ and $f_2(X)$ is merely the negation (N) of input x_2; the relationship between $f_2(X)$ and $f_3(X)$ is the permutation (P) of inputs x_1 and x_3, whilst the relationship between $f_3(X)$ and $f_4(X)$ is the overall function negation (N). Hence, all four linearly separable functions belong to the same NPN-algebraic classification group.

The a_i and d numerical values themselves, however, are not completely satisfactory for classification purposes, since overall function negation usually involves some numerical revision to the value of d. Whilst the

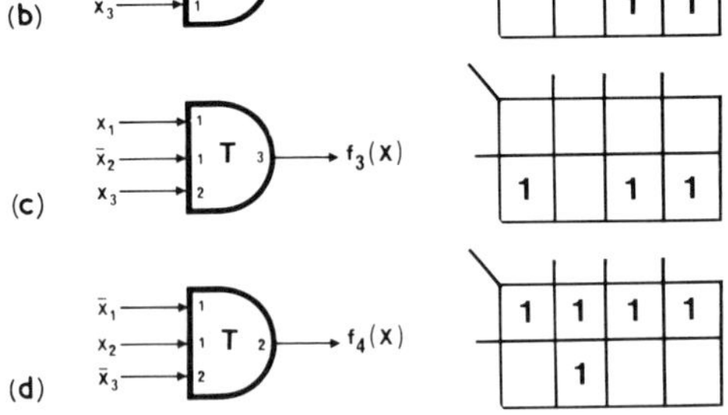

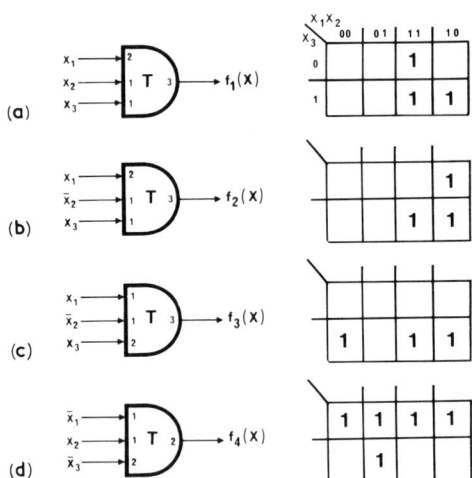

Fig. 3.3. NPN variants of a linearly separable function:

(a) $f_1(X) = x_1x_2 + x_1x_3$
(b) $f_2(X) = x_1\bar{x}_2 + x_1x_3$
(c) $f_3(X) = x_1x_3 + \bar{x}_2x_3$
(d) $f_4(X) = \bar{x}_3 + \bar{x}_1x_2$

numerical relationships involved in input variable and overall function negation are relatively trivial to compute,† nevertheless it remains that the realizing parameters themselves do not prove to be the most relevant or satisfactory means of classification. It may also be appreciated that there exists no linear mathematical relationship between the functional domain of Boolean algebra and the threshold domain of input summation and output discrimination.

However, the a_i input weights do constitute a measure of the "importance" of each input variable in determining the output state of the function, that is, the "order" of the variables, and the output discrimination value d constitutes some measure of how many true input minterms there are in the function $f(X)$. Hence the concept of $n + 1$ parameters involving the order of the variables and the number of true input minterms is present within the a_i, d parameters. Let us revert back to the functional domain to consider developments along this path.

The developments we shall briefly review are those based upon the original work of Chow,[13] and hence are generally referred to as Chow or modified Chow parameters.[4,5,8–12] The basic Chow parameters of a function $f(X)$ are defined as

$$\mathbf{CH}(X) \triangleq \mathrm{Ch}(x_1), \mathrm{Ch}(x_2), \ldots, \mathrm{Ch}(x_n); \mathrm{Ch}(x_0) \tag{3.7}$$

where $\mathrm{Ch}(x_i)$ $(i = 1 \text{ to } n) = \sum$ occurrence of x_i taken over all true minterms, and $\mathrm{Ch}(x_0)$ = total number of true minterms. For example, given $f(X) = x_1x_2 + x_1x_3$, we first expand the function into its three true minterms $x_1x_2\bar{x}_3 + x_1x_2x_3 + x_1\bar{x}_2x_3$, whence:

	$\mathrm{Ch}(x_1)$	$\mathrm{Ch}(x_2)$	$\mathrm{Ch}(x_3)$;	$\mathrm{Ch}(x_0)$
$x_1x_2\bar{x}_3$	1	1	—	
$x_1x_2x_3$	1	1	1	
$x_1\bar{x}_2x_3$	1	—	1	
$\therefore$ **CH**(X) =	3	2	2;	3

† The threshold realisation of $\overline{f(X)}$ is initially given by multiplying all the a_i and d parameter values of $f(X)$ by -1.0, without negating the x_i inputs themselves. However, negative weight and threshold values are not usually required, and hence further manipulation requires the x_i input variables to be complemented, input weights to be returned to their original sign, and the gate threshold to be adjusted to $\sum_{i=1}^{n} a_i - d + 1$. Details of all negation operations, which are based upon the arithmetic relationship of $\bar{x}_i = (1 - x_i)$, may be found.[10,11]

It will be apparent that each $\mathrm{Ch}(x_i)$ value corresponds to the number of times each literal x_i occurs in the minterm expansion, ignoring the presence of $\bar{x}_i$. A further interesting feature is that

$$\frac{1}{\mathrm{Ch}(x_0)}(\mathrm{Ch}(x_1), \ldots, \mathrm{Ch}(x_n)) \tag{3.8}$$

corresponds to the centre of gravity of all true minterms of $f(X)$ in the Euclidean n-space.

The ordering (importance) of the x_i input variables is related to the magnitude of the $\mathrm{Ch}(x_i)$ parameter values for any given function. Note, however, that the threshold a_i parameters and the Chow $\mathrm{Ch}(x_i)$ parameters, while possessing this same property, are not themselves related by any mathematical relationship; further, the Chow parameters so far defined are applicable to any function $f(X)$, whether linearly separable or not.

However, let us modify the basic Chow parameters as follows, in order to derive an integer-value vector that will finally prove more relevant for threshold function classification purposes.[5,6,14] For any function $f(X)$, we may define

$$\mathbf{B}(X) \triangleq b_1, b_2, \ldots, b_n; b_0 \tag{3.9}$$

where b_i, $(i = 1 \text{ to } n) = 2\mathrm{Ch}(x_i) - \mathrm{Ch}(x_0)$ and $b_0 = \mathrm{Ch}(x_0) - 2^{n-1}$. Note that the original Chow parameters were always positive integer values; the modified Chow parameters will still be integer, but not necessarily all positive.

Taking our previous example with the original Chow parameters $\mathbf{CH}(X) = 3, 2, 2; 3$, the modified Chow parameters become $\mathbf{B}(X) = 3, 1, 1; -1$. A particular and useful property of these modified parameters is that if a given function $f(X)$ is independent of a particular variable x_i, then the associated b_i parameter is always zero-valued. This is not the case with the original Chow parameters. However, the converse is not true, as may be shown by determining the Chow parameters for, say $f(X) = x_1 \oplus x_2$.

It may be simply shown that each linearly separable function has one unique non-zero set of Chow parameters, and that no other function possesses this same set of values.[5] Note that both magnitudes and signs are necessary in the modified Chow parameter set for uniqueness. Hence the Chow parameters, and more particularly the modified set, are particularly appropriate and convenient for cataloguing threshold functions.

Let us look a little deeper into the modified set introduced, and indeed modify them further so as to correspond to developments in Chapter 2. Taking the incomplete $2^n \times n$ binary orthogonal matrix considered in

Chapter 2, for the example function $f(X)$ we have

$$\begin{bmatrix} 1 & 1 & 1 & 1 & 1 & 1 & 1 & 1 \\ 1 & 1 & 1 & 1 & -1 & -1 & -1 & -1 \\ 1 & 1 & -1 & -1 & 1 & 1 & -1 & -1 \\ 1 & -1 & 1 & -1 & 1 & -1 & 1 & -1 \end{bmatrix} \begin{bmatrix} 1 \\ 1 \\ 1 \\ 1 \\ 1 \\ -1 \\ -1 \\ -1 \end{bmatrix} = \begin{bmatrix} +2 \\ +6 \\ +2 \\ +2 \end{bmatrix} \begin{matrix} s_0 \\ s_1 \\ s_2 \\ s_3 \end{matrix} \qquad (3.10)$$

$$\mathbf{T} \qquad \mathbf{Y} = \mathbf{S}$$

If we maintain 0, 1 coding for $f(X)$, we alternatively have

$$\begin{bmatrix} 1 & 1 & 1 & 1 & 1 & 1 & 1 & 1 \\ 1 & 1 & 1 & 1 & -1 & -1 & -1 & -1 \\ 1 & 1 & -1 & -1 & 1 & 1 & -1 & -1 \\ 1 & -1 & 1 & -1 & 1 & -1 & 1 & -1 \end{bmatrix} \begin{bmatrix} 0 \\ 0 \\ 0 \\ 0 \\ 0 \\ 1 \\ 1 \\ 1 \end{bmatrix} = \begin{bmatrix} 3 \\ -3 \\ -1 \\ -1 \end{bmatrix} \begin{matrix} r_0 \\ r_1 \\ r_2 \\ r_3 \end{matrix} \qquad (3.11)$$

$$\mathbf{T} \qquad \mathbf{Z} = \mathbf{R}$$

Considering the first of these alternatives, it will now be noticed that

(a) the modified Chow parameters b_i, $i = 1$ to n, introduced above are given by $\frac{1}{2}$ the resultant spectral coefficients s_i, $i = 1$ to n; and

(b) the modified Chow parameter b_0 is given by $-\frac{1}{2}$ times the spectral coefficient s_0.

Hence the b_i, $i = 0$ to n, modified Chow parameters (and the original Chow parameters) have exactly the same information content as the first $n + 1$ spectral coefficients. Note that theoretically we may multiply the n parameters b_1 to b_n ($\text{Ch}(x_1)$ to $\text{Ch}(x_n)$) by any positive or negative constant without altering their information content, and similarly, we may multiply the b_0 ($\text{Ch}(x_0)$) parameter by the same or any other constant. Thus there are an infinite range of Chow-parameter definitions theoretically possible, but clearly only a few are most relevant and sensible.

However, great care must be taken in reading published literature in order to appreciate the particular definition of the Chow-parameter values being used, particularly the factor of two in the b_1 to b_n parameters, and the sign convention for b_0.† In this chapter we shall continue to use the convention given by the transform $\mathbf{TY} = \mathbf{S}$ shown in (3.10), which is not the same as the definitions initially given in (3.9) for the modified Chow parameters. In addition we shall drop the qualification modified from here on, and refer to them merely as Chow parameters.

With our final chosen definition, we have the simple relationships:

(a) b_i ($i = 1$ to n) $= \sum_{m=0}^{2^n-1}$ {(no. of agreements between x_i and $f(X)$) − (no. of disagreements between x_i and $f(X)$)};
$\equiv$ spectral coefficient s_i;

(b) $b_0 = \sum_{m=0}^{2^n-1}$ {(no. of false minterms in $f(X)$) − (no. of true minterms in $f(X)$)};
$\equiv$ spectral coefficient s_0.

Hence the $n + 1$ realising weight and threshold parameters $a_1, \ldots, a_n$; d and the $n + 1$ characterising Chow parameters b_i, $i = 0$ to n, both uniquely define any given linearly separable function $f(X)$. The Chow parameters, however, constitute the most powerful set. Consider again the four linearly separable functions shown in Fig. 3.3. Evaluating their Chow parameters, we have

$$\begin{aligned} f_1(X) &= x_1x_2 + x_1x_3 \quad \text{(Boolean)} \\ &= 6, 2, 2; 2 \quad \text{(Chow)} \end{aligned}$$

$$\begin{aligned} f_2(X) &= x_1\bar{x}_2 + x_1x_3 \quad \text{(Boolean)} \\ &= 6, -2, 2; 2 \quad \text{(Chow)} \end{aligned}$$

$$\begin{aligned} f_3(X) &= x_1x_3 + \bar{x}_2x_3 \quad \text{(Boolean)} \\ &= 2, -2, 6; 2 \quad \text{(Chow)} \end{aligned}$$

$$\begin{aligned} f_4(X) &= \bar{x}_3 + \bar{x}_1x_2 \quad \text{(Boolean)} \\ &= -2, 2, -6; -2 \quad \text{(Chow)} \end{aligned}$$

Thus the negation of variable x_2 is shown by the sign change in the value of b_2 between $f_1(X)$ and $f_2(X)$; the permutation of $x_1 \leftrightarrow x_3$ is shown by the interchange of $b_1 \leftrightarrow b_3$ between $f_2(X)$ and $f_3(X)$, and the negation of the whole function is shown by the sign change of all $n + 1$ parameters between $f_3(X)$ and $f_4(X)$. Hence the three NPN-classification operators are implemented in these positional and sign changes. It may readily be proved that

† For example, in Hurst,[4] b_0 is defined with the opposite sign convention as here, b_1 to b_n being the same convention. It is not significant which is chosen.

the magnitudes $|b_i|$, $i = 1$ to n, of 6, 2, 2 and $|b_0| = 2$ exactly cover all NPN variants of function $f_1(X)$, and thus constitute a numerical classification for all functions in this NPN group.

However, full permutation across all $n + 1$ Chow parameters, including b_0, has not yet been invoked. Consider the permutation $b_1 \leftrightarrow b_0$ in function $f_1(X)$, giving the Chow parameters 2, 2, 2; 6. This function is the linearly separable function $f_5(X) = x_1x_2x_3$. Thus $|2|, |2|, |2|; |6|$ embraces a further NPN-classified group of functions. The full significance of the Chow-parameter classification, therefore, is that a given set of parameter values may be unrestrictedly permutated, with independent sign allocations for each parameter, all such allocations uniquely defining a particular function $f(X)$ which is always linearly separable.

The theoretical development of the Chow-parameter classification, and the relationship between the $n + 1$ numerical parameters that embrace more than one NPN-classification group and SD-algebraic classifications that likewise each embrace more than one NPN group, have been extensively studied.[5–9,12–14] However, when dealing exclusively with linearly separable functions, it is not necessary to follow in detail the whole of these algebraic developments, and hence we shall not review them here.†

The Chow-parameter classification table for all linearly separable functions of $n \leq 5$ is given in Table 3.4. Note that (conventionally) the b_i parameters are listed in descending magnitude order, with no fixed correspondence to $b_1, \ldots, b_n; b_0$. The principal reason for such tabulations, however, is not merely as a classification table, but much more

(a) as a means to determine whether any given function is linearly separable or not, and

(b) if a function is linearly separable, then to give the minimum-integer threshold realisation for the function.

Details of the full use of these classification tables may be found.[4,9,11] Here we shall briefly note:

(i) If the $n + 1$ Chow parameters for any given function are computed and arranged in numerically descending magnitude order, then if the function is not linearly separable the derived numbers will not appear in the classification tabulation.

(ii) If the derived numbers do appear, then the given function is linearly separable, and the $|a_i|$ values tabulated against the appropriate $|b_i|$ entry

† We shall, however, return to these Chow parameters and their numerical interchanges again in Sections 3.5 and 3.6, when we relate them to the general case covering all switching functions.

may be used to derive the minimum integer $a_1, \ldots, a_n; d$ threshold gate realising parameters. The relationship between the listed $|a_i|$ value and the final $a_1, \ldots, a_n; d$ gate parameters is a one-to-one mapping for $a_1, \ldots, a_n$ with a simple arithmetic relationship to give d.[4,11]

(iii) The entries for any n also appear in the subsequent tabulation for $n + 1$, but with all values multiplied by 2 in the latter, and with a further

Table 3.4. The Chow-parameter classification for all linearly separable functions of $n \leq 5$.[a]

	n	$\|b_i\|$						$\|a_i\|$					
$n \leq 3$	1	8	0	0	0			1	0	0	0		
	2	6	2	2	2			2	1	1	1		
	3	4	4	4	0			1	1	1	0		
$n \leq 4$	1	16	0	0	0	0		1	0	0	0	0	
	2	14	2	2	2	2		3	1	1	1	1	
	3	12	4	4	4	0		2	1	1	1	0	
	4	10	6	6	2	2		3	2	2	1	1	
	5	8	8	8	0	0		1	1	1	0	0	
	6	8	8	4	4	4		2	2	1	1	1	
	7	6	6	6	6	6		1	1	1	1	1	
$n \leq 5$	1	32	0	0	0	0	0	1	0	0	0	0	0
	2	30	2	2	2	2	2	4	1	1	1	1	1
	3	28	4	4	4	4	0	3	1	1	1	1	0
	4	26	6	6	6	2	2	5	2	2	2	1	1
	5	24	8	8	4	4	4	4	2	2	1	1	1
	6	24	8	8	8	0	0	2	1	1	1	0	0
	7	22	10	10	6	2	2	5	3	3	2	1	1
	8	22	10	6	6	6	6	3	2	1	1	1	1
	9	20	12	12	4	4	0	3	2	2	1	1	0
	10	20	12	8	8	4	4	4	3	2	2	1	1
	11	20	8	8	8	8	8	2	1	1	1	1	1
	12	18	14	14	2	2	2	4	3	3	1	1	1
	13	18	14	10	6	6	2	5	4	3	2	2	1
	14	18	10	10	10	6	6	3	2	2	2	1	1
	15	16	16	16	0	0	0	1	1	1	0	0	0
	16	16	16	12	4	4	4	3	3	2	1	1	1
	17	16	16	8	8	8	0	2	2	1	1	1	0
	18	16	12	12	8	8	4	4	3	3	2	2	1
	19	14	14	14	6	6	6	2	2	2	1	1	1
	20	14	14	10	10	10	2	3	3	2	2	2	1
	21	12	12	12	12	12	0	1	1	1	1	1	0

[a] The b_i values are as defined by development (3.10) and may be dissimilar but related in other publications; for the 135 $n \leq 6$ entries, see Hurst[4]; for the 2470 $n \leq 7$ entries, see Winder.[15]

zero-valued component. This directly corresponds to the doubling of the n-space between functions of n and $n + 1$ variables.

To conclude, the numerical classification procedure for linearly separable functions, which links directly with orthogonal transforms and spectral coefficients, is firmly established, being the principal means of threshold-logic synthesis for functions of up to, say, $n \leq 6$ without difficulty. For $n \leq 7$ the length of tabulation becomes an embarrassment, whilst above $n = 7$ no fully published classification tables are generally available.

3.4 WIDER CONSIDERATIONS

When we review the several possible classes of Boolean functions, we find that the linearly separable class constitutes the most restrictive class. Between this class and the unrestricted class of all 2^{2^n} Boolean functions lie several other classification areas, which we shall briefly consider. None, individually, have the importance of the linearly separable (threshold) class, and hence do not have the practical significance of the latter.

We may readily demonstrate that the $n + 1$ Chow parameters are inadequate to define unambiguously functions of unrestricted class—as an extreme example, all the Chow parameters of any Exclusive-OR or Exclusive-NOR function will be found to be zero, and hence cannot have a unique functional identity. Thus the use of only $n + 1$ numerical parameters for function classification purposes has obvious limitations.

3.4.1 Unate Functions

A unate function is one in which no x_i input variable appears in both complemented and uncomplemented form (x_i and $\bar{x}_i$) in a minimised sum-of-products expression for $f(X)$. For example, the function $x_1\bar{x}_2 + \bar{x}_2x_3$ is a unate function, but $x_1\bar{x}_2 + x_2x_3$ is a non-unate function. Care must be taken to ensure that the expression for $f(X)$ is appropriately minimised; for example, $f(X) = x_1 + \bar{x}_1\bar{x}_2$ is unate because there exists the minimised form $f(X) = x_1 + \bar{x}_2$.†

It may readily be shown that all linearly separable functions must be unate.[4,5] The converse, however, does not hold, as may be illustrated by the

† If in a unate function all variables appear in the uncomplemented form, then such a function may be further referred to as a positive unate function; if all variables appear in the complemented form, then such a function may be referred to as a negative unate function. We shall not require such distinction here.

simple unate but non-linearly separable function $f(X) = x_1x_2 + x_3x_4$. Unateness, therefore, is the weakest test that may be applied to a Boolean expression for linear separability.

3.4.2 Monotonicity

Monotonicity first requires the definition of function comparability (incomparability). Two functions $f_1(X)$ and $f_2(X)$ are said to be *comparable* if

(i) for all inputs where $f_1(X)$ is 1, $f_2(X)$ is also 1, written as $f_1(X) \subseteq f_2(X)$,

or

(ii) for all inputs where $f_2(X)$ is 1, $f_1(X)$ is also 1, written as $f_2(X) \subseteq f_1(X)$.†

Should both (i) and (ii) hold, we have equality, namely, $f_1(X) = f_2(X)$. If neither $f_1(X) \subseteq f_2(X)$ or $f_2(X) \subseteq f_1(X)$ holds, then $f_1(X)$ and $f_2(X)$ are said to be *incomparable*.

Consider now a given function $f(X) = f(x_1, \ldots, x_i, \ldots, x_n)$, and its decomposition into two disjoint functions $f_{\bar{x}_i}(X) = f(x_1, \ldots, 0, \ldots, x_n)$ and $f_{x_i}(X) = f(x_1, \ldots., 1, \ldots, x_n)$. If the comparability

$$f_{\bar{x}_i}(X) \subseteq f_{x_i}(X) \qquad \text{or} \qquad f_{x_i}(X) \subseteq f_{\bar{x}_i}(X)$$

holds for all n assignments to x_i, then $f(X)$ is said to be 1-monotonic. If we continue the decompositon of $f(X)$ into four disjoint functions about two variables x_i, x_j, and comparability holds between all four functions for each x_i, x_j decomposition and for all possible x_i, x_j assignments from $x_1, \ldots, x_n$, then $f(X)$ is said to be 2-monotonic.

Continuing, if comparability holds within each set of 2^k functions decomposed about k variables, and for all assignments of k out of n variables, then the given function is k-monotonic. Finally, if comparability holds for all possible k assignments, $1 \leq k \leq n - 1$, $f(X)$ is said to be *completely monotonic*. Further details of the algebraic developments of monotonicity may be found referenced,[5] and graphically illustrated for $n = 4$ for both completely monotonic and 1-monotonic only examples.[4]

Monotonicity provides a powerful check of linear separability. Function unateness may be shown to be equivalent to 1-monotonicity, which is the weakest test for linear separability. k-Monotonicity, $k = 2, 3, \ldots$, provides increasingly strong tests, and it was earlier considered that complete mono-

† The alternative designation, $f_1(X) \subseteq$ or $\supseteq f_2(X)$, may be used to express comparability of $f_1(X)$ and $f_2(X)$.

tonicity was a necessary and sufficient test for linear separability. However, sufficiency was disproved by a counterexample for $n = 9$, although it has been shown to hold for all functions of $n \leq 8$.[16,17]

3.4.3 Summability and Asummability

A function $f(X)$ is said to be *k-summable*, $k \geq 2$, if there exists any two sets of k input vectors, one set taken from the set of all true input vectors, $(f(X) = 1)$, the other set taken from the set of all false input vectors, $(f(X) = 0)$, such that the vector summation of each set is equal, that is,

$$\sum^{k} (\text{true vectors}) = \sum^{k} (\text{false vectors})$$

For example, given $f(X) = x_1 x_2 + x_3 x_4$, we may select the two true input vectors 0011 and 1100, and the two false input vectors 0101 and 1010, giving us

$$\sum^{k=2} (\text{true vectors}) = 1111$$

$$\sum^{k=2} (\text{false vectors}) = 1111$$

thus proving 2-summability of $f(X)$.

If k-summability cannot be realised, then the function is said to be *k-asummable*; 2-asummability may be shown to be equal to complete monotonicity.[5] Complete asummability, however, that is k-asummable for any k is a necessary and sufficient test for linear separability. All threshold logic functions, therefore, are completely asummable.

3.4.4 Dual-Comparable Functions

The dual $f^D(X)$ of function $f(X)$ was defined in Section 3.2. If with any given function $f(X)$ the comparability relationships

$$f^D(X) \subseteq f(X) \qquad \text{or} \qquad f(X) \subseteq f^D(X)$$

holds, then $f(X)$ is said to be dual comparable. The function $f(X) = x_1 + x_2$ is a trivial example of dual comparability.

The class of n-variable dual-comparable functions will be found to embrace all n-variable completely monotonic functions and, in addition, some (but not all) lower monotonic, unate, and non-unate functions. Thus this class of function does not lie between any bounds of asummability, monotonicity, or unateness, being a relatively weak function classification.

3.4.5 Continued Function Classification

The $n + 1$ numbers which constitute the Chow parameters have been noted as being ideal for the classification of linearly separable functions, but inadequate for the classification of all 2^{2^n} functions of n input variables. However, let us consider the classification of functions in the intermediate classification groups we have now listed.

Yajima and Ibaraki[18] in particular have investigated the range of functions that have unique Chow parameters, such that any function outside this range does not have the same parameter magnitudes as any function within the range. While the necessary asummability relationships can be employed to define such a classification, there is no known way of directly determining all functions that lie within this boundary. Clearly all the linearly separable (threshold) functions belong to this class, which we may term *Chow-parameter definable*, but exhaustive testing is required to determine the bounds of this classification.

It has been shown that for $n \leq 8$ the whole class of completely monotonic (2-asummable) functions are also Chow-parameter definable. However, for $n = 9$, at least one function (with its NPN variants) is known that does not belong to the completely monotonic class and yet remains Chow-parameter definable. The case for $n \geq 9$ has not been studied in detail, but the situation must be that the Chow-parameter definable class does not extend beyond the completely monotonic class of functions.

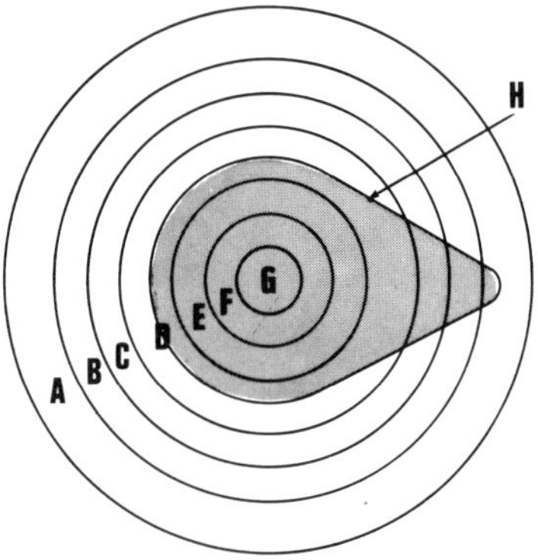

Fig. 3.4. The set inclusive relationships between classes of switching functions of n variables (not to scale): A: unrestricted class of 2^{2^n} functions; B: 1-monotonic functions, unate; C: 2-monotonic; D: higher monotonic; E: completely monotonic, 2-asummable; F: Chow-parameter definable; G: linearly separable (threshold) functions, completely asummable; and H: dual comparable.

Our function classification divisions are thus indicated by the set inclusive relationships illustrated in Fig. 3.4, with the Chow parameters being adequate to define and classify functions of the inner two regions F and G only. Of these two, only functions in G can be usefully defined. Beyond these regions more information than is contained in $n + 1$ numerical parameters becomes necessary in order to define and classify the increasing functional complexity.

3.5 SPECTRAL CLASSIFICATION

The number of functions with increasing n in the main classification areas is given in Table 3.5. The decreasing percentage of linearly separable functions from all 2^{2^n} functions as n increases is evident.

In Chapter 2 we saw how the operation

$$\mathbf{T}^n\mathbf{Y} = \mathbf{S} \tag{3.12}$$

transformed a functional domain vector $\mathbf{Y}$ into a spectral domain vector $\mathbf{S}$, the complete, orthogonal $2^n \times 2^n$ matrix $\mathbf{T}^n$ ensuring no loss of information between the two domains. $\mathbf{S}$ is thus uniquely related to $\mathbf{Y}$ for any given $\mathbf{Y}$. From this property, we shall now consider the use of the 2^n spectral coefficients in $\mathbf{S}$ for function classification purposes.

When the spectra of a number of functions of any given n are compared, it will be evident that identical sets of magnitudes occur in many spectra, albeit with position and sign changes. It is further clear that the NPN-classification operations are reflected in the spectral coefficients in a similar manner to that which we have seen in the Chow parameters. Consider the following four $n = 3$ non-linearly separable functions

$$f_1(X) = x_1x_2 + \bar{x}_2x_3 + x_2\bar{x}_3$$

$$f_2(X) = x_1\bar{x}_2 + x_2x_3 + \bar{x}_2\bar{x}_3 = \text{negation of } x_2 \text{ in } f_1(X)$$

$$f_3(X) = x_1\bar{x}_3 + x_2x_3 + \bar{x}_2\bar{x}_3 = \text{permutation } x_2 \leftrightarrow x_3 \text{ in } f_2(X)$$

$$f_4(X) = \bar{x}_2x_3 + \bar{x}_1x_2\bar{x}_3 = \text{negation of } f_3(X)$$

and their spectra

	s_0	s_1	s_2	s_3	s_{12}	s_{13}	s_{23}	s_{123}
$f_1(X)$:	-2	2	2	2	-2	-2	6	2
$f_2(X)$:	-2	2	-2	2	2	-2	-6	-2
$f_3(X)$:	-2	2	2	-2	-2	2	-6	-2
$f_4(X)$:	2	-2	-2	2	2	-2	6	2

From these four spectra, it is clear that the magnitudes

$$|6|, |2|, |2|, |2|, |2|, |2|, |2|, |2|$$

(but not their order) characterise these four NPN-related functions. However, full permutation of the positions of the spectral coefficients has not been invoked.

It has been shown that there are in total five invariance operations that may be applied to the full set of 2^n spectral coefficients, of which only three have

Table 3.5. Classification statistics for functions of $\leq n$ independent binary input variables, including the linea,ly separable (threshold) class.[a]

	n				
	2	3	4	5	6
Boolean functions					
(a) total	16	256	65536	$\simeq 4.3 \times 10^9$	$\simeq 1.8 \times 10^{19}$
(d) non-degenerate	10	218	64594	$\simeq 4.3 \times 10^9$	$\simeq 1.8 \times 10^{19}$
(c) degenerate	6	38	942	325262	2.58×10^{10}
Threshold functions					
(d) total	14	104	1882	94572	$\simeq 1.5 \times 10^7$
% of (a)	(88%)	(41%)	(2.9%)	(2.2×10^{-2}%)	(8×10^{-9}%)
(e) non-degenerate	8	72	1536	86080	$\simeq 1.4 \times 10^7$
% of (b)	(80%)	(33%)	(2.4%)	(2×10^{-2}%)	(8×10^{-9}%)
(f) degenerate	6	32	346	8492	541094
% of (c)	(100%)	(84%)	(37%)	(2.6%)	(2.1×10^{-3}%)
No. of entries in the NPN classification for all (a)	4	14	222	616126	$\simeq 2 \times 10^{14}$
No. of entries in the NPN classification for the unate functions only in (a)	3	6	17	112	8282
No. of entries in the SD classification for all (a)	3	7	83	109958	*b*
No. of entries in the NPN classification for all (d)	3	6	15	63	567
No. of entries in the Chow parameter classification for all (d)[c]	2	3	7	21	135

[a] See also Tables 3.6 and 3.9.
[b] Not computed.
[c] Table 3.4 for full details of $n \leq 5$, and Hurst[4] for $n \leq 6$.

been illustrated here so far. Here we shall merely state the operations; formal proofs may be found in Edwards[19] and elsewhere.[4,9,20] They are as follows:

(i) Permutation of any input variables x_i and x_j, $i \neq j$, $i, j \in 1$ to n: this requires the interchange of 2^{n-2} pairs of coefficient values

$$\begin{aligned} s_i &\leftrightarrow s_j \\ s_{ik} &\leftrightarrow s_{jk} \\ s_{ikl} &\leftrightarrow s_{jkl} \\ &\vdots \end{aligned} \tag{3.13}$$

Note that coefficients s_0, s_k, s_{ij}, ... remain unchanged.

(ii) Negation of any input variable x_i, $i \in 1$ to n: this requires the negation of 2^{n-1} spectral coefficient values

$$\begin{aligned} s_i &\to -s_i \\ s_{ij} &\to -s_{ij} \\ s_{ik} &\to -s_{ik} \\ &\vdots \end{aligned} \tag{3.14}$$

Note that coefficients s_0, s_j, s_{jk}, ... remain unchanged.

(iii) Negation of a network output: this requires the negation of all 2^n spectral coefficients

$$\begin{aligned} s_0 &\to -s_0 \\ s_1 &\to -s_1 \\ s_2 &\to -s_2 \\ &\vdots \\ s_{12\ldots n} &\to -s_{12\ldots n} \end{aligned} \tag{3.15}$$

(iv) Replacement of any variable x_i into a network with $x_i \oplus x_j$, $i \neq j$, $i, j \in 1$ to n: this requires the interchange of 2^{n-2} pairs of coefficient values

$$\begin{aligned} s_i &\leftrightarrow s_{ij} \\ s_{ik} &\leftrightarrow s_{ijk} \\ s_{ikl} &\leftrightarrow s_{ijkl} \\ &\vdots \end{aligned} \tag{3.16}$$

Note that coefficients s_0, s_j, s_{jk}, ... remain unchanged.

(v) Finally, modification of a network output from $f(X)$ to $f(X) \oplus x_i$, $i \in 1$ to n: this results in the interchange of 2^{n-1} pairs of spectral coefficients

$$\begin{aligned} s_i &\leftrightarrow s_0 \\ s_{ij} &\leftrightarrow s_j \\ s_{ijk} &\leftrightarrow s_{jk} \\ &\vdots \end{aligned} \tag{3.17}$$

Note that all 2^n coefficients are involved in this operation.

Figure 3.5 illustrates these five operations.† Note that the first three, the NPN-invariance operations, do not alter the order of any coefficient value; the fourth operation interchanges certain coefficient values from all orders except s_0, whilst the final operation brings the value of s_0 into the interchange. Hence operations (iv) and (v), repeated as necessary, allow any spectral coefficient value to be moved to any other order, including the zero-order position s_0. The magnitude of all the individual coefficients, however, remains invariant under any of these operations.

The canonic classification tables for functions of $\leq n$ variables is a listing of the coefficient values, in the normal identification order s_0; $s_1, s_2, \ldots, s_n$; $s_{12}, \ldots; s_{12\ldots n}$, but (conventionally) with the zero- and first-order coefficients s_0; $s_1, \ldots, s_n$ all positive integer values and arranged in decreasing magnitude order. Thus s_0 is conventionally the highest magnitude coefficient in these canonic classification tables, followed by $s_1, \ldots, s_n$. Table 3.6 lists the canonic classification entries for $n \leq 4$. Several points concerning Table 3.6 may be highlighted:

(a) With all the zero- and first-order spectral coefficients made positive, this does not mean that all the higher-order coefficients are likewise positive. It should be recalled that all valid spectra have the property that the sum of the spectral coefficients must always total $\pm 2^n$, and this must be true for the canonic functions listed in the classification tables.

(b) Every entry in any n also appears in the $n + 1$ classification list, but with all magnitudes multiplied by 2. This is as already encountered in the Chow-parameter classification tables for linearly separable functions.

(c) The compactness of this classification procedure is considerable; all 65,536 functions of $n \leq 4$ are contained in the eight $n \leq 4$ canonic classification entries. For the 4.3×10^9 functions of $n \leq 5$, the number of classification entries increases to 48, details of which will be found in Edwards[21] and elsewhere.[4,22]

† See Appendix A for details of the spectral coefficient map.

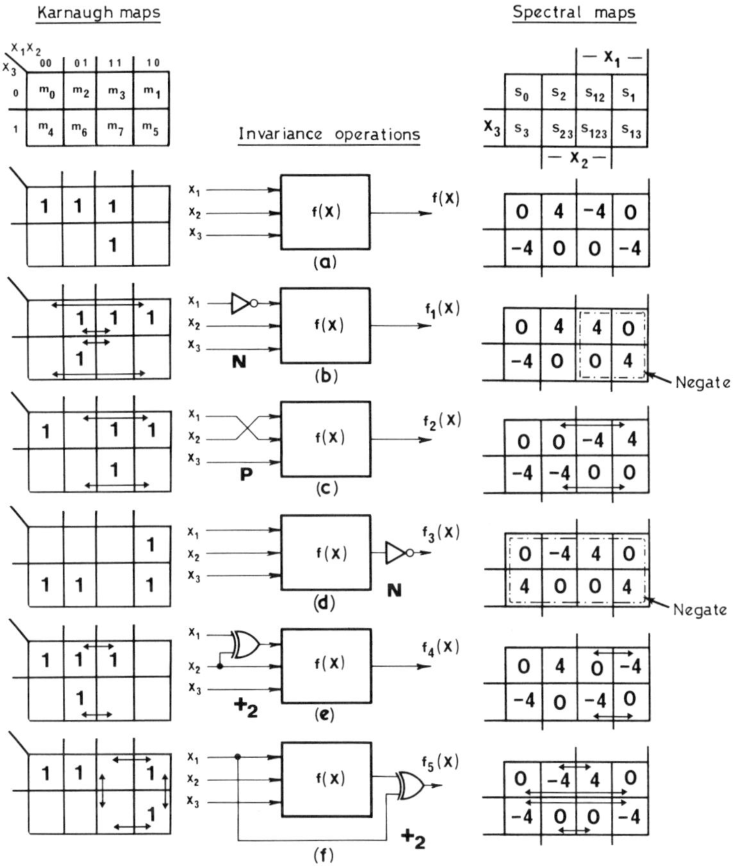

Fig. 3.5. The five spectral invariance operations of Eqs. (3.13)–(3.17):

(a) given function, $f(X) = x_1x_2 + \bar{x}_1\bar{x}_3$
spectrum 0; 0, 4, −4; −4, −4, 0; 0

(b) negation of x_1, $f_1(X) = \bar{x}_1x_2 + x_1\bar{x}_3$
spectrum 0; 0, 4, −4; 4, 4, 0; 0

(c) permutation $x_1 \leftrightarrow x_2$, $f_2(X) = x_1x_2 + \bar{x}_2\bar{x}_3$
spectrum 0; 4, 0, −4; −4, 0, −4; 0

(d) output negation, $f_3(X) = x_1\bar{x}_2 + \bar{x}_1x_3$
spectrum 0; 0, −4, 4; 4, 4, 0; 0

(e) input operation $x_1 \to x_1 \oplus x_2$, $f_4(X) = \bar{x}_1x_2 + \bar{x}_1\bar{x}_3 + x_2\bar{x}_3$
spectrum 0; −4, 4, −4; 0, 0, 0; −4

(f) output operation $f(X) \oplus x_1$, $f_5(X) = x_1\bar{x}_2 + \bar{x}_1\bar{x}_3$
spectrum 0; 0, −4, −4; 4, −4, 0; 0.

Table 3.6. The positive canonic spectral classification for all binary functions of $n \leq 4$ under the full spectral coefficient invariance operations.

	Spectral coefficient															
n	s_0;	s_1	s_2	s_3	s_4;	s_{12}	s_{13}	s_{14}	s_{23}	s_{24}	s_{34};	s_{123}	s_{124}	s_{134}	s_{234};	s_{1234}
≤2: (1)	4	0	0			0										
(2)	2	2	2			−2										
≤3: (1)	8	0	0	0		0	0		0			0				
(2)	6	2	2	2		−2	−2		−2			2				
(3)	4	4	4	0		−4	0		0			0				
≤4: (1)	16	0	0	0	0	0	0	0	0	0	0	0	0	0	0	0
(2)	14	2	2	2	2	−2	−2	−2	−2	−2	−2	2	2	2	2	−2
(3)	12	4	4	4	0	−4	−4	0	−4	0	0	4	0	0	0	0
(4)	10	6	6	2	2	−6	−2	−2	−2	−2	2	2	2	−2	−2	2
(5)	8	8	8	0	0	−8	0	0	0	0	0	0	0	0	0	0
(6)	8	8	4	4	4	−4	−4	−4	0	0	0	0	0	0	−4	4
(7)	6	6	6	6	6	−2	−2	−2	−2	−2	−2	−2	−2	−2	−2	6
(8)	4	4	4	4	4	4	4	−4	−4	4	−4	−4	−4	4	4	−4

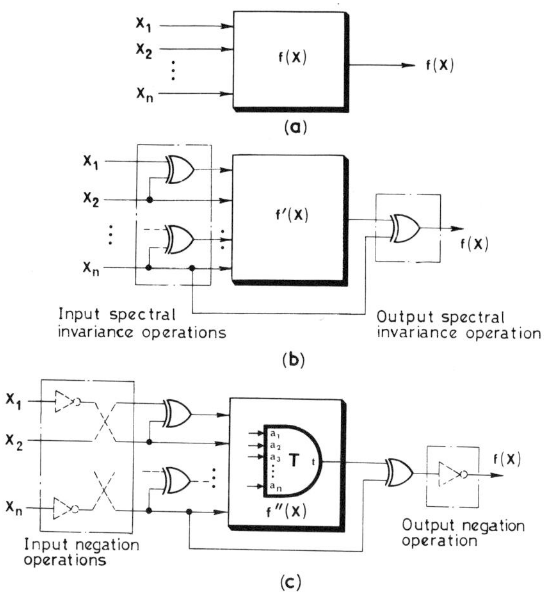

Fig. 3.6. Function synthesis using embedded threshold-logic core function (not always possible for $n \geq 4$). (a) Required function $f(X)$, as given not linearly separable; (b) $+\text{mod}_2$ spectral invariance operations to linearise the core function; (c) final NPN operations to realise $f(X)$ from the canonic linearly separable core function.

Comparing this complete classification for all functions with the linearly separable classification of Table 3.4, it will be noted that the entries for $n \leq 3$ in the Chow-classification tables are identical to the entries in the unrestricted class of functions given in Table 3.6, with the exception that Table 3.6 gives the higher-order coefficients in addition to the zero- and first-order (Chow) ones.† For $n \leq 4$, seven of the eight entries of Table 3.6 appear in the Chow list of Table 3.4, and hence there is only one further unique entry, namely, the "all-fours" canonic function (8) in the unrestricted class. For $n \leq 5$, it will be found that of the 48 spectral classification entries, 21 are the linearly separable entries of Table 3.4, leaving 27 additional entries in the unrestricted class.[4,21]

What this implies is that, considering $n \leq 4$ functions, all functions *except* those covered by the final classification entry in Table 3.6 may be realised by a threshold-logic gate, appropriately prefaced and/or followed by Exclusive-OR gates that perform the required invariance operations. This is illustrated in Fig. 3.6. The functions covered by the final entry in Table 3.6,

† We noted in Section 3.3 that there were sign and magnitude variants on the definition of the Chow parameters. It will now be appreciated that the definition we chose in Section 3.3 was such as to give conformity with the full set of spectral coefficients that we are now considering.

Table 3.7. Eight invariance operations on $f(X) = \bar{x}_1 x_2(\bar{x}_3 + \bar{x}_4) + \bar{x}_2(\bar{x}_3\bar{x}_4 + x_3x_4) + x_2\bar{x}_3x_4$ to generate the positive canonic spectrum of the final function.

	Spectral coefficients															
	s_0	s_1	s_2	s_3	s_4	s_{12}	s_{13}	s_{14}	s_{23}	s_{24}	s_{34}	s_{123}	s_{124}	s_{134}	s_{234}	s_{1234}
Given function $f(X)$	0	−4	0	−4	0	4	0	−4	4	0	−4	0	4	0	−12	0
(1) Replace x_2 by $x_2 \oplus x_3$, $= f_1(X)$	0	−4	4	−4	0	0	0	−4	0	−12	−4	4	0	0	0	4
(2) Replace x_4 by $x_4 \oplus x_2$, $= f_2(X)$	0	−4	4	−4	−12	0	0	0	0	0	0	4	−4	4	−4	0
(3) Modify output to $f_2(X) \oplus x_4$, $= f_3(X)$	−12	0	0	0	0	−4	4	−4	−4	4	−4	0	0	0	0	4
(4) Replace x_1 by $x_1 \oplus x_2$, $= f_4(X)$	−12	−4	0	0	0	0	0	0	−4	4	−4	4	−4	4	0	0
(5) Replace x_2 by $x_2 \oplus x_4$, $= f_5(X)$	−12	−4	4	0	0	−4	0	0	0	0	−4	0	0	4	−4	4
(6) Replace x_3 by $x_3 \oplus x_4$, $= f_6(X)$	−12	−4	4	−4	0	−4	4	0	−4	0	0	4	0	0	0	0
(7) Negate $f_6(X)$, $= f_7(X)$	12	4	−4	4	0	4	−4	0	4	0	0	−4	0	0	0	0
(8) Negate x_2, $= f_8(X)$	12	4	4	4	0	−4	−4	0	−4	0	0	4	0	0	0	0

and similar non-Chow entries in $n \geq 5$ spectral classification tables, cannot be realised by the embedded threshold function topology shown in this figure. Whether this is a realistic topology to adopt for realising a range of initially non-linearly separable functions is open to debate, since the practical cost of the Exclusive-OR gates in addition to the threshold-logic core is a serious factor. Further details of function synthesis using embedded threshold-logic gates have been published.[4,9,19,20]

As a final example of this classification procedure for the unrestricted range of functions, consider the $n = 4$ function

$$f(X) = \bar{x}_1 x_2(\bar{x}_3 + \bar{x}_4) + \bar{x}_2(\bar{x}_3\bar{x}_4 + x_3 x_4) + x_2\bar{x}_3 x_4$$

We evaluate its spectral coefficients by the normal transform $\mathbf{T}^n\mathbf{Y} = \mathbf{S}$, giving

$$\begin{array}{cccccccccccc} s_0; & s_1 & s_2 & s_3 & s_4; & s_{12} & s_{13} & s_{14} & s_{23} & s_{24} & s_{34}; \\ 0 & -4 & 0 & -4 & 0 & 4 & 0 & -4 & 4 & 0 & -4 \end{array}$$

$$\begin{array}{ccccc} s_{123} & s_{124} & s_{134} & s_{234}; & s_{1234} \\ 0 & 4 & 0 & -12 & 0 \end{array}$$

One can see by inspection that this function must belong to classification (3) in the $n \leq 4$ tabulation of Table 3.6. The individual steps that can transform this particular spectrum into the positive canonic classification entry are detailed in Table 3.7. Note that

(a) no permutation (P) operation is required in this worked example to produce the final positive canonic order, only input and output negation from the three NPN-invariance operations being used;

(b) alternative sequences of invariance operations from those shown in Table 3.7 may be chosen to perform the same final overall transformation; the intermediate functions will of course be dissimilar from those shown if some other sequence is adopted; and finally,

(c) the operations listed in Table 3.7 are those that may be implemented if it was required to realise the given function $f(X)$ using the canonic classification function as a core function (compare Fig. 3.6).

3.6 FURTHER CLASSIFICATION COMMENTS

We have seen how the 2^n spectral coefficients for any given function $f(X)$ may be used as a ready means of function classification. The Chow-parameter classification for the class of linearly separable functions has been shown to be a special case of spectral classification, the particular properties of linearly separable functions being such that only the first $n + 1$ of the full set of 2^n

coefficients are necessary for classification purposes. It should be appreciated that we may always determine the function $f(X)$ corresponding to a given set of 2^n coefficients by evaluating the inverse transform from the spectral domain to the functional domain, but it is *not possible* to perform any meaningful inverse transformation from the reduced set of $n + 1$ Chow coefficients to the functional domain, even for the linearly separable class of functions.†

However, there are two related features of academic interest. First, we have seen in Fig. 3.4 that there are certain functions other than the linearly separable class that may be defined explicitly by the first $n + 1$ (Chow) parameters, and, second, the full set of 2^n spectral coefficients always contains redundancy, in that not all the 2^n coefficients are essential in order to define any given function. The last comment may be expressed in an informal way by saying that if we know the majority of spectral coefficients for any given function, then we are not free to allocate unrestricted magnitudes and signs to the remainder, since there are relationships between the coefficient values, and their total must always sum to $\pm 2^n$.

These features together suggest that it may be possible to derive a numerical classification procedure for functions, using a set of coefficients whose number varies from $n + 1$ (minimum) to $\leq 2^n$ (maximum), the actual number necessary increasing as one moves from the most restrictive class of functions to the unrestricted class of all 2^{2^n} functions. We know that the $n + 1$ (Chow) parameters suffice for the linearly separable class, class G in Fig. 3.4, and also for the next slightly larger class F, but there is no known work published to date which considers the increasingly wider function classification groups E, D, C, B, A of Fig. 3.4, and how each may be unambiguously classified by some minimum set of positive canonic spectral coefficients. It may be of interest to emphasize the build-up in the number of spectral coefficients with n, which follows the familiar Pascal triangle structure shown in Table 3.8, and it is intriguing to conjecture whether the incorporation of increasingly higher-ordered coefficients may in some way be employed in the increasingly wider function classification groups of Fig. 3.4.

As a final point of consideration in this area of function classification, let us briefly link back to the algebraic SD classification procedure considered in Section 3.2 and see how the subsequent spectral classifications relate to this earlier algebraic procedure.

Consider the Chow-parameter classification for the positive linearly separable function

$$f(X) = x_1x_2x_3 + x_4(x_1 + x_2)$$

† Note that none of the linearly separable class of functions of $n \geq 2$ have second- and higher-order coefficients which are all zero-valued in the full canonic spectral classification. If this situation did exist, than all the function information would be contained in the first $n + 1$ coefficients, and the inverse transform $[\mathbf{T}^n]^{-1}$ would reconstruct the function. However, this situation never arises.

Table 3.8. The increase in number of spectral coefficients with increasing n.[a]

n													Total number
0							1						1
1						1		1					2
2					1		2		1				4
3				1		3		3		1			8
4			1		4		6		4		1		16
5		1		5		10		10		5		1	32
⋮													⋮
n	1	n		$\binom{n}{2}$			$\cdots$			$\binom{n}{2}$	n	1	2^n

[a] Left-hand entry, the zero-order coefficient s_0; next entry, the primary or first-order coefficients s_i, $i = 1$ to $n, \ldots$; and the right-hand entry, the highest-order coefficient $s_{12\ldots n}$.

The Chow parameters are

$$\begin{matrix} b_0; & b_1 & b_2 & b_3 & b_4 \\ 2 & 6 & 6 & 2 & 10 \end{matrix}$$

which immediately shows that the function belongs to the Chow positive canonic classification 10, 6, 6, 2, 2. It will be recalled that permutations within the b_i's, $i = 1$ to n, merely correspond to input permutation (P) operations, but permutations involving b_0, which are freely permissible, must involve Exclusive-OR operations as introduced during the unrestricted spectral classification procedure.

To convert the Chow parameters of the given function (or indeed the full 2^n spectral parameters) into the positive canonic order, the following operations may be performed

	b_0 (s_0)	b_1 (s_1)	b_2 (s_2)	b_3 (s_3)	b_4 (s_4)	s_{14}	s_{24}	s_{34}
(a) Given function $f(X)$:	2	6	6	2	10			
(b) Operation $f(X) \oplus x_4$:	10	*	*	*	2	6	6	2
(c) Operation $x_1 \rightarrow (x_1 \oplus x_4)$:	10	6	*	*	2	*	6	2
(d) Operation $x_2 \rightarrow (x_2 \oplus x_4)$:	10	6	6	*	2	*	*	2
(e) Operation $x_3 \rightarrow (x_3 \oplus x_4)$:	10	6	6	2	2	*	*	*

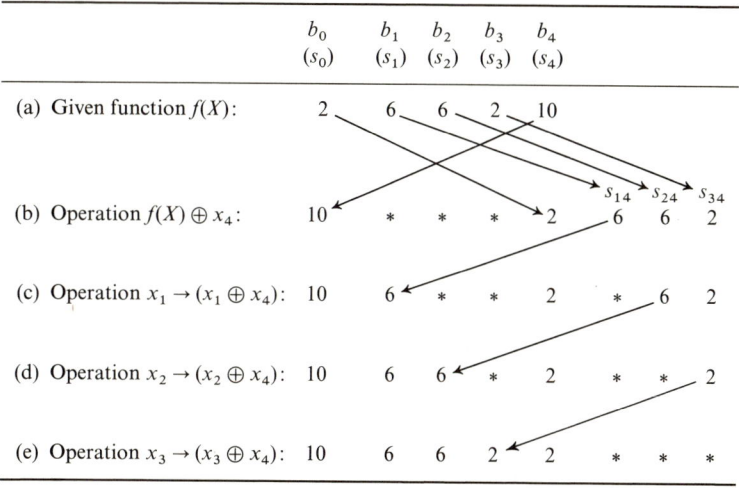

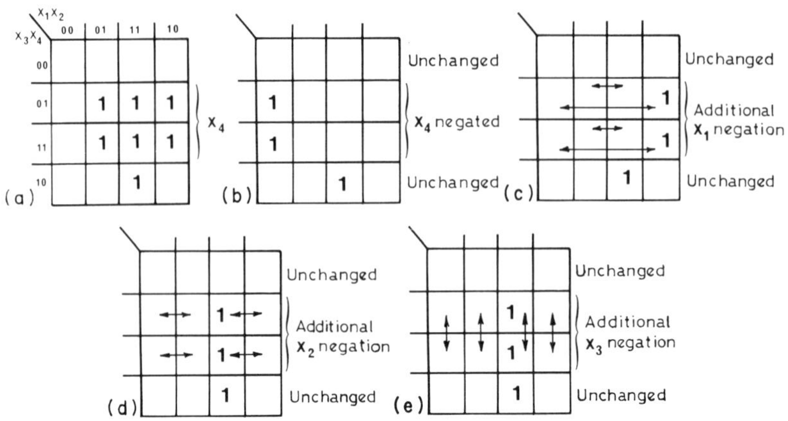

Fig. 3.7. The invariance operations in the function domain to interchange $b_0 \leftrightarrow b_4$, leaving b_1, b_2, b_3 unchanged. (a) Given function $f(X)$; (b) $f(X) \oplus x_4$; (c) $x_1 \rightarrow x_1 \oplus x_4$; (d) $x_2 \rightarrow x_2 \oplus x_4$; (e) $x_3 \rightarrow x_3 \oplus x_4$, giving the final positive canonic function $x_1x_2(x_3 + x_4)$, $= 10$; 6, 6, 2, 2; . . . .

The Karnaugh map plot of these invariance operations is shown in Fig. 3.7. Note that we need not trouble to evaluate the spectral coefficient values marked * for the purpose of this discussion.

What we have done in translating the spectral coefficient of value 10 down from the b_4 (s_4) position into the zero-order position is initially to perform the disjoint spectral translation operation $f(X) \oplus x_4$, but this operation simultaneously transforms all remaining first-order coefficients to second-order positions, outside the Chow-parameter range. In order to return each of these coefficients to its original position, such that we finally have interchanged b_0 and b_4 only, it is necessary to perform the further $n - 1$ invariance operations of $x_i \rightarrow (x_i \oplus x_4)$, $i = 1, 2, 3$.†

This procedure may be followed in Fig. 3.7. The sum total of the operations has

(a) left the original function $f(X)$ unchanged in the $\bar{x}_4$ area of the function, and

(b) negated the original function in the x_4 area, followed by negation of each of the remaining x_i inputs within this area,

giving the final positive canonic function

$$f'(X) = \bar{x}_4 f(X) + x_4 f^{D}\{f(X) \oplus x_4\} \tag{3.18}$$

Hence the final function $f'(X)$ with the spectrum 10; 6,6,2,2; ...is the posi-

† Note that it may not always be essential to perform all these further operations, should the first-order values become maximised at some interim stage. However the full set of $n - 1$ operations will always restore the original first-order coefficients.

tive canonic classification function to which $f(X)$ belongs under the spectral (and Chow-parameter) classification procedure.

We obtain the same final result by the algebraic SD classification, but the steps involved are indirect since the procedure involves the introduction of the higher-order self-dualized function. In the unrestricted spectral domain, the generation of $f^{\mathrm{SD}}(X_{n+1})$ is as follows.

As will be covered in Section 4.2.1, given any function $f(X)$ and its Shannon decomposition

$$f(X) = \{\bar{x}_n f(x_1, \ldots, x_{n-1}, 0) + x_n f(x_1, \ldots, x_{n-1}, 1)\}$$

(see Eq. (4.2)), then the spectrum $\mathbf{S}$ of $f(X)$ is given by the ordered vector additions and subtractions

$$\mathbf{S} = \begin{bmatrix} \mathbf{S}_0 + \mathbf{S}_1 \\ \mathbf{S}_0 - \mathbf{S}_1 \end{bmatrix} \tag{3.19}$$

where $\mathbf{S}_0$ is the spectrum of $f(x_1, \ldots, x_{n-1}, 0)$ and $\mathbf{S}_1$ is the spectrum of $f(x_1, \ldots, x_{n-1}, 1)$, where $\mathbf{S}$, $\mathbf{S}_0$ and $\mathbf{S}_1$ are in Hadamard order.

Let x_p be the further variable introduced in $f^{\mathrm{SD}}(X_{n+1})$ additional to $x_1, \ldots, x_n$, giving us

$$f^{\mathrm{SD}}(X_{n+1}) = \{\bar{x}_p f^{\mathrm{D}}(X) + x_p f(X)\}$$

Let the spectrum $\mathbf{S}^{\mathrm{D}}$ of $f^{\mathrm{D}}(X)$ and the spectrum $\mathbf{S}$ of $f(X)$ be distinguished as follows:

$$\mathbf{S}^{\mathrm{D}} = \begin{bmatrix} s_0^{\mathrm{d}} \\ s_1^{\mathrm{d}} \\ s_2^{\mathrm{d}} \\ s_{12}^{\mathrm{d}} \\ \vdots \\ s_{12\ldots n}^{\mathrm{d}} \end{bmatrix}, \quad \mathbf{S} = \begin{bmatrix} s_0^{\mathrm{f}} \\ s_1^{\mathrm{f}} \\ s_2^{\mathrm{f}} \\ s_{12}^{\mathrm{f}} \\ \vdots \\ s_{12\ldots n}^{\mathrm{f}} \end{bmatrix}$$

Then by Eq. (3.19), the spectrum of $f^{\mathrm{SD}}(X_{n+1})$ is given by

$$\mathbf{S}^{\mathrm{SD}} = \begin{bmatrix} s_0 \\ s_1 \\ s_2 \\ \vdots \\ s_{12\ldots n} \\ s_p \\ s_{1p} \\ s_{2p} \\ \vdots \\ s_{12\ldots np} \end{bmatrix} = \begin{bmatrix} s_0^{\mathrm{d}} + s_0^{\mathrm{f}} \\ s_1^{\mathrm{d}} + s_1^{\mathrm{f}} \\ s_2^{\mathrm{d}} + s_2^{\mathrm{f}} \\ \vdots \\ s_{12\ldots n}^{\mathrm{d}} + s_{12\ldots n}^{\mathrm{f}} \\ s_0^{\mathrm{d}} - s_0^{\mathrm{f}} \\ s_1^{\mathrm{d}} - s_1^{\mathrm{f}} \\ s_2^{\mathrm{d}} - s_2^{\mathrm{f}} \\ \vdots \\ s_{12\ldots n}^{\mathrm{d}} - s_{12\ldots n}^{\mathrm{f}} \end{bmatrix}$$

However, recalling that by definition $f^{D}(X) = \overline{f(\bar{X})}$, this algebraic relationship requires that all the spectral coefficients in $\mathbf{S}^{D}$ and $\mathbf{S}$ must have the same *magnitudes*, but zero- and all even-order coefficients must have *opposite* signs in $\mathbf{S}^{D}$ and $\mathbf{S}$, and all odd-order coefficients must have the *same* signs. Hence the above spectrum for $f^{SD}(X_{n+1})$ becomes as follows, when reassembled in spectral-order row format:

$$\begin{array}{lccccccccc} S^{SD} = s_0; & s_1 & s_2 & \cdots s_n & s_p; & s_{12} & s_{13} \cdots; & s_{123} & \cdots \\ 0 & 2s_1^d & 2s_2^d & 2s_n^d & 2s_0^d & 0 & 0 & 2s_{123}^d & \end{array}$$

Applying these relationships to the previous example $f(X) = x_1x_2x_3 + x_4(x_1 + x_2)$, whose full spectrum is

$$\begin{array}{ccccccccccc} s_0; & s_1 & s_2 & s_3 & s_4; & s_{12} & s_{13} & s_{14} & s_{23} & s_{24} & s_{34}; \\ 2 & 6 & 6 & 2 & 10 & 2 & -2 & -2 & -2 & -2 & 2 \end{array}$$

$$\begin{array}{ccccc} s_{123} & s_{124} & s_{134} & s_{234}; & s_{1234} \\ 2 & -6 & -2 & -2 & 2 \end{array}$$

the self-dualized function $f^{SD}(X_{n+1})$ with the self-dualizing variable x_5 has the spectrum

$$\begin{array}{cccccccccccccccc} s_0; & s_1, & s_2, & s_3, & s_4, & s_5; & s_{12}, & s_{13}, & s_{14}, & s_{15}, & s_{23}, & s_{24}, & s_{25}, & s_{34}, & s_{35}, & s_{45}; \\ 0 & 12 & 12 & 4 & 20 & -4 & 0 & 0 & 0 & 0 & 0 & 0 & 0 & 0 & 0 & 0 \end{array}$$

$$\begin{array}{cccccccccc} s_{123}, & s_{124}, & s_{125}, & s_{134}, & s_{135}, & s_{145}, & s_{234}, & s_{235}, & s_{245}, & s_{345}; \\ 4 & -12 & -4 & -4 & 4 & 4 & -4 & 4 & 4 & -4 \end{array}$$

$$\begin{array}{cccccc} s_{1234}, & s_{1235}, & s_{1245}, & s_{1345}, & s_{2345}; & s_{12345} \\ 0 & 0 & 0 & 0 & 0 & -4 \end{array}$$

The zero-magnitude s_0 value confirms that the self-dualized function has an equal number of true and false minterms; all even-order coefficients are also zero.

The Shannon decomposition of this spectrum $\mathbf{S}^{SD}$ about any x_i, $i = 1$ to 5, is readily available from the converse of Eq. (3.19), that is,

$$\begin{aligned} \mathbf{S}_0 &= \tfrac{1}{2}(\mathbf{S}^0 + \mathbf{S}^1) \\ \mathbf{S}_1 &= \tfrac{1}{2}(\mathbf{S}^0 - \mathbf{S}^1) \end{aligned} \tag{3.20}$$

where $\mathbf{S}^0$, $\mathbf{S}^1$ are the ordered halves of the spectrum $\mathbf{S}^{SD}$ in Hadamard order. (See also Eq. (4.6).)

Clearly, if we perform this operation about x_5, we reconstruct our starting example function $f(X)$ and $f^{D}(X)$, but if we decompose about x_4, corresponding to the highest-value primary spectral coefficient, we obtain

	s_0;	s_1, s_2, s_3, s_5;				$s_{12}, s_{13}, s_{15}, s_{23}, s_{25}, s_{35}$;					
$f_{x_4=0}(X)$:	10	6	6	2	−2	−6	−2	2	−2	2	−2
$f_{x_4=1}(X)$:	−10	6	6	2	−2	6	2	−2	2	−2	2

$s_{123}, s_{125}, s_{135}, s_{235}$;				s_{1235}
2	−2	2	2	−2
2	−2	2	2	2

Inspection confirms that this decomposition is a function $f(X)$ and its dual $f^{D}(X)$ as algebraic theory requires.

Negation operations on (either of) these functions will construct the positive canonic classification function of Fig. 3.7e, with its spectrum of 10; 6,6,2,2; .... Hence the SD classification procedure implements (within function and variable negation) the spectral classification procedure of the interchange of a first-order spectral coefficient with the zero-order coefficient, but in order to do so the published algebraic procedure has introduced the additional $(n + 1)$th variable. However, it may be noted that the equivalent algebraic procedure could be defined without recourse to the self-dualized function, but not so succinctly, by generalization of the algebra of Eq. (3.18).

Thus, to conclude:

(a) Since all linearly separable functions may be classified by consideration of the first $n + 1$ spectral coefficients (Chow parameters) only, and

Table 3.9. Concluding statistics on function classification.[a]

		n			
		≤ 2	≤ 3	≤ 4	≤ 5
(a)	Total no. of functions	16	256	65536	$\simeq 4.3 \times 10^9$
	No. of entries in the SD classification	3	7	83	109958
	No. of entries in the spectral classification	2	3	8	48
(b)	No. of linearly separable functions	14	104	1882	94572
	No. of entries in the SD classification	2	3	7	21
	No. of entries in the spectral (Chow) classification	2	3	7	21

[a] See also Tables 3.5 and 3.7.

since the spectral operation of $b_i \leftrightarrow b_0$ corresponds to the SD-algebraic classification, then the SD-algebraic classification for threshold functions contains precisely the same number of classification entries as the Chow-parameter classification tables.

(b) If the spectral translation operations were confined to $b_i \leftrightarrow b_0$, $i = 1$ to n, then spectral classification and SD classification would remain the same for all functions. However, the full spectral translation operations permit invariance operations between *all orders* of spectral coefficients, for example, as illustrated in Table 3.7, and hence the unrestricted spectral classification is more efficient than the algebraic procedures.

Table 3.9 indicates this comparison. It is left as an informative exercise for the reader to compute the spectrum of each of the seven self-dualized canonic SD-representative functions for $n \geq 3$ listed in Section 3.2, and to confirm that all seven are covered by the three spectral classification entries (1), (3) and (5) of Table 3.6.

3.7 UNIVERSAL LOGIC ELEMENTS

We have noted in, for example, Fig. 3.1 how a single representative function[†] may be used in the realisation of all functions which are represented by this function, the appropriate invariance operations having to be implemented in order to realise any specific function required. However, this development has not pursued the possibility of one single core function from which all 2^{2^n} functions of a given n can be realised.

A single function that has the capability of realising all 2^{2^n} functions of a given n is termed a universal logic gate (ULG). or alternatively, a universal logic module (ULM) or universal logic circuit (ULC). We shall employ the terminology ULG from henceforth. A ULG circuit that has the capability of realising all functions of two input variables is termed a ULG.2; for the general case the terminology ULG.n is used. Note that a ULG.n circuit is always capable of realising all functions of less than n variables, and thus has $\leq n$ universality.

The external operations that are normally involved in order to commit a given ULG.n module for a particular function realisation $f(x_1, \ldots, x_n)$ are

(i) negation of the x_i input variables,
(ii) permutation of the x_i input variables,

[†] The various terminologies, canonic function, representative function, prototype function, and core function, are used interchangeably in this subject area.

(iii) negation of the circuit output, and
(iv) use of a constant input.

Thus the three NPN-invariance operations considered in Section 3.2 are relevant. The pre- and post-Exclusive-OR operations of Fig. 3.6 are not normally considered as part of the commitment requirements of a ULG circuit, but, as may be seen, such relationships may be involved inside the perimeter of a ULG circuit.

If with a given ULG.n circuit it is necessary to evoke all operations (i)–(iv) in order to realise all possible functions of n input variables, then such a circuit is termed an

NPN-complete ULG.n circuit

However, if external negation of the output is unnecessary, then we have an

NP-complete ULG.n circuit

If output negation but no input negation are required, we have a

PN-complete ULG.n circuit

Finally, if no external negations at input or output are required, we have a

P-complete ULG.n circuit

Thus permutation of the x_i inputs in order to generate specific output functions is the most significant operation with ULG circuits. It is this that distinguishes them from more familiar Boolean AND, OR, NAND and NOR primitives, which have complete functional symmetry in their input connections, and perform the same duty no matter how their inputs are permutated.†

Figure 3.8 illustrates the basic concept. Points of note are:

(a) The input terminals of the ULG.n circuit are labelled $y_1, y_2, \ldots, y_m$, where $m \geq n$.

(b) The y_i input connections are taken from the set $\{0, 1, x_1, \bar{x}_1, \ldots, x_n, \bar{x}_n\}$.

(c) The number of possible different permutations of $\{0, 1, x_1, \bar{x}_1, \ldots, x_n, \bar{x}_n\}$ to the m inputs must be $\geq \frac{1}{2}(2^{2^n})$ in order to realise all possible functions of n input variables, assuming output negation is also invoked, or $\geq 2^{2^n}$ without output negation.

Clearly the most significant parameter in the design of any ULG.n is the minimisation of m for any given n.

† Note that the term "polypheck" may also be encountered for fully functional primitives such as NAND and NOR gates. However, this terminology is mainly found in multiple-valued logic applications, rather than in binary applications.[31,32]

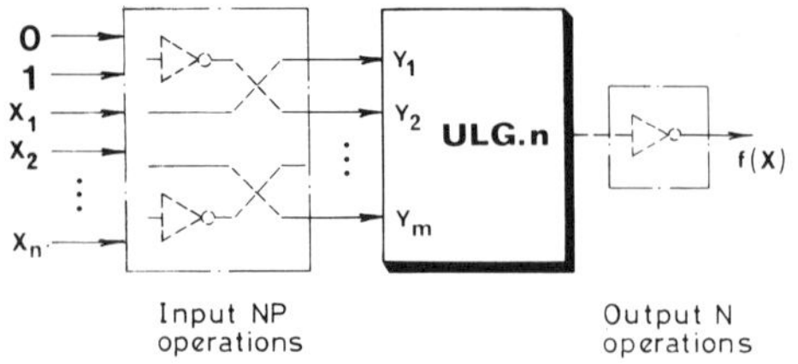

Fig. 3.8. The basic concept of the ULG.*n*, where *f*(*X*) = any function of $\leq n$ variables.

Investigations into the formulation of ULG specifications have been extensively pursued.[4,23–28] It has been shown that for $n = 2$, the minimum number of input terminals necessary is $m = 3$ for both NPN-complete and NP-complete circuits, whilst for $n = 3$ the minimum value is $m = 5$, again both for NPN- and NP-complete circuits. For $n = 4$, the situation has not been exhaustively reported, but $m = 7$ has been noted.[25] The original developments in this area were based upon algebraic considerations; later work, however, is based upon the consideration of spectral classification data, and it is this more relevant area that we shall briefly discuss herewith.

Should we have available all the NPN (or NP)-representative functions for any *n*, then we may realise any combinatorial function of *n* (or fewer) variables by appropriate NPN (or NP) hardwiring. If we can now determine one circuit from which all these representative functions may be realised, then this one function is a universal NPN (or NP)-complete ULG.*n* circuit. Our search, therefore, is to determine a function that subsumes all the required representative functions as follows:

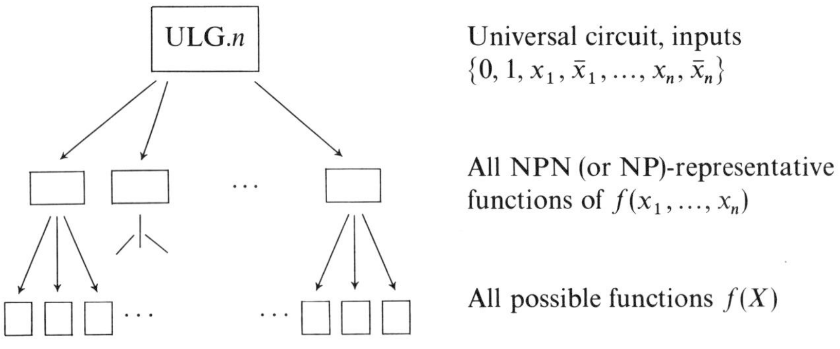

The derivation of ULG.*n* candidates using spectral data involves four further spectral operations, which are required in addition to the negation and permutation invariance operations detailed in Eqs. (3.13) to (3.15), these being

(i) replacement of any input variable x_i, $i = 1$ to n, by logic 0: this results in a compaction involving all spectral coefficients of

$$\begin{aligned} s_0' &= \tfrac{1}{2}(s_0 + s_i) \\ s_j' &= \tfrac{1}{2}(s_j + s_{ij}) \\ &\vdots \\ s_{jk}' &= \tfrac{1}{2}(s_{jk} + s_{ijk}) \\ &\vdots \end{aligned} \tag{3.21}$$

$i \neq j \neq k, \ldots, i, j, k \in 1$ to n, and where $s_0', s_j', \ldots$ are the resulting spectral coefficients of the network with one fewer independent input variable;

(ii) replacement of any input variable x_i, $i = 1$ to n, by logic 1: this results in a compaction involving all spectral coefficient of

$$\begin{aligned} s_0' &= \tfrac{1}{2}(s_0 - s_i) \\ s_j' &= \tfrac{1}{2}(s_j - s_{ij}) \\ &\vdots \\ s_{jk}' &= \tfrac{1}{2}(s_{jk} - s_{ijk}) \\ &\vdots \end{aligned} \tag{3.22}$$

$i, j, k, \ldots, s_0', s_j', \ldots$ as in (i);

(iii) replacement of any input variable x_i by variable x_j, $i \neq j$: this results in the compaction involving all spectral coefficients of

$$\begin{aligned} s_0' &= \tfrac{1}{2}(s_0 + s_{ij}) \\ s_j' &= \tfrac{1}{2}(s_j + s_i) \\ s_k' &= \tfrac{1}{2}(s_k + s_{ijk}) \\ &\vdots \\ s_{jk}' &= \tfrac{1}{2}(s_{jk} + s_{ik}) \\ &\vdots \end{aligned} \tag{3.23}$$

$i, j, k, \ldots, s_0', s_j', \ldots$ as in (i);

(iv) finally, replacement of any input variable x_i by variable $\bar{x}_j$, $i \neq j$: this results in the compaction involving all spectral coefficients of

$$\begin{aligned} s'_0 &= \tfrac{1}{2}(s_0 - s_{ij}) \\ s'_j &= \tfrac{1}{2}(s_j - s_i) \\ s'_k &= \tfrac{1}{2}(s_k - s_{ijk}) \\ &\vdots \\ s'_{jk} &= \tfrac{1}{2}(s_{jk} - s_{ik}) \\ &\vdots \end{aligned} \tag{3.24}$$

$i, j, k, \ldots, s'_0, s'_j, \ldots$ as in (i).

Each of these operations reduces the dimensions of the resulting network from n to $n - 1$ input variables, input variable x_i being discarded, the factor of $\frac{1}{2}$ in the spectral additions or subtractions reflecting this reduction in dimension. These spectral operations may be clearly illustrated on spectral coefficient maps, as shown in Fig. 3.9.

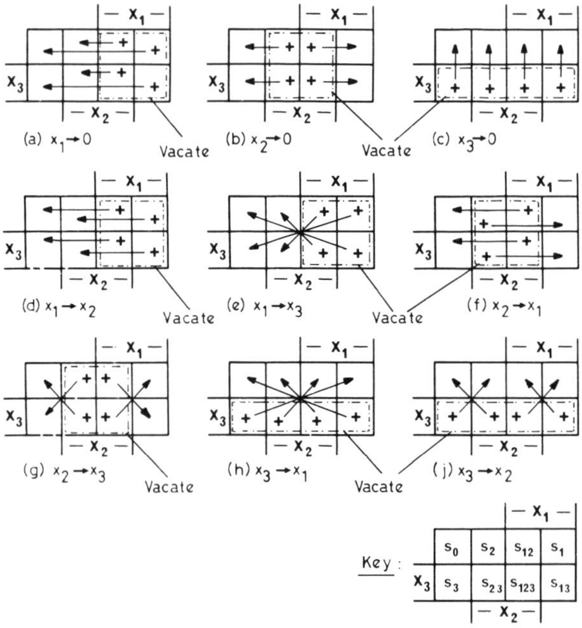

Fig. 3.9. The spectral operations that reduce the number of variables from n to $n - 1$. (a)–(c) Setting x_i to 0, addition of spectral coefficient values, leaving x_i area void; for x_i set to 1, subtract s_i's instead of add. (d)–(j) Setting x_i to x_j, addition of spectral coefficient values, leaving x_i area void; for x_i set to $\bar{x}_j$, subtract s_i's instead of add. (Note that the factor of $\frac{1}{2}$ is necessary on all coefficient pair additions or subtractions.)

Table 3.10. The spectral classification of all NPN canonic functions of $n \leq 3$.[a]

No. of independent input variables n	NPN classification of functions of ≤ 3 input variables								
	Spectral coefficients								Algebraic expression
	s_0;	s_1	s_2	s_3;	s_{12}	s_{13}	s_{23};	s_{123}	
0	1								$f_A = 0$
1	0	2							$f_B = x_1$
2	2	2	2		-2				$f_C = x_1x_2$
	0	0	0		4				$f_D = x_1 \oplus x_2$
3	0	0	0	0	0	0	0	8	$f_E = x_1 \oplus x_2 \oplus x_3$
	6	2	2	2	-2	-2	-2	2	$f_F = x_1x_2x_3$
	2	6	2	2	-2	-2	2	-2	$f_G = x_1(x_2 + x_3)$
	2	2	2	2	-6	2	2	2	$f_H = x_1x_2 + \bar{x}_1\bar{x}_2x_3$
	2	2	2	2	2	2	2	-6	$f_I = \bar{x}_1x_2x_3 + x_1\bar{x}_2x_3 + x_1x_2\bar{x}_3$
	4	4	0	0	0	0	4	-4	$f_J = x_1x_2\bar{x}_3 + x_1\bar{x}_2x_3$
	4	0	0	0	4	-4	4	0	$f_K = \bar{x}_1x_2\bar{x}_3 + x_1\bar{x}_2x_3$
	0	0	4	4	-4	4	0	0	$f_L = x_1x_2 + \bar{x}_1x_3$
	0	0	0	4	0	4	4	-4	$f_M = x_1x_2\bar{x}_3 + \bar{x}_2x_3 + \bar{x}_1x_3$
	0	4	4	4	0	0	0	-4	$f_N = x_1x_2 + x_1x_3 + x_2x_3$

[a] Obtainable from Table 3.6 by appropriate spectral translation operations on the $n \leq 3$ data.

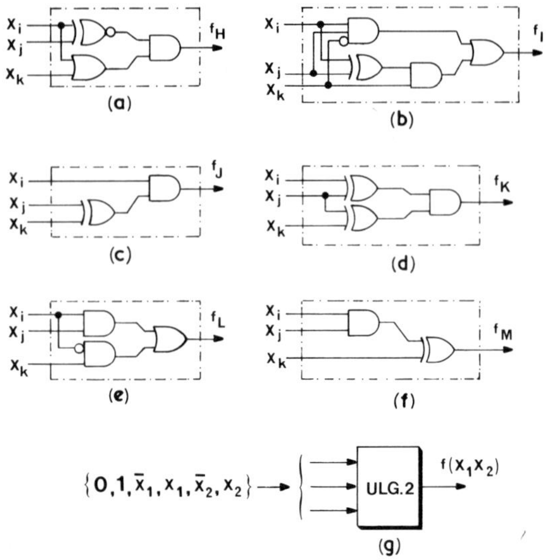

Fig. 3.10. The six possible ULG.2 circuit configurations with three input terminals. (Note that negation of inputs and outputs may be freely made without impairing universality.)

(a) $f_{\mathrm{H}} = x_i x_j + \bar{x}_i \bar{x}_j x_k = (x_i \overline{\oplus} x_j)(x_i + x_k)$;
(b) $f_{\mathrm{I}} = \bar{x}_i x_j x_k + x_i \bar{x}_j x_k + x_i x_j \bar{x}_k = (x_i \oplus x_j) x_k + x_i x_j \bar{x}_k$;
(c) $f_{\mathrm{J}} = x_i x_j \bar{x}_k + x_i \bar{x}_j x_k = x_i (x_j \oplus x_k)$;
(d) $f_{\mathrm{K}} = \bar{x}_i x_j \bar{x}_k + x_i \bar{x}_j x_k = (x_i \oplus x_j)(x_j \oplus x_k)$;
(e) $f_{\mathrm{L}} = x_i x_j + \bar{x}_i x_k$;
(f) $f_{\mathrm{M}} = x_i x_j \bar{x}_k + \bar{x}_j x_k + \bar{x}_i x_k, = x_i x_j \oplus x_k$;
(g) general schematic of all configurations.

We are now in a position to consider the spectral coefficients of, say an $n + 1$ variable function, and how by (possibly repeated) operations (i) to (iv) this function may be reduced to the spectrum of all the representative NPN (or NP) functions of less than $n + 1$ variables. For compactness, let us first illustrate the procedure for the ULG.2 case.

We know from simple arithmetic considerations[25,27] that a minimum of three input terminals ($m = 3$) is necessary on a ULG.2 circuit in order to provide sufficient input permutation capacity. Our procedure, therefore, is to examine the spectrum of all three-variable NPN canonic functions, and determine which (if any) subsume the spectra of all two-variable NPN canonic functions. Any three-variable candidate that does so is an NPN-complete ULG.2 candidate.†

† Note that the NPN class includes the NP class; hence if we perform our investigation on NPN data, we may afterwards extract NP-complete candidates from the wider class of NPN-complete.

The full spectral details of all NPN canonic functions of $n \leq 3$ are given in Table 3.10. We now examine each of the spectral entries f_E to f_N to determine whether the operations of Eqs. (3.21)–(3.24) enable all spectral entries of $n = 2$ to be generated.

An examination of Table 3.10 will show that the spectra of the canonic functions f_E, f_F, f_G and f_N do not possess appropriate spectral coefficient values to enable the spectra of both f_C and f_D to be obtained under any of the available spectral operations. This leaves us with the six possible candidates f_H–f_M. It is a relatively simple search to show that each is able to realise both f_C and f_D; an example of the spectral operations to do so follows (*, void):

	s_0	s_1	s_2	s_3	s_{12}	s_{13}	s_{23}	s_{123}
f_H:	2	2	2	2	−6	2	2	2
Replace $x_3 \to 0$:	$\frac{1}{2}\{4$	4	4	*	−4	*	*	$*\}$
Replace $x_3 \to 1$:	$\frac{1}{2}\{0$	0	0	*	−8	*	*	$*\}$
f_I:	2	2	2	2	2	2	2	−6
Replace $x_3 \to 0$:	$\frac{1}{2}\{4$	4	4	*	−4	*	*	$*\}$
Replace $x_3 \to 1$:	$\frac{1}{2}\{0$	0	0	*	8	*	*	$*\}$
f_J:	4	4	0	0	0	0	4	−4
Replace $x_3 \to 0$:	$\frac{1}{2}\{4$	4	4	*	−4	*	*	$*\}$
Replace $x_1 \to 1$:	$\frac{1}{2}\{0$	*	0	0	*	*	8	$*\}$
f_K:	4	0	0	0	4	−4	4	0
Replace $x_2 \to 0$:	$\frac{1}{2}\{4$	4	*	4	*	−4	*	$*\}$
Replace $x_3 \to x_1$:	$\frac{1}{2}\{0$	0	0	*	8	*	*	$*\}$
f_L:	0	0	4	4	−4	4	0	0
Replace $x_3 \to 0$:	$\frac{1}{2}\{4$	4	4	*	−4	*	*	$*\}$
Replace $x_3 \to \bar{x}_2$:	$\frac{1}{2}\{0$	0	0	*	−8	*	*	$*\}$
f_M:	0	0	0	4	0	4	4	−4
Replace $x_3 \to 0$:	$\frac{1}{2}\{4$	4	4	*	−4	*	*	$*\}$
Replace $x_2 \to 1$:	$\frac{1}{2}\{0$	0	*	0	*	8	*	$*\}$

These results are within NPN permutations of the two canonic NPN functions f_C and f_D. Other operations on the spectra of f_H to f_M will produce similar results.

Hence we are in a position to state that there are six and only six $m = 3$ candidates to adopt as possible ULG.2 circuits. These are illustrated in Fig. 3.10. However, it should be appreciated that there are several variants possible on each of these six candidates, since negation of inputs and outputs may be freely made without impairing in any way their $n \leq 2$ universality.

A detailed discussion of the evaluation of these six basic candidates and their 42 variants and use are available.[27,29,30] Points of interest are:

(a) Candidates f_H, f_L and f_M prove to be NP-complete, not requiring output negation to be invoked in the realisation of any $n \leq 2$ function; candidates f_I, f_J and f_K however are only NPN-complete.

(b) Candidate f_L will be noted as being a standard Shannon expansion multiplexer configuration, whilst candidate f_M will be noted as being the realisation of the Exclusive-OR Reed–Muller canonic expansion; the remaining four candidates, however, do not correspond to any standard algebraic expansion.

(c) There is no possible N-complete ULG.2 circuit with the minimum $m = 3$ input structure.

When this procedure is extended to search all $n = 4$ canonic spectra for candidates that subsume all $n \leq 3$ spectra, then it will be found that no $n = 4$ candidate will cover all $n \leq 3$ entries. A minimum of $m = 5$ is required, at which value both NP-complete and NPN-complete candidates will be found. Note that in these procedures it is only necessary to show failure to realise any one of the $n \leq 3$ entries in order to discard the $n = 4$ candidate under consideration. Also, rather than implementing the $\frac{1}{2}$ factor of Eqs. (3.21)–(3.24), it generally proves more convenient to retain the spectral dimensions of the functions that are being examined, rather than the reduced dimensions of the functions that are subsumed. This will be noted in the published developments.[27]

Thus spectral coefficients have proved to be particularly useful in this universal-logic area, since although algebraic developments that parallel the spectral procedures are available, the numerical format of the spectral domain is significantly more convenient.

3.8 CHAPTER SUMMARY

In this chapter we have considered the classification of combinatorial logic functions, both in the function domain using an algebraic approach, and also in the spectral domain using operations on the spectral coefficients. The algebraic approach is of greater antiquity, but all procedures are mirrored by available operations in the spectral domain, which in general prove to be more convenient and compact than the function domain procedures.

We have based all these spectral classification discussions on spectra obtained using the Hadamard or Rademacher–Walsh orthogonal transformation, that is, where each spectral coefficient is a global parameter of the given

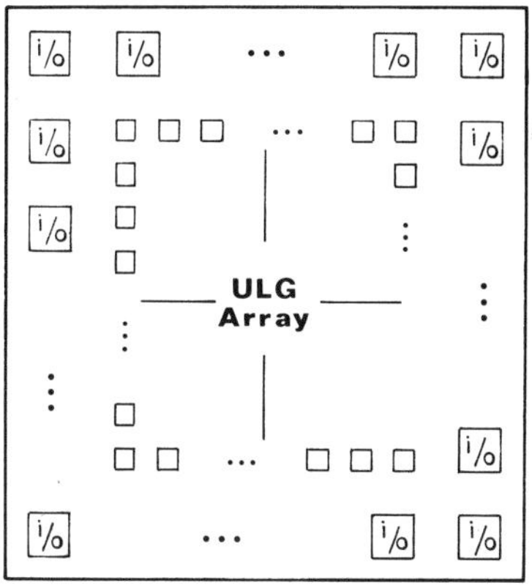

Fig. 3.11. The principle of the master-chip gate array, containing a matrix of standard cells with peripheral I/O buffer circuits.

function $f(X)$. This corresponds to the situation in all known publications to date. It may well, however, be possible to compile a similar classification structure and usage using the spectra obtained by the Haar transformation, although it is not intuitively obvious that any advantages would accrue with this alternative approach.

Without doubt, the greatest success to date of the spectral domain data has been in function classification, leading to the formulation of universal logic gates and hence to their potential adoption in master-chip gate arrays (see Fig. 3.11). The advantages of master-chip gate arrays to original equipment manufacturers is outside the scope of this volume; it suffices here to say that gate arrays represent a major growth area of present-day digital electronics.[33–38]

References

1. Hellerman, L., A catalogue of three-variable OR-invert and AND-invert logic circuits, *Trans. IEEE* **EC-12,** 198–223 (1963).
2. Staff of the Harvard Computational Laboratory, Synthesis of Electronic Computing Systems, Harvard University Press, Cambridge, Massachusetts, 1957.
3. Golomb, S. W., On the classification of Boolean functions, *Trans. IRE* **CT-6** (Suppl.) 176–186 (May 1959).
4. Hurst, S. L., "The Logic Processing of Digital Signals." Crane-Russak, New York, and Edward Arnold, London, 1978.
5. Muroga, S., "Threshold Logic and Its Application." Wiley, New York, 1971.
6. Goto, E., and Takahashi, H., Some theorems useful in threshold logic for enumerating Boolean functions, *Proc. IFIP Congress*, 747–751 (August 1962).

7. Toda, I., On the number of types of self-dual logic functions, *Trans. IRE* **EC-11,** 282–284 (1962).
8. Winder, R. O., Fundamentals of threshold logic, Air Force Cambridge Research Laboratory Report No. 1, Contract AFCRC-68-0066, January 1968.
9. Dertouzos, M. L., "Threshold Logic: A Synthesis Approach." MIT Press, Cambridge, Massachusetts, 1965.
10. Lewis, P. M., and Coates, C. L., "Threshold Logic." Wiley, New York, 1967.
11. Hurst, S. L., "Threshold Logic: An Engineering Survey." Mills and Boon, London, 1971 (translated "Schwellwertlogik." UTB, Heidelberg, 1974).
12. Sheng, C. L., "Threshold Logic." Academic Press, New York, 1969.
13. Chow, C. K., On the characterization of threshold functions, *SCTLD*, IEEE Special Publication No. S134, 34–38 (September 1961).
14. Minnick, R. C., Linear input logic, *Trans. IRE* **EC-10,** 6–16 (1961).
15. Winder, R. O., Threshold functions through $n = 7$, Scientific Report No. 7, Air Force Cambridge Research Laboratory, Contract AF19(604)-8423, October 1969.
16. Gabelman, I. J., The synthesis of Boolean functions using a single threshold element, *Trans. IRE* **EC-11,** 639–642 (1962).
17. Muroga, S., Tsuboi, T., and Baugh, C. A., Enumeration of threshold functions of eight variables, *Trans. IEEE* **C-19,** 818–825 (1970).
18. Yajima, S., and Ibaraki, T., Realization of arbitrary logic functions by completely monotonic functions and its application to threshold logic, *Trans. IEEE* **C-17,** 328–351 (1968).
19. Edwards, C. R., The application of the Rademacher–Walsh transform to Boolean function classification and threshold-logic synthesis, *Trans. IEEE* **C-24,** 48–62 (1975).
20. Karpovsky, M. G., "Finite Orthogonal Series in the Design of Digital Devices." Wiley, New York, 1976.
21. Edwards, C. R., Characterisation of threshold functions under the Walsh transform and linear translation, *IEE Electron. Lett.* **11,** 563–565 (1975).
22. Lechner, R. J., Harmonic analysis of switching functions, *in* "Recent Developments in Switching Theory" (A. Mukhopadhyay, ed.). Academic Press, New York, 1971.
23. Yau, S. S., and Tang, C. K., Universal logic modules and their application, *Trans. IEEE* **C-19,** 141–149 (1970).
24. Preparata, F. P., On the design of universal Boolean functions, *Trans. IEEE* **C-20,** 418–423 (1971).
25. Stone, H. S., Universal logic modules, *in* "Recent Developments in Switching Theory" (A. Mukhopadhyay, ed.). Academic Press, New York, 1971.
26. Edwards, C. R., A special class of universal logic gate and its evaluation under the Walsh transform, *Internat. J. Electron.* **44,** 49–59 (1978).
27. Chen, X., and Hurst, S. L., A consideration of the minimum number of input terminals on universal logic gates and their realisation, *Internat. J. Electron.* **50,** 1–13 (1981).
28. Chen, X., and Wu, X., Derivation of universal logic modules for $n \geq 3$ by algebraic means, *Proc. IEE* **E 128,** 205–211 (1981).
29. New, A. M., The statistical efficiency of universal-logic elements in the realisation of logic functions, *Proc. IEE* **E 129,** 93–99 (1982).
30. Chen, X., and Hurst, S. L., A comparison of universal-logic-module realizations and their application in the synthesis of combinatorial and sequential logic networks, *Trans. IEEE* **C-31,** 140–147 (1982).
31. Tanaka, S., and Tahara, M., Functional completeness and polyphecks in three-valued logic, *Trans. IECE, Japan* **53C,** 111–117 (1970).
32. Yanagita, M., Fuduka, N., Miyoshi, Y., Nakashima, K., and Yamato, K., Synthesis methods for ternary logic functions, based upon NAND-type polyphecks, *Proc. IEEE Internat. Symp. Multiple-Valued Logic, Japan*, 172–174 (1983).

33. Hartman, R. F., Design and market potential for gate arrays, *Lambda* **1**(3), 55–59 (1980).
34. Posa, J. G., Gate arrays, *Electronics*, 145–158 (September 1980).
35. Huffman, G. D., Gate array logic, *EDN* **26**(19), 86–96 (1981).
36. Hurst, S. L., Custom lsi design: The universal-logic-module approach, *Proc. IEEE Internat. Conf. Circuits Computers, New York*, (ICCC 80), 1116–1119 (1980).
37. Braeckelmann, W., Custom-made integrated circuits, *Proc. Internat. Conf. New Trends Integrated Circuits, Paris*, 99–107 (1981).
38. Hurst, S. L., "Custom-Specific Integrated Circuits: Design and Fabrication." Marcel Dekker, New York, 1985.

4

Synthesis of Completely Specified Binary Functions Using Spectral Techniques

4.1 INTRODUCTION

In this chapter we shall consider the use of the spectral coefficients for network synthesis, as an alternative to the various conventional design methods using function domain (Boolean) data.[1-5] We shall not be specifically concerned with the generation of the coefficients, but instead we shall assume that the appropriate transform procedure has been executed on the given functional domain data.

In general, we shall employ the spectra **S** generated from the $\{+1, -1\}$ recoding of the binary data, rather than the alternative spectra **R** generated from the $\{0, 1\}$ data. Whether **S** has been generated by the Hadamard or the Rademacher–Walsh ordered forward transform, or indeed any other alternative order of the orthogonal transform row vectors, is largely immaterial for our purpose, since we shall be primarily concerned with their information content rather than the order of their initial generation. When writing them in horizontal row form, we shall generally group them in the conventional spectral meaning order, namely,

$$s_0; \quad s_1, s_n; \quad s_{12}, \ldots, s_{n-1n}; \quad s_{123}, \ldots; \quad s_{12\ldots n},$$

where s_0 is the zero-order coefficient, s_1 to s_n are the first-order coefficients, ..., up to the final nth order coefficient $s_{12\ldots n}$. However, when executing any transform multiplication procedure we may be required to reorder the above row ordering into the Hadamard column-vector order, namely,

$$\begin{matrix} s_0 \\ s_1 \\ s_2 \\ s_{12} \\ s_3 \\ s_{13} \\ \vdots \end{matrix}$$

A further feature of this chapter is that (a) we shall mainly concern ourselves with the synthesis of completely specified Boolean functions, and (b) we shall not consider synthesis techniques based upon the detection and exploitation of symmetries in the function being synthesised. The reason for these exceptions is that the properties of symmetries in Boolean functions will form the subject matter of Chapter 5, and in the exploitation of symmetries for design purposes don't-care conditions play a significant part. Thus both these aspects will follow later.

There is the possibility of the synthesis of required functions directly from the spectral data by a hardware execution of the inverse transform $[\mathbf{T}^n]^{-1}\mathbf{S} = \mathbf{Y}$. With the Hadamard order of $\mathbf{T}^n$, $\mathbf{S}$, we have $(1/2^n)\mathbf{T}^n\mathbf{S} = \mathbf{Y}$. Thus for each input minterm condition, the appropriate output minterm value is generated by the specific vector multiplication

$$\frac{1}{2^n}\{\mathbf{T}^n_{j*}\mathbf{S}\} = y_j$$

where y_j is the value of $f(X)$ at minterm m_j, $y_j \in \{+1, -1\}$. Equally if we take the spectra $\mathbf{R}$ rather than $\mathbf{S}$, we have

$$(1/2^n)\{\mathbf{T}^n_{j*}\mathbf{R}\} = z_j$$

where z_j is the value of $f(X)$ at minterm m_j, $z_j \in \{0, 1\}$, whence

$$\frac{1}{2^n}\left\{\sum_{\mu=0}^{2n-1}\{t_{j\mu}r_\mu\}\right\} = z_j \tag{4.1}$$

where μ is the row entry in vector $\mathbf{T}^n_{j*}$ and the matching column entry in vector $\mathbf{R}$. Hence a possible realisation technique is as shown in Fig. 4.1.

It is clear that although the input and output data are binary, the hardware implementation of Fig. 4.1 implies operations of a non-binary nature, and hence would involve either precise analogue hardware or recoding of the data in appropriate binary-coded form. In either case noticeable component complexity of the hardware may accrue in comparison with more conventional techniques, such as we shall consider later in this chapter.

It should, however, be noted that if we have unrestricted capability in the arithmetic blocks of Fig. 4.1, then **R** can be the spectrum covering two or more output functions $f(X)$, the final output being the appropriate encoding of the two or more functions for each input word $x_1x_2 \ldots x_n$. This arises since $\mathbf{T}^n\mathbf{Y} = \mathbf{R}$ applies for any values in **Y**; hence we may encode **Y** to be, say, $0, 1, 2, 3$, where $0 = 0, 0$, $1 = 0, 1$, $2 = 1, 0$ and $3 = 1, 1$. Thus the final output $f(X)$ of Fig. 4.1 would be 0, 1, 2 or 3 as appropriate for each input condition.

Thus this arithmetic form of realisation has many attractive possibilities if the premium for the arithmetic operations could be accommodated. It may be noted that the adoption of Harr transformations rather than Hadamard/Rademacher–Walsh variants may provide some simplification in the total arithmetic requirements. For a comprehensive exposition, see Karpovsky.[6] However, we shall continue with network syntheses using more conventional digital devices (gates), using the spectral domain data for feature recognition in these synthesis techniques.

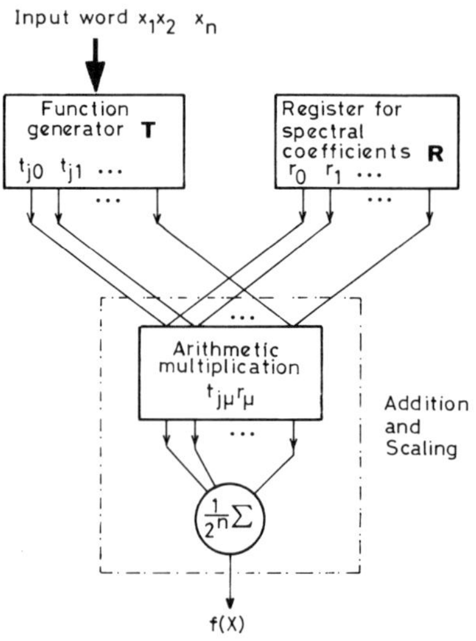

Fig. 4.1. Arithmetic synthesis of $f(X)$ by Eq. (4.1).

4.2 SYNTHESIS BY SPECTRAL DECOMPOSITION

Evaluation of the spectrum of the sum, product, or mod 2 addition (Exclusive-OR) of any two functions in terms of their individual spectra has been considered earlier in Section 2.5. However, for synthesis purposes we normally seek the decomposition of a given function $f(X)$, which is the converse to our previous considerations. Therefore we shall now consider the spectral approach to function decomposition.

4.2.1 Shannon Decompositions

Algebraic techniques for the decomposition of given binary functions have been extensively investigated.[7–10] Here we shall consider the Shannon decomposition of a given function $f(X)$:

$$\begin{aligned} f(X) &= \{\bar{x}_n f(x_1, \ldots, x_{n-1}, 0) + x_n f(x_1, \ldots, x_{n-1}, 1)\} \\ &= \{\bar{x}_n f_0(x_1, \ldots, x_{n-1}) + x_n f_1(x_1, \ldots, x_{n-1})\} \end{aligned} \tag{4.2}$$

By relabelling the x_i inputs, x_n can be any original input x_1 through to x_n. Further decompositions

$$\begin{aligned} f(X) = \{&\bar{x}_{n-1}\bar{x}_n f_0(x_1, \ldots, x_{n-2}) + x_{n-1}\bar{x}_n f_1(x_1, \ldots, x_{n-2}) \\ &+ \bar{x}_{n-1} x_n f_2(x_1, \ldots, x_{n-2}) + x_{n-1} x_n f_3(x_1, \ldots, x_{n-2})\} \end{aligned} \tag{4.3}$$

etc., follow. Function decomposition is illustrated in Fig. 4.2.

A general theorem concerning the relationship between the spectrum of a given function and the spectra of all subfunctions involved in a general Shannon decomposition of the given function may be developed. Consider the decomposition of Eq. (4.2). Let $\mathbf{S}$, $\mathbf{S}_0$ and $\mathbf{S}_1$ be the spectra for functions $f(X)$, $f_0(x_1, \ldots, x_{n-1})$, $f_1(x_1, \ldots, x_{n-1})$, respectively. The full spectrum $\mathbf{S}$ is given by

$$\mathbf{S} = \begin{bmatrix} \mathbf{T}^{n-1} & \mathbf{T}^{n-1} \\ \mathbf{T}^{n-1} & -\mathbf{T}^{n-1} \end{bmatrix} \begin{bmatrix} \mathbf{Y}_0 \\ \mathbf{Y}_1 \end{bmatrix} \tag{4.4}$$

where $\mathbf{Y}_0$ and $\mathbf{Y}_1$ represent the column truthtable vectors for $f_0(x_1, \ldots, x_{n-1})$ and $f_1(x_1, \ldots, x_{n-1})$, respectively. Hence,

$$\mathbf{S} = \begin{bmatrix} \mathbf{T}^{n-1}\mathbf{Y}_0 + \mathbf{T}^{n-1}\mathbf{Y}_1 \\ \mathbf{T}^{n-1}\mathbf{Y}_0 - \mathbf{T}^{n-1}\mathbf{Y}_1 \end{bmatrix} \tag{4.5}$$

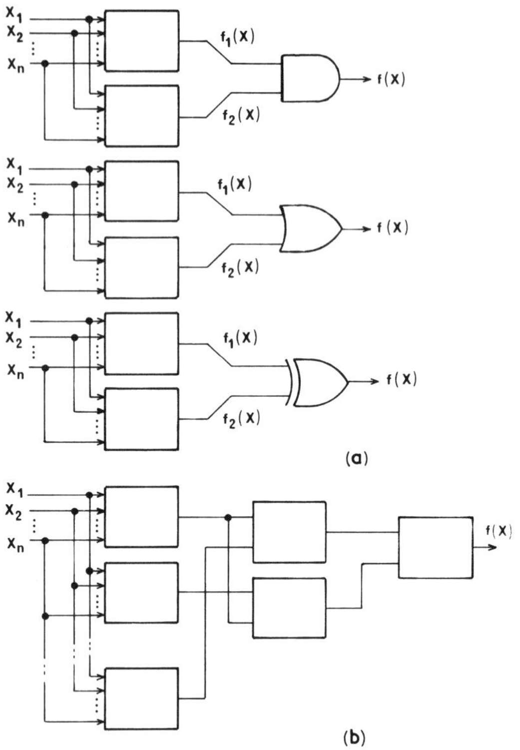

Fig. 4.2. The composition of a function *f*(*X*): (a) two-level AND, OR and Exclusive-OR compositions; (b) general three-level composition.

However, $\mathbf{T}^{n-1}\mathbf{Y}_0 = \mathbf{S}_0$ and $\mathbf{T}^{n-1}\mathbf{Y}_1 = \mathbf{S}_1$, whence it follows that if we partition $\mathbf{S}$ into two proper halves, such that

$$\mathbf{S} = \begin{bmatrix} \mathbf{S}^0 \\ \mathbf{S}^1 \end{bmatrix}$$

we have

$$\mathbf{S}^0 = \mathbf{S}_0 + \mathbf{S}_1 \qquad \text{and} \qquad \mathbf{S}^1 = \mathbf{S}_0 - \mathbf{S}_1$$

Rearranging, we therefore have

$$\mathbf{S}_0 = \tfrac{1}{2}(\mathbf{S}^0 + \mathbf{S}^1) \qquad \text{and} \qquad \mathbf{S}_1 = \tfrac{1}{2}(\mathbf{S}^0 - \mathbf{S}^1) \tag{4.6}$$

To clarify Eq. (4.6), consider the Shannon decomposition

$$f(X) = \bar{x}_3 f_0(x_1, x_2) + x_3 f_1(x_1, x_2)$$

with spectrum

$$\mathbf{S} = \begin{bmatrix} \mathbf{S}^0 \\ \\ \mathbf{S}^1 \end{bmatrix} = \begin{bmatrix} s_0 \\ s_1 \\ s_2 \\ s_{12} \\ s_3 \\ s_{13} \\ s_{23} \\ s_{123} \end{bmatrix}$$

Hence the spectrum $\mathbf{S}_0$ for $f_0(x_1, x_2)$ and the spectrum $\mathbf{S}_1$ for $f_1(x_1, x_2)$ are given by

$$\mathbf{S}_0 = \begin{bmatrix} \frac{1}{2}(s_0 + s_3) \\ \frac{1}{2}(s_1 + s_{13}) \\ \frac{1}{2}(s_2 + s_{23}) \\ \frac{1}{2}(s_{12} + s_{123}) \end{bmatrix}, \quad \mathbf{S}_1 = \begin{bmatrix} \frac{1}{2}(s_0 - s_3) \\ \frac{1}{2}(s_1 - s_{13}) \\ \frac{1}{2}(s_2 - s_{23}) \\ \frac{1}{2}(s_{12} - s_{123}) \end{bmatrix}$$

Equation (4.6) may be rewritten as

$$\begin{bmatrix} \mathbf{S}_0 \\ \mathbf{S}_1 \end{bmatrix} = \frac{1}{2}\left\{\begin{bmatrix} \mathbf{I}^{n-1} & \mathbf{I}^{n-1} \\ \mathbf{I}^{n-1} & -\mathbf{I}^{n-1} \end{bmatrix} \begin{bmatrix} \mathbf{S}^0 \\ \mathbf{S}^1 \end{bmatrix}\right\} \tag{4.7}$$

or,

$$[\mathbf{S}_0 \quad \mathbf{S}_1] = \tfrac{1}{2}\{[\mathbf{S}^0 \quad \mathbf{S}^1]\mathbf{T}^1\}$$

where $\mathbf{T}^1$ is the first-order Hadamard transform $\left[\begin{smallmatrix} 1 & 1 \\ 1 & -1 \end{smallmatrix}\right]$. This may readily be extended to higher orders; for example, the decomposition of Eq. (4.3) yields

$$[\mathbf{S}_0 \quad \mathbf{S}_1 \quad \mathbf{S}_2 \quad \mathbf{S}_3] = \tfrac{1}{4}\{[\mathbf{S}^0 \quad \mathbf{S}^1 \quad \mathbf{S}^2 \quad \mathbf{S}^3]\mathbf{T}^2\} \tag{4.8}$$

where $\mathbf{S}^0, \mathbf{S}^1, \mathbf{S}^2, \mathbf{S}^3$ is the ordered partition of the spectrum $\mathbf{S}$ of $f(X)$. The general result, a formal proof of which has been given by Tokmen,[11] is

$$[\mathbf{S}_0 \quad \mathbf{S}_1 \quad \cdots \quad \mathbf{S}_\beta] = \frac{1}{2^{n-m}}\{[\mathbf{S}^0 \quad \mathbf{S}^1 \quad \cdots \quad \mathbf{S}^\beta]\mathbf{T}^{n-m}\} \tag{4.9}$$

where $\beta = 2^{n-m} - 1$. There are $n - m$ detached variables in the Shannon decomposition corresponding to Eq. (4.9); $\mathbf{S}^0, \mathbf{S}^1, \ldots, \mathbf{S}^\beta$ again is the ordered partition of the given spectrum $\mathbf{S}$ of $f(X)$.

Equation (4.9) may be appreciated as being a partial application of the inverse transform. In the extreme case of $m = 0$, the right-hand side of Eq.

(4.9) becomes the normal inverse transform on the individual spectral coefficients, in row rather than in column order, and the left-hand side becomes a minterm row vector representing $f(X)$.

An immediate use of these spectral relationships is when a Shannon sum-of-products decomposition for a given function is required with a given level of decomposition. A practical aspect, however, may be that some optimum choice of the variables of decomposition (the "control" variables) may be desired in order that the residual subfunctions $f_0(\ldots)$, $f_1(\ldots)$, ..., may be as simple as possible, or as many identical to each other as possible.

If no optimisation is required, then the direct application of the above spectral relationships may be implemented. For example, if the Shannon second-level decomposition of $f(X)$ about variables x_1, x_2 is sought, then with the appropriate reordering and partitioning of **S** into $\mathbf{S}^0$, $\mathbf{S}^1$, $\mathbf{S}^2$ and $\mathbf{S}^3$, Eq. (4.8) will generate the spectra of the four subfunctions $f_0(x_3, \ldots)$, $f_1(x_3, \ldots)$, $f_2(x_3, \ldots)$ and $f_3(x_3, \ldots)$. The inverse transformation on each resulting spectrum $\mathbf{S}_0, \mathbf{S}_1, \mathbf{S}_2, \mathbf{S}_3$ will give the Boolean data for each decomposition.

For this particular case, Eq. (4.8) becomes

$$[\mathbf{S}_0 \quad \mathbf{S}_1 \quad \mathbf{S}_2 \quad \mathbf{S}_3] = \frac{1}{4} \begin{bmatrix} s_0 & s_1 & s_2 & s_{12} \\ s_3 & s_{13} & s_{23} & s_{123} \\ s_4 & s_{14} & s_{24} & s_{124} \\ s_{34} & s_{134} & s_{234} & s_{1234} \\ s_5 & s_{15} & s_{25} & s_{125} \\ s_{35} & s_{135} & s_{235} & s_{1235} \\ \vdots & \vdots & \vdots & \vdots \end{bmatrix} \mathbf{T}^2 \tag{4.8a}$$

Note also that the spectrum column vectors $\mathbf{S}_0, \mathbf{S}_1, \mathbf{S}_2, \mathbf{S}_3$ are produced in Hadamard spectral order rather than in zero-order, first-order,..., nth-order.

Should some optimum decomposition be sought so as to optimise the subfunctions in some sense, there is no guaranteed procedure short of exhaustive examination of all possibilities. Certain work has been pursued on summing subsets of spectral coefficients in **S** to give a first indication of "good" decompositions,[12] this work being based upon the concept that high values for appropriate subset summations indicate above average correlation of the Shannon decomposition product terms, for example, $\bar{x}_1\bar{x}_2$, $\bar{x}_1x_2$, $x_1\bar{x}_2$ and x_1x_2, with $f(X)$, and hence more simple subfunctions $f_0(\ldots)$, $f_1(\ldots)$, ... While this approach provides a guide to the problem, it will not cater for certain conditions, such as identical subfunctions being present, which itself may provide some optimum overall decomposition.

We shall return to this area of logic synthesis later in this chapter when we consider multiplexer realisations, since it may be appreciated that an optimum multiplexer realisation for a given function $f(X)$ is precisely the problem that we have addressed in these developments.

4.2.2 Disjoint Spectral Decompositions

Spectral methods have been pursued for the disjoint decomposition of any given function $f(X)$. Consider a simple disjoint decomposition of $f(X)$, as shown in Fig. 4.3a. The function output may be expressed as

$$f(X) = h(g(x_1, \ldots, x_m), x_{m+1}, \ldots, x_n) \tag{4.10}$$

where $x_1, \ldots, x_m$ and $x_{m+1}, \ldots, x_n$ is a partition of X. Both g and h may be any logical function.

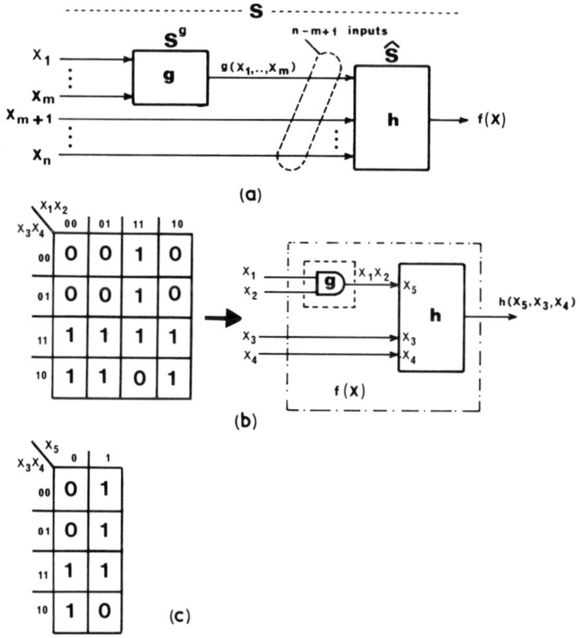

Fig. 4.3. Disjoint synthesis of $f(X)$: (a) general disjoint decomposition, $\mathbf{S}$ = spectrum of $f(X)$, 2^n entries, $\mathbf{S}^g$ = spectrum of function g, 2^m entries; $\hat{\mathbf{S}}$ = spectrum of function h, 2^{n-m+1} entries; (b) specific function $f(X)$ with the disjoint synthesis $f(X) = (x_1x_2)\bar{x}_3 + (\overline{x_1x_2})x_3 + x_3x_4$; (c) function h of example (b), minterms = 0, 1, 0, 1, 1, 0, 1, 1, and function = $x_5\bar{x}_3 + \bar{x}_5x_3 + x_3x_4$.

If we fix each of the latter variables x_{m+1} at u_i, $u_i \in \{0, 1\}$, $1 \leq i \leq n - m$, we obtain a particular subfunction $f_u(x_1, \ldots, x_m)$, where

$$f_u(x_1, \ldots, x_m) = f(x_1, \ldots, x_m, u_1, \ldots, u_{n-m})$$

where $u = \sum_{j=1}^{n-m} u_j 2^{j-1}$.

Recall Eq. (4.9),

$$[\mathbf{S}_0 \quad \mathbf{S}_1 \quad \ldots \quad \mathbf{S}_\beta] = \frac{1}{2^{n-m}}\{[\mathbf{S}^0 \quad \mathbf{S}^1 \quad \ldots \quad \mathbf{S}^\beta]\mathbf{T}^{n-m}\} \tag{4.9}$$

where $\beta = 2^{n-m} - 1$, whence $\mathbf{S}_u$ is the spectrum of $f_u(x_1, \ldots, x_m)$, $0 \leq u \leq \beta$.

If we separately consider the final function h as an $(n - m + 1)$ input function with spectrum $\hat{\mathbf{S}}$ (see Fig. 4.3a), we may again consider the 2^{n-m} subfunctions that result in fixing the inputs x_{m+1} through x_n. This gives

$$[\hat{\mathbf{S}}_0 \quad \hat{\mathbf{S}}_1 \quad \ldots \quad \hat{\mathbf{S}}_\beta] = \frac{1}{2^{n-m}}\{[\hat{\mathbf{S}}^0 \quad \hat{\mathbf{S}}^1 \quad \ldots \quad \hat{\mathbf{S}}^\beta]\mathbf{T}^{n-m}\} \tag{4.11}$$

where $\hat{\mathbf{S}}^0, \ldots, \hat{\mathbf{S}}^\beta$ are the ordered partitions of spectrum $\hat{\mathbf{S}}$.

To clarify Eqs. (4.9) and (4.11) in this context, consider the four-variable function $f(X)$ shown in Fig. 4.3b. The decomposition of f about x_3, x_4 gives

$$f_0(x_1, x_2) = x_1 x_2, \qquad \mathbf{S}_0 = 2; 2, 2; -2$$

$$f_1(x_1, x_2) = x_1 x_2, \qquad \mathbf{S}_1 = 2; 2, 2; -2$$

$$f_2(x_1, x_2) = \overline{x_1 x_2}, \qquad \mathbf{S}_2 = -2; -2, -2; 2$$

$$f_3(x_1, x_2) = 1, \qquad \mathbf{S}_3 = -4; 0, 0; 0$$

For the decomposition of h about x_3, x_4, (see Fig. 4.3c) we have

$$h_0(x_5) = x_5, \qquad \hat{\mathbf{S}}_0 = 0; 2$$

$$h_1(x_5) = x_5, \qquad \hat{\mathbf{S}}_1 = 0; 2$$

$$h_2(x_5) = \bar{x}_5, \qquad \hat{\mathbf{S}}_2 = 0; -2$$

$$h_3(x_5) = 1, \qquad \hat{\mathbf{S}}_3 = -2; 0$$

where x_5 is the input from g to h.

If there exists a disjoint decomposition of $f(X)$, it is clear that there must exist a valid relationship between the spectra of the logic functions f, g, and h. A full discussion of these relationships may be found in Tokmen,[11,13] the original development being for many-valued functions, with binary $(p = 2)$ functions as a special case. A close connection between these considerations and the detection of function symmetries considered in Chapter 5 may also be found.

Here we shall formally state without proof the final relationship for a disjoint decomposition of a binary function $f(X)$:

An n-variable binary function $f(X)$ has a disjoint decomposition of the form

$$f(X) = h(g(x_1, \ldots, x_m), x_{m+1}, \ldots, x_n)$$

if the following relationship is satisfied:

$$\frac{1}{2}\begin{bmatrix} [\mathbf{S}^G] & & 0 \\ & \ddots & \\ 0 & & [\mathbf{S}^G] \end{bmatrix} \begin{bmatrix} \\ \hat{\mathbf{S}} \\ \\ \end{bmatrix} = \begin{bmatrix} \\ \mathbf{S} \\ \\ \end{bmatrix} \tag{4.12}$$

where $\hat{\mathbf{S}}$ is the spectrum of the logic function h, $\mathbf{S}$ is the spectrum of the given function $f(X)$, and $\mathbf{S}^G$ is a 2×2^m matrix given by

$$[\mathbf{S}^G] = \begin{bmatrix} 2^m & \\ 0 & \\ 0 & \mathbf{S}^g \\ \vdots & \end{bmatrix} \tag{4.13}$$

where $\mathbf{S}^g$ is the spectrum of $g(x_1, \ldots, x_m)$. Note that in Eqs. (4.12) and (4.13) all spectra are in Hadamard order.

By illustration, let us apply this procedure to confirm the function $f(X)$ previously illustrated in Fig. 4.3. We take g as the AND function x_1x_2 as before, and confirm the function h necessary to realise $f(X)$.

Computation of the spectra of $f(X)$ and $g(x_1, x_2)$ gives

$$\begin{aligned} \mathbf{S} = {} & s_0; \quad s_1 \; s_2 \; s_3 \; s_4; \quad s_{12} \; s_{13} \; s_{14} \; s_{23} \; s_{24} \; s_{34}; \\ & -2 \quad 2 \; 2 \; 10 \; 2 \quad -2 \; 6 \; -2 \; 6 \; -2 \; -2 \\ & s_{123} \; s_{124} \; s_{134} \; s_{234}; \quad s_{1234} \\ & -6 \; 2 \; 2 \; 2 \quad -2 \end{aligned}$$

and

$$\begin{aligned} \mathbf{S}^g = {} & s_0^g; \quad s_1^g \; s_2^g; \quad s_{12}^g \\ & 2 \quad 2 \; 2 \quad -2 \end{aligned}$$

Rearranging to the required Hadamard ordering yields the following matrix formulation for the spectrum of h:

$$\frac{1}{2}\begin{bmatrix} \begin{bmatrix} 4 & 2 \\ 0 & 2 \\ 0 & 2 \\ 0 & -2 \end{bmatrix} & & & 0 \\ & \begin{bmatrix} 4 & 2 \\ 0 & 2 \\ 0 & 2 \\ 0 & -2 \end{bmatrix} & & \\ & & \begin{bmatrix} 4 & 2 \\ 0 & 2 \\ 0 & 2 \\ 0 & -2 \end{bmatrix} & \\ 0 & & & \begin{bmatrix} 4 & 2 \\ 0 & 2 \\ 0 & 2 \\ 0 & -2 \end{bmatrix} \end{bmatrix} \begin{bmatrix} \hat{s}_0 \\ \hat{s}_5 \\ \hat{s}_3 \\ \hat{s}_{53} \\ \hat{s}_4 \\ \hat{s}_{54} \\ \hat{s}_{34} \\ \hat{s}_{534} \end{bmatrix} = \begin{bmatrix} -2 \\ 2 \\ 2 \\ -2 \\ 10 \\ 6 \\ 6 \\ -6 \\ 2 \\ -2 \\ -2 \\ 2 \\ -2 \\ 2 \\ 2 \\ -2 \end{bmatrix}$$

Matrix based upon the spectrum of g | Spectrum of h $= \hat{\mathbf{S}}$ | Spectrum of f $= \mathbf{S}$

Note that x_5 is the input from g to h, as previously.

Hence, we have

$$\tfrac{1}{2}(4\hat{s}_0 + 2\hat{s}_5) = -2$$
$$\tfrac{1}{2}(0 + 2\hat{s}_5) = 2$$
$$\tfrac{1}{2}(0 + 2\hat{s}_5) = 2$$
$$\tfrac{1}{2}(0 - 2\hat{s}_5) = -2$$

$$\tfrac{1}{2}(4\hat{s}_3 + 2\hat{s}_{53}) = 10$$
$$\tfrac{1}{2}(0 + 2\hat{s}_{53}) = 6$$
$$\tfrac{1}{2}(0 + 2\hat{s}_{53}) = 6$$
$$\tfrac{1}{2}(0 - 2\hat{s}_{53}) = -6$$

$$\tfrac{1}{2}(4\hat{s}_4 + 2\hat{s}_{54}) = 2$$
$$\tfrac{1}{2}(0 + 2\hat{s}_{54}) = -2$$
$$\tfrac{1}{2}(0 + 2\hat{s}_{54}) = -2$$
$$\tfrac{1}{2}(0 - 2\hat{s}_{54}) = 2$$

$$\tfrac{1}{2}(4\hat{s}_{34} + 2\hat{s}_{534}) = -2$$
$$\tfrac{1}{2}(0 + 2\hat{s}_{534}) = 2$$
$$\tfrac{1}{2}(0 + 2\hat{s}_{534}) = 2$$
$$\tfrac{1}{2}(0 - 2\hat{s}_{534}) = -2$$

A solution for $\hat{\mathbf{S}}$ clearly exists, being

$$\hat{\mathbf{S}} = \begin{matrix} \hat{s}_0 & \hat{s}_5 & \hat{s}_3 & \hat{s}_{53} & \hat{s}_4 & \hat{s}_{54} & \hat{s}_{34} & \hat{s}_{534} \\ -2 & 2 & 2 & 6 & 2 & -2 & -2 & 2 \end{matrix}$$

Applying the inverse transform yields

$$\hat{\mathbf{Y}} = +1, -1, -1, +1, +1, -1, -1, -1]^t$$

which is the Boolean function shown in Fig. 4.3c. Thus the realisation for $f(X)$ of

$$f(X) = (x_1x_2)\bar{x}_3 + (\overline{x_1x_2})x_3 + x_3x_4$$

is confirmed.

There are considerable advantages but some disadvantages in the above procedure. First, the advantages:

(i) There are no restrictions on the logical complexity of the functions g and h in the disjoint decomposition topology. The AND function for g just considered is a trivial case, but *no additional computational complexity* arises in incorporating any function in Eq. (4.12), such flexibility being completely absent when working in the conventional Boolean domain. This flexibility may prove particularly advantageous when syntheses using complex functions, such as are now available in CMOS technology,[†] are required.

(ii) Whilst the consideration of 2^n simultaneous equations in total is necessary to confirm the disjoint decomposition, it is usually possible to detect

[†] Due to CMOS silicon layout advantages, complex logic functions such as $\overline{x_1x_2 + (x_3 + x_4)x_5}$ are readily fabricated with compact utilisation of silicon area.

failure very rapidly, since the incompatibility of any one subset of simultaneous equations will show failure. Further, the very simple relationships required within each subset of coefficients of **S** for any given function of g, and the equality of the subsets, is readily observable, as demonstrated in the example cited.

The possible disadvantages are:

(i) If a disjoint synthesis of a given function $f(X)$ using given functions g and/or h is required, it may not be immediately apparent how the x_i input variables should be partitioned to achieve a possible realisation. In the example shown, we deliberately chose x_1, x_2 as the input variables to g, with x_3, x_4 as the disjoint input variables to h. If we attempt to interchange $x_2 \leftrightarrow x_3$, giving x_1, x_3 as the input variables to the AND function, we shall find that no disjoint synthesis is now possible for $f(X)$. This may be demonstrated by reordering the coefficients in **S** to the alternative (Hadamard) order

$$\begin{array}{c} s_0 \\ s_1 \\ s_3 \\ s_{13} \\ s_2 \\ \vdots \end{array}$$

the first subset of simultaneous equations that now results being

$$\tfrac{1}{2}(4\hat{s}_0 + 2\hat{s}_5) = -2$$
$$\tfrac{1}{2}(0 + 2\hat{s}_5) = 2$$
$$\tfrac{1}{2}(0 + 2\hat{s}_5) = 10$$
$$\tfrac{1}{2}(0 - 2\hat{s}_5) = 6$$

This clearly has no solution, and hence indicates an impossible decomposition.

Theoretically it is therefore necessary to reorder **S** $n!$ times to ensure that no disjoint decomposition exists for a given **S** and g, although again failure of the first subset of relationships provides a rapid no-go check.

(ii) The second disadvantage is merely one of maintaining the proper order and interpretation of the spectral coefficient vectors **S**, $\mathbf{S}^g$ and $\hat{\mathbf{S}}$. This is merely a housekeeping problem, but very important, particularly where repeated application of disjoint decomposition synthesis is sought.

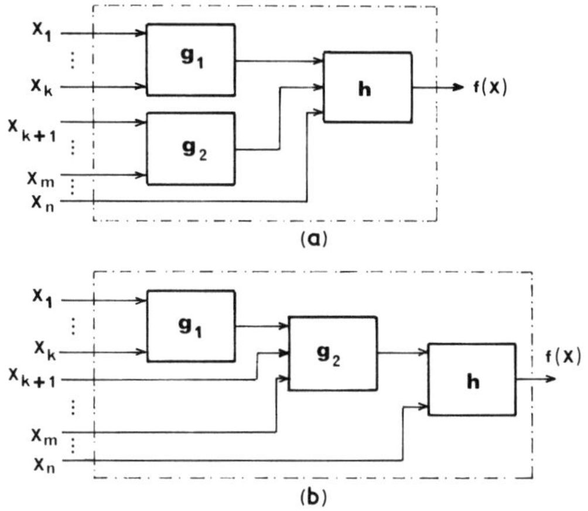

Fig. 4.4. Repeated application of disjoint synthesis:

(a) $f(X) = h\{g_1(x_1, \ldots, x_k), g_2(x_{k+1}, \ldots, x_m), \ldots, x_n\}$;
(b) $f(X) = h\{g_2(g_1(x_1, \ldots, x_k), x_{k+1}, \ldots, x_m), \ldots, x_n\}$.

Other features of the subgroups of simultaneous equations may be found.[11,13] The procedure may be extended to further decomposition topologies such as shown in Fig. 4.4, but clearly careful housekeeping of the spectral coefficients becomes necessary when three or more decomposition functions are present. The principles leading to the strongly structured relationships of Eq. (4.12), however, remain unchanged.

The final problem of synthesis by disjoint decompositions is, of course, where no such decomposition exists for a given set of conditions. Less restrictive realising topologies then become necessary, such as those considered in the following sections.

4.3 SYNTHESIS BY PRIME IMPLICANT EXTRACTION

The classic method of synthesis of any given combinatorial function is to search for some minimum cover of prime implicant terms, redundant prime implicants being deleted unless race hazard requirements dictate otherwise. Such classic methods include geometric techniques based upon Karnaugh maps. and algorithmic techniques based upon Quine–McClusky and other algorithms.[1–5,10] Exceptions to these general methods are when a realisation in ROM form is required, which implies the individual generation of every true minterm of the function (or every false minterm with final output negation), or where a realisation in some other specific topology is required, such as a multiplexer or other universal-logic-module realisation.

There are 2^{n+1} trivial cases where the spectrum of $f(X)$ immediately gives a synthesis, this being when one coefficient is maximum-valued, the remaining coefficients being zero-valued. These functions range from $f(X) = 0$ or 1 to $f(X)$ or $\overline{f(X)} = x_1 \oplus x_2 \oplus \cdots \oplus x_n$; note that all except the last are degenerate functions of less than n variables. However, in the following we shall pursue the case where two (or more) non-zero coefficients sum to $\pm 2^n$. It will be shown that such summations between appropriate coefficients indicate prime implicant terms of $f(X)$, and hence may be employed in the synthesis of the function. We shall assume in the following that we are dealing with non-degenerate functions, and hence minimum-length spectra **S**.

4.3.1 Spectral Summations

Consider the pairwise addition of the corresponding elements of any two row vectors, row α and row β, of $\mathbf{T}^n$, giving the row vector

$$\mathbf{T}^n_{(\alpha+\beta)*} \triangleq t_{\alpha k} + t_{\beta k}, \qquad \alpha \neq \beta, \quad k = 0 \text{ to } 2^n - 1$$

This addition will always result in half the elements of the resulting vector being zero, and the other half ± 2.

Should we now multiply this row vector by the truthtable vector **Y**, we obtain

$$\mathbf{T}^n_{(\alpha+\beta)*}\mathbf{Y} = s^+$$

There is now no information in s^+ concerning half the minterm values in **Y**, whilst the remaining minterms will each contribute ± 2 to the value of s^+. A similar development follows should we consider the pairwise subtraction of the corresponding elements of any two rows of $\mathbf{T}^n$, resulting in a value s^-, which now depends upon the other half of the minterms in **Y** for any given row α and row β.

Hence s^+ (s^-) is the summation of 2^{n-1} terms whose values are each ± 2. The maximum value of s^+ (s^-) is therefore $\pm 2 \times 2^{n-1}, = \pm 2^n$. In this special case, the minterms of **Y** that are effective in contributing to s^+ (s^-) must fully agree or fully disagree with the relevant elements in row α and row β of $\mathbf{T}^n$. Further, it will be appreciated that

$$\mathbf{T}^n_{(\alpha\pm\beta)*}\mathbf{Y} \equiv \mathbf{T}^n_{\alpha*}\mathbf{Y} \pm \mathbf{T}^n_{\beta*}\mathbf{Y},$$

and hence s^+ (s^-) merely represents the addition (subtraction) of two of the normal spectral coefficients of **S**, corresponding to the rows α and β of $\mathbf{T}^n$.

As an example, consider the function $f(X) = \bar{x}_1 x_2 + x_1 \bar{x}_2 + x_2 x_3$. The normal Hadamard transformation is

$$\begin{bmatrix} 1 & 1 & 1 & 1 & 1 & 1 & 1 & 1 \\ 1 & -1 & 1 & -1 & 1 & -1 & 1 & -1 \\ 1 & 1 & -1 & -1 & 1 & 1 & -1 & -1 \\ 1 & -1 & -1 & 1 & 1 & -1 & -1 & 1 \\ 1 & 1 & 1 & 1 & -1 & -1 & -1 & -1 \\ 1 & -1 & 1 & -1 & -1 & 1 & -1 & 1 \\ 1 & 1 & -1 & -1 & -1 & -1 & 1 & 1 \\ 1 & -1 & -1 & 1 & -1 & 1 & 1 & -1 \end{bmatrix} \begin{bmatrix} 1 \\ -1 \\ -1 \\ 1 \\ 1 \\ -1 \\ -1 \\ -1 \end{bmatrix} = \begin{bmatrix} -2 \\ +2 \\ +2 \\ +6 \\ +2 \\ -2 \\ -2 \\ +2 \end{bmatrix} \begin{matrix} s_0 \\ s_1 \\ s_2 \\ s_{12} \\ s_3 \\ s_{13} \\ s_{23} \\ s_{123} \end{matrix}$$

$$\mathbf{T}^n \qquad \mathbf{Y} = \mathbf{S}$$

Selecting and adding, say, rows 2 and 3 of $\mathbf{T}^n$ gives

$$\mathbf{T}^n_{(2+3)*} = 2 \ 0 \ 0 \ -2 \ 2 \ 0 \ 0 \ -2$$

which if multiplied by $\mathbf{Y}$ gives[†]

$$s^+ = 2 + 0 + 0 - 2 + 2 + 0 + 0 + 2 = +4 = s_1 + s_2$$

Similarly, subtracting rows 2 and 3 gives

$$\mathbf{T}^n_{(2-3)*} = 0 \ -2 \ 2 \ 0 \ 0 \ -2 \ 2 \ 0$$

which produces

$$s^- = 0 + 2 - 2 + 0 + 0 + 2 - 2 + 0 = 0 = s_1 - s_2$$

However, selecting and adding, say, rows 3 and 4 of $\mathbf{T}^n$ gives

$$\mathbf{T}^n_{(3+4)*} = 2 \ 0 \ -2 \ 0 \ 2 \ 0 \ -2 \ 0$$

which produces

$$s^+ = 2 + 0 + 2 + 0 + 2 + 0 + 2 + 0 = +8 = s_2 + s_{12}$$

The latter case illustrates that the four minterms in $\mathbf{Y}$ not multiplied by zero fully agree with the corresponding elements in both row 3 and row 4 of $\mathbf{T}^n$, indicated by the maximum value for s^+.

This procedure can be extended to the summation of more than two rows of $\mathbf{T}^n$, which will lead to the consideration of the summation of more than two coefficients of $\mathbf{S}$. However, the choice of rows must be such as to complete an orthogonal subset of rows from $\mathbf{T}^n$. We must therefore use 2, 4, 8, ... rows

[†] Note that for convenience we here refer to the rows of $\mathbf{T}^n$ as row 1, row 2, ..., rather than the formal matrix notation $\mathbf{T}_{j*}$, $j = 0$ to $2^n - 1$.

of $\mathbf{T}^n$, which implies the eventual summation of 2, 4, 8, ... coefficients of **S**. Also, the larger the number of rows and coefficients considered, the fewer will be the minterms of $f(X)$ that may be uniquely identified.

Any two rows of $\mathbf{T}^n$ complete an orthogonal subset, and hence in our previous consideration we are free to take any two rows (coefficients of **S**) without restriction. For four rows, however, we may take any three rows α, β, γ, but the fourth row must be the mod 2 addition

$$\mathbf{T}^n_{\delta *} = t_{\alpha k} \oplus t_{\beta k} \oplus t_{\gamma k}, \qquad \alpha \neq \beta \neq \gamma, \quad k = 0 \text{ to } 2^n - 1,$$

that is the row representing the Exclusive-OR of the functions constituting rows α, β, γ. This is readily identified by the subscripts of the coefficients in **S**; for example, the three rows of $\mathbf{T}^n$ that generate, say, s_1, s_2 and s_3 must have s_{123} as the fourth component.

For eight rows from any $\mathbf{T}^n$, we have

$$\left.\begin{array}{l} \text{row } \alpha \\ \text{row } \beta \\ \text{row } \gamma \\ \text{row } \delta \end{array}\right\} \text{“basis” rows—any four}$$

$$\begin{aligned} \text{row } \alpha\beta\gamma &= \text{row } \alpha + \text{row } \beta + \text{row } \gamma \\ \text{row } \alpha\beta\delta &= \text{row } \alpha + \text{row } \beta + \text{row } \delta \\ \text{row } \alpha\gamma\delta &= \text{row } \alpha + \text{row } \gamma + \text{row } \delta \\ \text{row } \beta\gamma\delta &= \text{row } \beta + \text{row } \gamma + \text{row } \delta \end{aligned}$$

In general, for any subset of 2^v orthogonal rows from $\mathbf{T}^n$, $0 \leq v \leq n$, we need to choose $(v + 1)$ basis rows and from these define the remaining $2^v - (v + 1)$ rows to complete the orthogonal subset. Note also that we may apply any $\pm$ multiplication to the basis rows, which corresponds to adding or subtracting the rows and hence the corresponding spectral coefficients, but this freedom does not extend to the remaining rows derived from the basis set. Further considerations of this orthogonality may be found.[14,15]

Referring back to our previous example, suppose we choose the first, second and fifth rows of $\mathbf{T}^n$, corresponding to spectral coefficients s_0, s_1 and s_3; we now must select the sixth row, corresponding to s_{13}, to complete the orthogonal subset. Suppose we add rows 2 and 5, subtract row 1, and therefore subtract row 6; this gives

$$\mathbf{T}^n_{(-1+2+5-6)*} = 0\,0\,0\,0\,0\; -4\,0\; -4$$

Multiplication by **Y** gives

$$s^{\pm} = 0 + 0 + 0 + 0 + 0 + 4 + 0 + 4 = +8,$$
$$= -s_0 + s_1 + s_3 - s_{13}$$

This maximum resultant value of $+8$ shows that the minterms of $f(X)$ that are multiplied by the non-zero terms of the summation row vector must

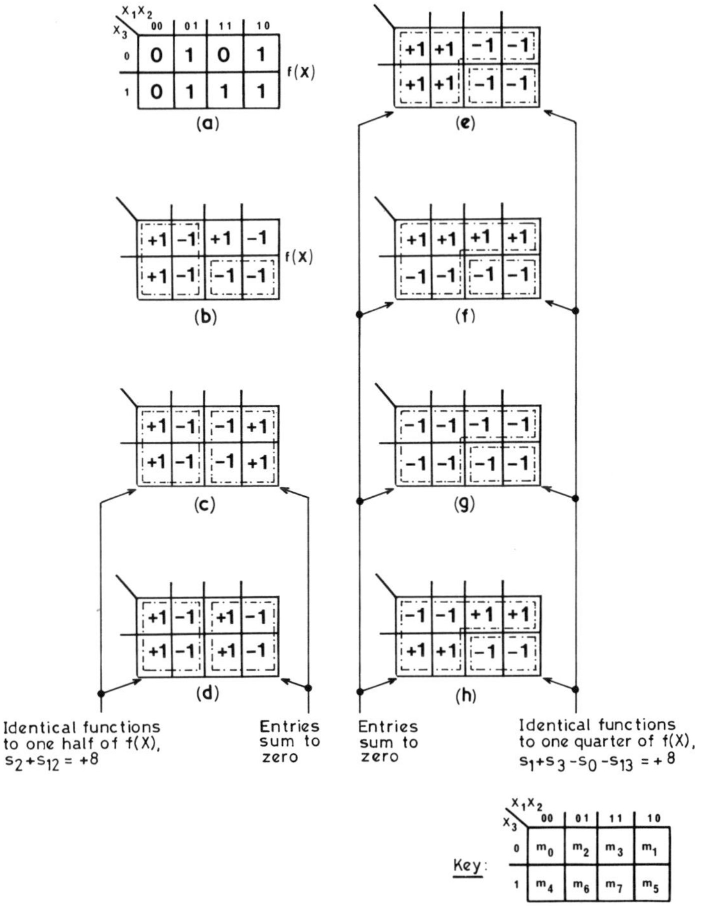

Fig. 4.5. Examples of summation to $\pm 2^n$ of spectral coefficients of $f(X)$: (a) given function $f(X) = \bar{x}_1x_2 + x_1\bar{x}_2 + x_2x_3 = -2; 2\ 2\ 2; 6\ -2\ -2; 2$; (b) $f(X)$ recoded $\{+1, -1\}$, $=$ **Y**; (c) third row of $\mathbf{T}^n$, producing s_2; (d) fourth row of $\mathbf{T}^n$, producing s_{12}; (e) second row of $\mathbf{T}^n$, producing s_1; (f) fifth row of $\mathbf{T}^n$, producing s_3; (g) first row of $\mathbf{T}^n \times -1$, producing $-s_0$; (h) sixth row of $\mathbf{T}^n \times -1$, producing $-s_{13}$.

agree with the corresponding entries in the rows 2 and 5 of $\mathbf{T}^n$, and disagree with the same entries in rows 1 and 6 of $\mathbf{T}^n$, that is, one quarter of $f(X)$ has been defined by this test. Figure 4.5 illustrates what has been developed in these two examples.

In the extreme case, we may consider the $\pm$ summation of all the 2^n rows of $\mathbf{T}^n$. It is a trivial exercise to show that while maintaining the proper sign relationships between the basis rows and the remaining rows, we may select 2^{n+1} different sign relationships, which will each sum to either $+2^n$ or -2^n in individual element positions of the summation vector, all remaining entries being zero-valued. Hence each such case corresponds to a $\pm$ summation of all 2^n spectral coefficients of $\mathbf{S}$, and uniquely identifies the value of one individual minterm of $f(X)$.

4.3.2 Prime Implicant Extraction by Spectral Summation

While the developments above were based upon the consideration of row summations from the transform $\mathbf{T}^n$, it is seen that this leads to a consideration of the proper summation of 2^v spectral coefficients of $\mathbf{S}$, with the target of achieving $\sum^{2^v}(\pm s_{\alpha\beta\ldots}) = \pm 2^n$, $v = 0, 1, \ldots, n$. The problems remaining however, are the selection of the coefficients to be considered for any given function $f(X)$, and the precise meaning of each achievable summation to $\pm 2^n$.

It will be noted from Fig. 4.5 that a maximum-value summation can define a decomposition of $f(X)$ containing both 0- and 1-valued minterms and not exclusively same-valued minterms as in conventional synthesis techniques. Consider first the summation of two rows from $\mathbf{T}^n$ without any negation of either row, and with the summation of the two corresponding s_i coefficients equal to $+2^n$. For example, take the second and third rows of an $n = 4$ transform, as shown in Fig. 4.6. Should we now find that $s_1 + s_2 = +16$ for a given function, then the minterms of $f(X)$ that are defined by this summation must be as shown, that is,

(i) the 1-valued minterms = area within the logical AND of the two row vector functions, and

(ii) the 0-valued minterms = area outside the logical OR of the two row vector functions.

Hence these 0- and 1-valued areas of the complete function $f(X)$ may be realised as shown in Fig. 4.6e, leaving a remainder function $f'(X)$ to be considered. Note that no information whatsoever is available from this summation concerning the remainder function, that is, areas $x_1\bar{x}_2$ and $\bar{x}_1x_2$ in this example.

This part-synthesis procedure may be found to be perfectly general for 2, 4, 8, ... spectral coefficient summations to $+2^n$. This is illustrated in Fig. 4.7a, with an alternative logically equivalent realisation in Fig. 4.7b. However, there remain further aspects, namely:

(i) If the spectral coefficient summation is to -2^n rather than $+2^n$, then negation of *all* the row vector functions of Fig. 4.7a or b is required.

(ii) If proper negation of one or more of the individual spectral coefficients is necessary in order to sum to $+2^n$, then negation of the corresponding one (or more) individual row vector function is required.

(iii) If the zero-order coefficient s_0 is involved in a summation to $+2^n$, then with $+s_0$ we have the constant function 0 as one of the AND and OR gate inputs. Similarly, with $-s_0$ involved in the summation to $+2^n$, we have the constant function 1 as one of the inputs. In the former case, it is clear that

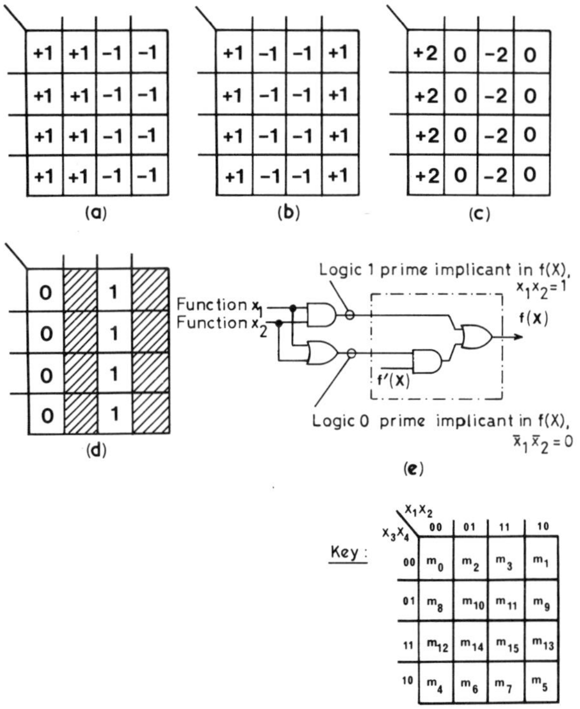

Fig. 4.6. Meaning of spectral summations to 2^n: (a) second row of $\mathbf{T}^n$, row function x_1, producing s_1; (b) third row of $\mathbf{T}^n$, row function x_2, producing s_2; (c) addition of (a) and (b), zero entries in $\bar{x}_1x_2$ and $x_1\bar{x}_2$ areas; (d) function minterms of $f(X)$ defined by $s_1 + s_2 = +16$, cross-hatched areas not defined by this summation; (e) part realisation of $f(X)$, realising the $\bar{x}_1\bar{x}_2$ and x_1x_2 areas.

the AND gate of Fig. 4.7a and b becomes redundant, and in the latter case, the OR gate becomes redundant. This is illustrated in Fig. 4.7c and d. Note that with s_0 included, we synthesise groups of 0-valued or 1-valued minterms only, and not two groups, one 0-valued and the second 1-valued, as before.

(iv) Finally, consider the case where we have a summation involving first-order coefficients and a higher-order coefficient, for example, s_1, s_2, s_4 and s_{124}. This group of coefficients implies the OR function $x_1 + x_2 + x_4 + (x_1 \oplus x_2 \oplus x_4)$, and the AND function $x_1x_2x_4(x_1 \oplus x_2 \oplus x_4)$. However, these functions will be recognised as merely $x_1x_2x_4$ and $x_1 + x_2 + x_4$, respectively, the Exclusive term being redundant. Note that should we have had x_1, x_2 and $\bar{x}_4$ present, then the fourth term of the orthogonal subset must be $x_1 \oplus x_2 \oplus \bar{x}_4$, which again is redundant in both the AND and the OR connective.

As an example illustrating these features, suppose the following summation is determined:

$$s_0 + s_1 + s_2 - s_3 + s_{12} - s_{13} - s_{23} - s_{123} = -2^n$$

Changing all signs, we have

$$-s_0 - s_1 - s_2 + s_3 - s_{12} + s_{13} + s_{23} + s_{123} = 2^n$$

giving the OR and AND input functions:

$-s_0$: logic 1, thus making the OR gate redundant (see Fig. 4.7d)

$-s_1$: $\bar{x}_1$

$-s_2$: $\bar{x}_2$

s_3: x_3

$$\left.\begin{array}{rl} -s_{12}: & \overline{x_1 \oplus x_2} \\ s_{13}: & x_1 \oplus x_3 \\ s_{23}: & x_2 \oplus x_3 \\ s_{123}: & x_1 \oplus x_2 \oplus x_3 \end{array}\right\} \text{all redundant in the resulting realisation}$$

Hence $\bar{x}_1\bar{x}_2x_3$ is a prime implicant factor of $f(X)$, giving the partial synthesis shown in Fig. 4.7e.

Clearly in the search for coefficients that sum to $\pm 2^n$, it is desirable to include s_0 and the first-order coefficients in preference to concentration upon the higher-order coefficients. The most desirable situation is the summation of the zero-order coefficient and one first-order coefficient to $\pm 2^n$, since this implies only one input variable to the OR or AND gate of Fig. 4.7c or d, hence making even this single gate redundant.

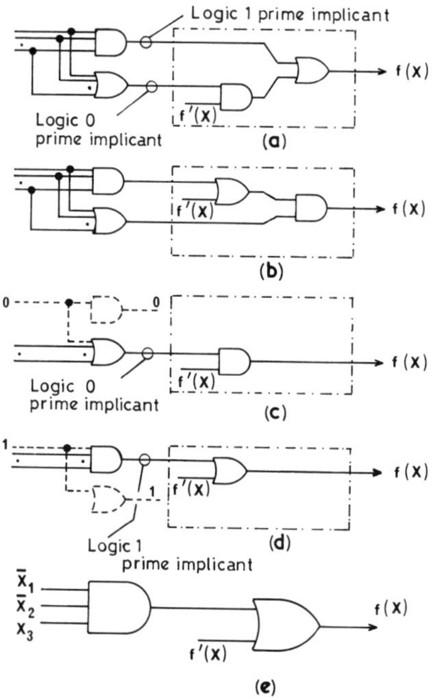

Fig. 4.7. General realisation topologies: (a) general case of Fig. 4.6e, where s_0 is not involved; (b) equivalent alternative to (a); (c) particular case of (a) and (b) when $+s_0$ is involved in the spectral summation to $+2^n$, AND gate redundant; (d) particular case of (a) and (b) when $-s_0$ is involved in the spectral summation to $+2^n$, OR gate redundant; (e) partial synthesis of $f(X)$, given $s_0 + s_1 + s_2 - s_3 + s_{12} - s_{13} - s_{23} - s_{123} = -2^n$, that is, $-s_0 - s_1 - s_2 + s_3 - s_{12} + s_{13} + s_{23} + s_{123} = +2^n$.

4.3.3 Further Considerations

It is a straightforward but unrewarding task to specify all the possible spectral summations that define one-half of the minterms of $f(X)$, for example,

$$\{s_0 + s_{\alpha\beta\ldots}\} = 2^n\text{: all minterms in area } x_\alpha \oplus x_\beta \oplus \cdots \text{ zero-valued}$$

$$\{s_0 - s_{\alpha\beta\ldots}\} = 2^n\text{: all minterms in area } \overline{x_\alpha \oplus x_\beta \oplus \cdots} \text{ zero-valued}$$

$$\{-s_0 + s_{\alpha\beta\ldots}\} = 2^n\text{: all minterms in area } x_\alpha \oplus x_\beta \oplus \cdots \text{ one-valued}$$

$$\{-s_0 - s_{\alpha\beta\ldots}\} = 2^n\text{: all minterms in area } \overline{x_\alpha \oplus x_\beta \oplus \cdots} \text{ one-valued}$$

but this listing becomes excessive if we define all possible one-quarter, one-eighth, ... decompositions of $f(X)$. Visual inspection of the spectra of simple functions is frequently adequate to determine a complete function cover, but in these circumstances a conventional Karnaugh-map synthesis would often be as relevant.†

Two particular features, however, may be observed in this synthesis procedure. First, the *larger* the prime implicant that is being realised at a particular stage of synthesis, then the *fewer* the spectral coefficients that have to be considered. This is in complete contrast to Quine–McClusky-based prime-implicant extraction algorithms, which require more computational effort to detect large prime implicants than smaller prime implicants, which is a serious consideration for CAD programs. Second, the remainder function $f'(X)$, which results after a partial synthesis has been executed, can take all the minterms of $f(X)$ that have been synthesised as "don't-care" terms, if desired. Examination of the realising topology of Figs. 4.6 and 4.7 will confirm that this freedom is always available, if required.

A suggested design philosophy based upon these developments is shown in Table 4.1. Certain features of this flow chart are:

(a) If a first-order coefficient $|s_\alpha|$ is present that with $|s_0|$ sums to 2^n, which we may reexpress as $\{a_0 s_0 + a_\alpha s_\alpha\} = +2^n$, where $a_0, a_\alpha \in \{+1, -1\}$, then the x_i or $\bar{x}_i$ half of $f(X)$ may be immediately realised. Variable x_i is now redundant in the remainder function $f'(X)$, and the spectrum for $f'(X)$ may be reduced from 2^n to 2^{n-1} coefficients.

The coefficient values in the reduced spectrum $\mathbf{S}'$ for $f'(X)$ are given by

(i) all coefficients in $\mathbf{S}$ containing α in their subscript identification are deleted from $\mathbf{S}'$;

(ii) the value of s_0' in $\mathbf{S}'$ is given by

$$s_0' = s_0 - a_0 2^{n-1}$$

where a_0 is $+1$ or -1 required by the previous spectral summation to $+2^n$; and

(iii) all remaining $2^{n-1} - 1$ coefficients of $\mathbf{S}'$ remain unchanged from $\mathbf{S}$.

(b) If a single higher-order coefficient $|s_{\alpha\beta\ldots}|$ is present that with $|s_0|$ sums to 2^n, then again half of $f(X)$ may be immediately synthesized. However, this involves an Exclusive-OR function $s_\alpha \oplus s_\beta \oplus \cdots$ to be generated and used as a function input for this part-realisation. If this is not acceptable, then we may return to search for further non-exclusive relationships involving two (or more) primary x_i inputs.

If this Exclusive-OR part-realisation is accepted, then we may reduce the spectrum for $f'(X)$ to one half of $\mathbf{S}$ by precisely the same rules as detailed

† Note that there is also the initial extreme of this summation technique, where one spectral coefficient alone $= \pm 2^n$, i.e., the summation of 2^v terms where $v = 0$. Clearly this is the trivial well-known case where all the function is fully defined by one spectral coefficient only.

Table 4.1. A spectral summation design procedure.

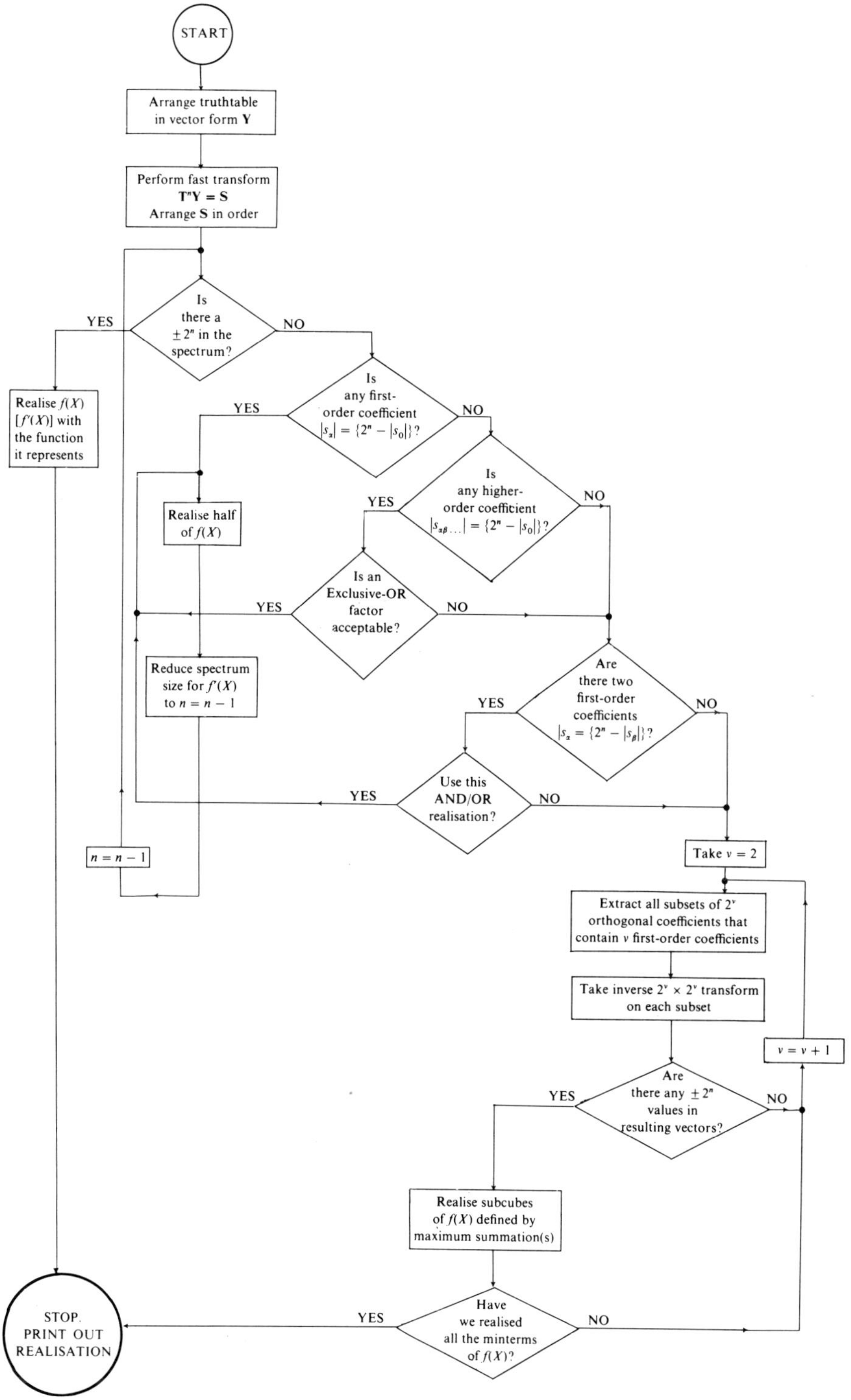

in (a) above. However, we now have the choice of deleting any one, but not more than one, of the x_i variables present in the Exclusive-OR function.

(c) When $\{|s_0| + |\text{any other single spectral coefficient}|\} = 2^n$ ceases to be available, we may consider searching for any other two coefficients of **S**, excluding s_0, that may appropriately sum to 2^n. However, it is debatable whether other than two first-order coefficients need be considered, since higher-order coefficients alone would necessitate Exclusive-OR input functions to the part-realisation. Table 4.1 thus suggests a search between pairs of first-order coefficients only before continuing with larger spectral groups.

(d) In the larger spectral groups of 4, 8, ... coefficients, it is advantageous to include s_0 as one of the coefficients. Hence each orthogonal subset of 2^v coefficients should include s_0 and v coefficients taken from the first-order coefficients as the basis set, from which the remaining necessary $2^v - (v + 1)$ coefficients are defined.

(e) The problem of determining which (if any) permitted sign permutations of a subset of 2^v coefficients sum to 2^n may be solved by applying a $2^v \times 2^v$ inverse transform to the subset. For example, should we be considering the following subset of eight coefficients extracted from a six-variable spectrum **S**

$$\begin{array}{cccccccc} s_0 & s_1 & s_4 & s_6 & s_{14} & s_{16} & s_{46} & s_{146} \\ 10 & 18 & -10 & 6 & 14 & -34 & 2 & 10 \end{array}$$

then we have

$$\begin{bmatrix} 1 & 1 & 1 & 1 & 1 & 1 & 1 & 1 \\ 1 & -1 & 1 & -1 & 1 & -1 & 1 & -1 \\ 1 & 1 & -1 & -1 & 1 & 1 & -1 & -1 \\ 1 & -1 & -1 & 1 & 1 & -1 & -1 & 1 \\ 1 & 1 & 1 & 1 & -1 & -1 & -1 & -1 \\ 1 & -1 & 1 & -1 & -1 & 1 & -1 & 1 \\ 1 & 1 & -1 & -1 & -1 & -1 & 1 & 1 \\ 1 & -1 & -1 & 1 & -1 & 1 & 1 & -1 \end{bmatrix} \begin{bmatrix} 10 \\ 18 \\ -10 \\ 14 \\ 6 \\ -34 \\ 2 \\ 10 \end{bmatrix} \begin{array}{l} s_0 \\ s_1 \\ s_4 \\ s_{14} \\ s_6 \\ s_{16} \\ s_{46} \\ s_{146} \end{array} = \begin{bmatrix} 16 \\ 0 \\ -16 \\ 64 \\ 48 \\ -64 \\ 64 \\ 32 \end{bmatrix}$$

whence from the maximum resulting values of $\pm 2^6$ we determine

$$s_0 - s_1 - s_4 + s_{14} + s_6 - s_{16} - s_{46} + s_{146} = +64$$

$$-s_0 + s_1 - s_4 + s_{14} + s_6 - s_{16} + s_{46} - s_{146} = +64$$

$$s_0 + s_1 - s_4 - s_{14} - s_6 - s_{16} + s_{46} + s_{146} = +64$$

Hence it follows that

$$\bar{x}_1 + \bar{x}_4 + x_6, \qquad x_1\bar{x}_4x_6 \qquad \text{and} \qquad x_1 + \bar{x}_4 + \bar{x}_6$$

respectively, are factors in the given function (see Fig. 4.8a).

The remainder of the flow diagram of Table 4.1 should be self-explanatory. Note that when an inverse transform is executed on a subset of 2^v spectral coefficients, the order of the coefficients must agree with the row order of the inverse transform being used. In the above example we employed the Hadamard transform, and hence the spectral coefficients were reassembled in the usual Hadamard order.

It is possible to suggest other strategies within this spectral-summation synthesis technique. For example, in the preceding procedure we have not taken advantage of the ability to allocate "don't cares" to the minterms within areas of $f(X)$ that have already been synthesized. Also spectral translation techniques, as used in Section 3.5, can be applied in order to move coefficients from one order to another, which may result in simpler summation subsets being formed. Details of further considerations may be found,[15] but considerable complexity and housekeeping may build up if an exhaustive procedure to determine an optimum realisation is pursued.

Finally it will be appreciated that should a gate in a synthesis be specified by the appropriate coefficient subset summation to $\pm 2^n$, then any logic fault on this gate will destroy this particular summation. Hence the orthogonal subsets of coefficients used during the design phase may also serve as test signatures for fault diagnosis purposes.

As a concluding example, consider the given five-variable spectrum:

s_0;	s_1	s_2	s_3	s_4	s_5;	s_{12}	s_{13}	s_{14}	s_{15}	s_{23}	s_{24}	s_{25}	s_{34}	s_{35}	s_{45};
-10	6	6	6	6	2	-2	-2	-2	2	-2	-2	-6	-2	-6	2

s_{123}	s_{124}	s_{125}	s_{134}	s_{135}	s_{145}	s_{234}	s_{235}	s_{245}	s_{345};
-2	-2	2	-2	2	-6	-2	-22	2	2

s_{1234}	s_{1235}	s_{1245}	s_{1345}	s_{2345};	s_{12345}
6	-6	2	2	-6	2

The Table 4.1 procedure is executed as follows:

(a) No single maximum-valued coefficient present; therefore proceed:

(b) No first-order coefficient $|s_\alpha| = \{32 - |s_0|\}$ present; therefore proceed:

(c) Third-order coefficient $s_{235} = -22$, $\{-s_0 - s_{235}\} = +32$; accept $\bar{x}_2 \oplus \bar{x}_3 \oplus \bar{x}_5 = \overline{x_2 \oplus x_3 \oplus x_5}$ as a factor of $f(X)$. Half of $f(X)$ now realised (see Fig. 4.8b); proceed:

(d) Reduce size of spectrum to four variables by eliminating x_2 or x_3 or x_5; choose, say x_5, giving $n = 4$ spectrum

s_0;	s_1	s_2	s_3	s_4;	s_{12}	s_{13}	s_{14}	s_{23}	s_{24}	s_{34};	s_{123}	s_{124}	s_{134}	s_{234};	s_{1234}
6	6	6	6	6	-2	-2	-2	-2	-2	-2	-2	-2	-2	-2	6

proceed:

(e) Repeat (a), (b), (c), target summation $\pm 2^4 = 16$; nothing present; therefore proceed:

(f) Considering all cases of $v = 2$, we find

$$s_0 + s_1 + s_2 + s_{12} = +16$$

$$s_0 + s_1 + s_3 + s_{13} = +16$$

$$s_0 + s_1 + s_4 + s_{14} = +16$$

$$s_0 + s_2 + s_3 + s_{23} = +16$$

$$s_0 + s_2 + s_4 + s_{24} = +16$$

$$s_0 + s_3 + s_4 + s_{34} = +16$$

The search may be terminated here since all minterms in $f'(X)$ are defined by this list. The final realisation is shown in Fig. 4.8c.

If we wished to investigate further $f'(X)$ realisations, we could have inspected the $v = 3$ subsets. The several complete cover subset summations available include

$$-s_0 + s_1 + s_2 + s_3 - s_{12} - s_{13} - s_{23} + s_{123} = +16$$

$$-s_0 + s_1 + s_2 + s_4 - s_{12} - s_{14} - s_{24} + s_{124} = +16$$

$$-s_0 + s_1 + s_3 + s_4 - s_{13} - s_{14} - s_{34} + s_{134} = +16$$

$$-s_0 + s_2 + s_3 + s_4 - s_{23} - s_{24} - s_{34} + s_{234} = +16$$

which give the alternative realisation shown in Fig. 4.8d. The choice of $v = 4$ would, of course, have defined the individual minterms of $f'(X)$.

Finally, an alternative philosophy to that previously discussed should be noted, using, however, the same underlying principle of the summation of a subset of spectral coefficients to achieve a maximum value of $\pm 2^n$. This approach is based upon a prime implicant detection theorem of Lechner,[16] and developed in detail by Besslich[17–20].

Unlike the design procedure given in Table 4.1, where deliberately chosen subsets of spectral coefficients were progressively investigated, the method of Besslich proposes an exhaustive search using the spectral domain data for all possible 3^n implicants (product terms) of a given function $f(X)$, this search being organised by appropriate matrix operations on the spectral data. Following the identification of all prime implicants, reduction to a minimal covering set finally has to be implemented. Since individual implicants are being sought, the zero-order coefficient s_0 is always involved in the summation searches, unlike our previous considerations where subsets of coefficients not necessarily including s_0 were considered.

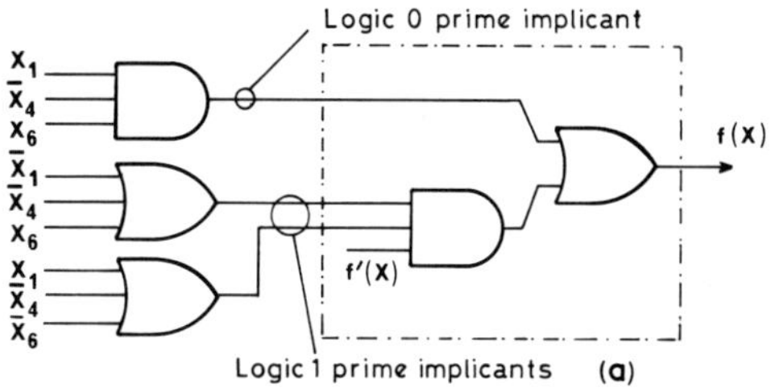

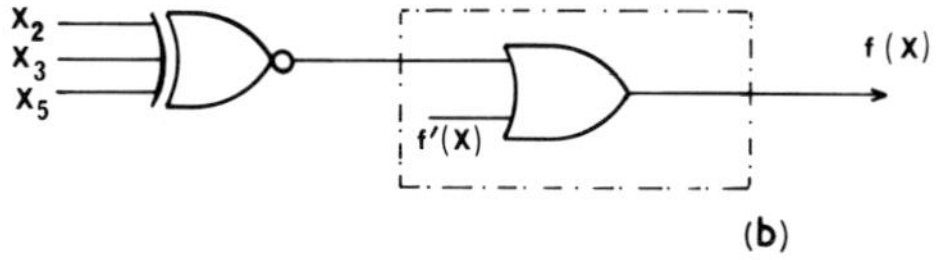

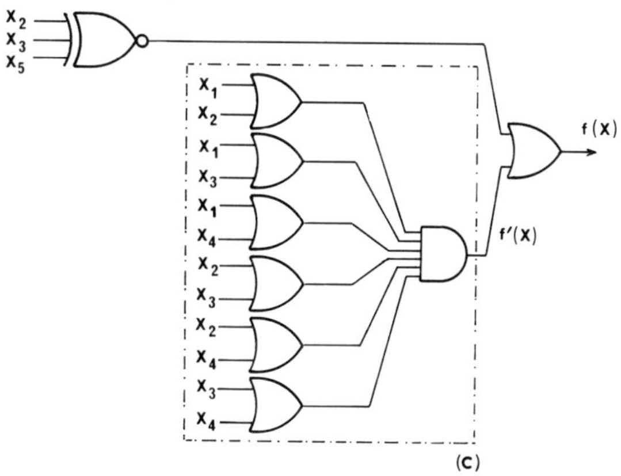

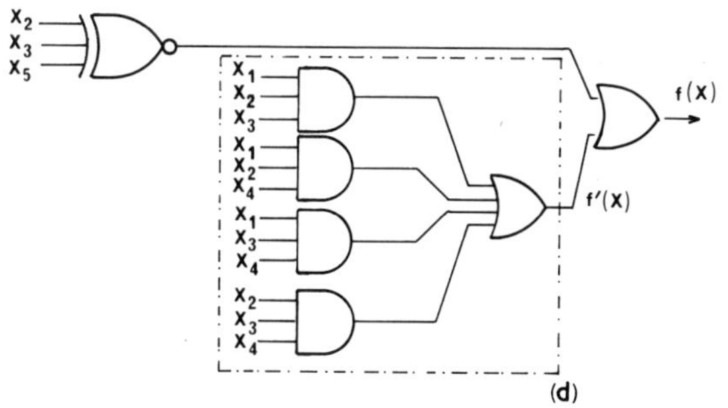

Fig. 4.8. Example synthesis by spectral summation: (a) part-synthesis from a given subset of eight spectral coefficients; (b) intitial synthesis of half of $f(X)$ of given five-variable function, compare Fig. 4.7d; (c) remaining synthesis of $f(X)$, s_0 positive in summation to $+2^n$; (d) alternative to (c), s_0 negative in summation to $+2^n$.

Referring back to Fig. 4.7d, it will be recalled that when $-s_0$ is involved in a summation to $+2^n$ (or, equivalently, $+s_0$ in a summation to -2^n), then an AND term of $f(X)$ has been identified, *with no corresponding OR term present*. Hence this term is an implicant of $f(X)$. To determine all possible implicants of $f(X)$, each true (logic 1) minterm of the given function is selected in turn, and all product terms that involve this minterm are then determined. This determination takes the form of a matrix operation on **S**, using a $2^n \times 2^n$ matrix operator $\mathbf{G}^n$, the resultant vector being inspected for maximum values.[17,18] As an example, consider the determination of all possible implicants containing minterm $\bar{x}_1\bar{x}_2\bar{x}_3\bar{x}_4$ of a four-variable function $f(X)$. We have, using the Rademacher–Walsh ordering for **S**,

$$\underbrace{\begin{bmatrix}
1&1&1&1&1&1&1&1&1&1&1&1&1&1&1&1\\
1&1&1&1&0&1&1&0&1&0&0&1&0&0&0&0\\
1&1&1&0&1&1&0&1&0&1&0&0&1&0&0&0\\
1&1&1&0&0&1&0&0&0&0&0&0&0&0&0&0\\
1&1&0&1&1&0&1&1&0&0&1&0&0&1&0&0\\
1&1&0&1&0&0&1&0&0&0&0&0&0&0&0&0\\
1&1&0&0&1&0&0&1&0&0&0&0&0&0&0&0\\
1&1&0&0&0&0&0&0&0&0&0&0&0&0&0&0\\
1&0&1&1&1&0&0&0&1&1&1&0&0&0&1&0\\
1&0&1&1&0&0&0&0&1&0&0&0&0&0&0&0\\
1&0&1&0&1&0&0&0&0&1&0&0&0&0&0&0\\
1&0&1&0&0&0&0&0&0&0&0&0&0&0&0&0\\
1&0&0&1&1&0&0&0&0&0&1&0&0&0&0&0\\
1&0&0&1&0&0&0&0&0&0&0&0&0&0&0&0\\
1&0&0&0&1&0&0&0&0&0&0&0&0&0&0&0\\
1&0&0&0&0&0&0&0&0&0&0&0&0&0&0&0
\end{bmatrix}}_{\mathbf{G}^n}
\underbrace{\begin{bmatrix} s_0\\ s_1\\ s_2\\ s_3\\ s_4\\ s_{12}\\ s_{13}\\ s_{14}\\ s_{23}\\ s_{24}\\ s_{34}\\ s_{123}\\ s_{124}\\ s_{134}\\ s_{234}\\ s_{1234} \end{bmatrix}}_{\mathbf{S}}
= \begin{bmatrix} \Sigma^0 \\ \vdots \\ \Sigma^\beta \end{bmatrix} = \Sigma_{m_0}$$

Clearly the right-hand summation vector can also be generated directly from the given function vector **Y** by combining the operations $\mathbf{G}^n\mathbf{T}^n\mathbf{Y} = \Sigma_{m_0}$.

Examination of this transform, or reference to Besslich, will reveal that it is performing the summation of 2^n orthogonal subsets of **S** that include s_0, which, if maximum-valued, indicate the presence of the appropriate AND term. For example, the seventh row of $\mathbf{G}^n$ sums s_0, s_1, s_4 and s_{14}, which would indicate the presence of $\bar{x}_1\bar{x}_4$ should the summation be -2^n.† Note

† Care is needed in sign conventions in various published works. In Besslich, the sign notation employed for the recoding to the function vector **Y** is logic $0 \rightarrow -1$, logic $1 \rightarrow +1$, which is opposite to that employed in this text. Hence his search for spectral summations involving $+s_0$ is to $+2^n$, rather than to -2^n considered here.

that the first summation term $\sum^0$ will always be maximum-valued, merely proving the existence of the true minterm itself.

For the determination of the implicants involving true minterms other than $\bar{x}_1\bar{x}_2\bar{x}_3\bar{x}_4$, the signs of the non-zero entries of $\mathbf{G}^n$ have to be permutated, or (more conveniently) the signs in $\mathbf{S}$. For each x_i variable changed from $\bar{x}_i$ to x_i, the signs of all coefficients in $\mathbf{S}$ containing i in their subscript identification require sign-changing should the latter policy be adopted; for example, the sign modifications necessary in $\mathbf{S}$ for the determination of implicants involving the true minterm $x_1\bar{x}_2\bar{x}_3x_4$ require the following coefficients in $\mathbf{S}$ to be changed in sign:

$$s_1, s_4; \quad s_{12}, s_{13}, s_{24}, s_{34}; \quad s_{123}, s_{234}$$

The seventh row of $\mathbf{G}^n$, for example, now sums s_0, $-s_1$, $-s_4$, s_{14}, which would indicate the presence of implicant x_1x_4 should the summation be -2^n.

It should be appreciated that the sign changing in $\mathbf{S}$ is merely to implement our previous considerations of $\{s_0 \pm s_i \pm \cdots\}$; the modified spectrum $\mathbf{S}$ with such sign changes no longer represents the given function $f(X)$. It may further be appreciated that these sign changes correspond to element-by element multiplication of $\mathbf{S}$ by an appropriate column vector taken from the Hadamard transform $\mathbf{T}^n$, the 2^n column vectors of $\mathbf{T}^n$ thus providing the sign modifications in $\mathbf{S}$ to cover all possible minterms m_0 to m_{2^n-1} of $f(X)$. Finally, it is possible to combine all these operations from the truth-table vector $\mathbf{Y}$ onwards into a single fast transform procedure for the determination of the implicants associated with each minterm, the final column vector being searched for maximum values.

Further details and examples of this approach, including the simple means of determining the prime implicants from all implicants detected at each minterm procedure, and the final determination of an irredundant prime implicant cover for $f(X)$, may be found in the published literature.[17–20] It may be seen to represent the equivalent in the spectral domain of the classic Quine–McClusky minimisation algorithm in the functional domain, but without the excessive housekeeping difficulties that accrue with the latter when dealing with large functions. However, it is confined to produce a two-level sum-of-products realisation for $f(X)$ and does not allow the possible introduction of Exclusive-OR relationships that our other spectral techniques may allow.

4.4 SYNTHESIS BY SPECTRAL TRANSLATION

We have already encountered a feature of the spectra of Boolean functions which is employed in the synthesis procedure that we shall now consider.

This feature concerns the significance of the zero-, first- and higher-order spectral coefficients in **S**. We may express this in a general way as follows:

(a) If the large-magnitude spectral coefficients of **S** are concentrated in the zero- and first-order group $s_0; s_1, \ldots, s_n$, then the function $f(X)$ has a simple realisation using vertex gates in a classical AND-OR synthesis.

(b) However, if the large-magnitude coefficients are concentrated towards the higher-order end of the spectrum, then a simple AND-OR realisation will not be possible.

We have encountered, although not emphasized, this in Chapter 3 when considering Boolean function classification. For example, Table 3.7 covers eight functions within the same spectral classification, ranging from the given function $f(X)$ with a particularly high-magnitude coefficient s_{234} to $f_8(X)$ that is maximally valued within the zero- and first-order group of coefficients. The latter function is the simple positive canonic function $x_1x_2x_3$.

There have been certain mathematical considerations on how functional complexity is reflected and may be quantified in the spectral domain, particularly as this may have subsequent relevance to network testability. Some early non-spectral work of Shannon[7] and Yablonski[21] may be cited, but quantification using spectral data has been the later work of Karpovsky[6] and Miller.[22]

A fundamental problem, however, is the precise meaning and definition of "complexity." The estimate of functional complexity studied by Miller is based on counting the number of minterms at a Hamming distance of one surrounding each minterm m_j that take the same function value ($f(X) = 0$ or 1) as that of m_j, this count being summed over all 2^n minterms m_0 to m_{2^n-1} to give a final function complexity value $C(f)$. Each count may range from zero, where the chosen minterm is surrounded by unlike-valued minterms, to n, where the chosen minterm is entirely surrounded by like-valued minterms. This procedure counts each matching minterm-pair twice, giving results as shown in Fig. 4.9.

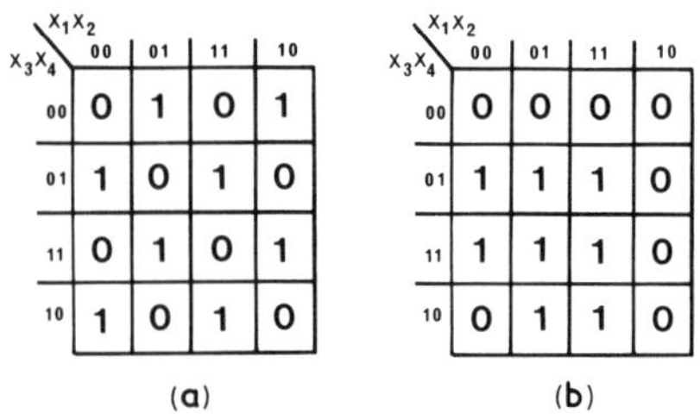

Fig. 4.9. Example functions to illustrate the complexity factor $C(f)$; (a) function $x_1 \oplus x_2 \oplus x_3 \oplus x_4$, $C(f) = 0$; (b) function $\bar{x}_1x_4 + x_2(x_3 + x_4)$, $C(f) = 40$.

This complexity factor $C(f)$ clearly increases as the 0 and 1 minterms of $f(X)$ become more grouped on a Karnaugh map, and hence is a useful measure of complexity if the function realisation is to be in classical AND-OR form.† However, it is less useful if Exclusive-OR gates are freely available, as Fig. 4.9a readily shows.

The development of an equation for $C(f)$ can be undertaken in the spectral domain,[22,39] giving the result

$$C(f) = n2^n - \frac{1}{2^{n-2}}\left\{\sum_{v=0}^{2^n-1} \|v\| r_v^2\right\} \tag{4.14}$$

where r_v are the spectral coefficients of **R**, and $\|v\|$ is a weighting factor representing the order of the coefficient r_v: $\|v\| = 0$ for r_0, $\|v\| = 1$ for r_1 to $r_n, \ldots, \|v\| = n$ for $r_{12\ldots n}$. Note that the first term $n2^n$ of Eq. (4.14) is the total number of adjacent minterms in $f(X)$, each pair counted twice‡; the second term accounts for the number of different-valued minterm pairs. Recall also that the spectra **R** and **S** are related by $r_0 = \frac{1}{2}\{2^n - s_0\}$, r_i, $i \neq 0$, $= -(1/2)s_i$, and hence Eq. (4.14) may be reexpressed as

$$C(f) = n2^n - \frac{1}{2^n}\left\{\sum_{v=0}^{2^n-1} \|v\| s_v^2\right\} \tag{4.14a}$$

Hence Eq. (4.14) provides mathematical support for the initial general statement that a "simpler" function has its large-magnitude spectral coefficients concentrated towards the lower-order end of the spectrum (smaller $\|v\|$ weightings), with less-simple functions having more prominent higher-order coefficients (higher $\|v\|$ weightings).

Whilst this development gives the same numerical result as our original definition for $C(f)$, it may be noted that if we ignore the constant $n2^n$ term, then the remaining expression decreases for simple AND OR functions, which may be considered more appropriate (see first footnote). Hence, an alternative definition of function complexity could be

$$C'(f) = \frac{1}{2^n}\left\{\sum_{v=0}^{2^n-1} \|v\| s_v^2\right\} \tag{4.14b}$$

However, in the following we shall continue to use where required our original definition for $C(f)$.

† It is possibly confusing that the magnitude of $C(f)$ increases for simple AND OR functions; the terminology "complexity factor" is therefore somewhat misleading.

‡ Adjacent minterms are minterms at a Hamming distance of one from each other.

4.4.1 Linear Translation Operations

Adopting the above concept of function simplicity (complexity), the principle of function synthesis by spectral translation is as follows:

(i) determine the spectrum **S** of the required function $f(X)$;

(ii) translate all large-magnitude spectral coefficients that are present in higher-order positions of **S** down to lower-order positions, so as to maximise the coefficient magnitudes towards the zero- and first-order groups;

(iii) realise $f(X)$ with the simple function defined by the latter spectrum, together with the necessary pre- and postoperators prefacing and following this simple core function, corresponding to the operations involved in the reordering of the spectrum of $f(X)$ into the latter form.

This will be appreciated as being the same procedure as encountered in Chapter 3, when considering Chow parameters and embedded threshold-logic core functions (see Fig. 3.6), but here we shall not insist upon the core function being necessarily a linearly separable function. Equally we may not insist upon finally ordering all the zero- and first-order coefficients in a positive, canonic order by applying appropriate NPN operations, as was necessary when canonic classification procedures were under consideration.

The general block schematic of this synthesis technique therefore is given in Fig. 4.10, which will be recognised as the generalisation of that previously shown in Fig. 3.6b. Since the pre- and postoperations which reorganise the spectrum of the core function are linear operators, the terminology "linear prefilter" and "linear postfilter" may be encountered for these two blocks in this general schematic.

The five spectral invariance operations that may be invoked in order to reorder a given spectrum into a canonic positive order have previously been given (see Chapter 3, Eqs. (3.13)–(3.17) inclusive). The last two are those that translate spectral coefficients from one order to another, and hence will be the two that now concern us. The remaining three correspond to NPN

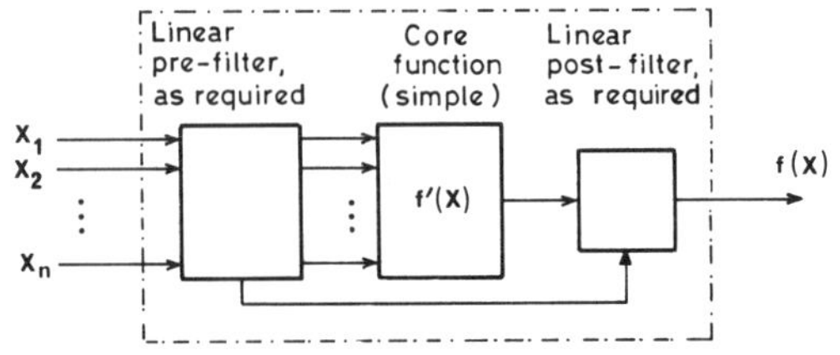

Fig. 4.10. Synthesis by spectral translation.

operations, which concern negations and variable permutations only. For convenience, we restate Eqs. (3.16) and (3.17) in a form relevant to us herewith.

Equation (3.16), *restated* The spectral domain interchange of 2^{n-2} pairs of coefficient values

$$\begin{aligned} s_i &\leftrightarrow s_{ij} \\ s_{ik} &\leftrightarrow s_{ijk} \\ s_{ikl} &\leftrightarrow s_{ijkl} \\ &\vdots \end{aligned} \tag{4.15}$$

which involves all coefficient orders except the zero-order and all coefficients containing i in their subscript identification, corresponds to the functional domain operation of replacing input x_i into the network with $\{x_i \oplus x_j\}$, $i \neq j$, that is, a linear prefilter implication.

Equation (3.17), *restated* The spectral domain interchange of 2^{n-1} pairs of coefficient values

$$\begin{aligned} s_i &\leftrightarrow s_0 \\ s_{ij} &\leftrightarrow s_j \\ s_{ijk} &\leftrightarrow s_{jk} \\ &\vdots \end{aligned} \tag{4.16}$$

which involves all coefficient orders including the zero-order coefficient, corresponds to the functional domain operation of modifying the network output to $\{f'(X) \oplus x_i\}$, that is, a linear postfilter implication. The reader should appreciate that because of the linearity of these operations, they may be applied either to keep the central "core" function constant and modify the overall resulting function (see Fig. 3.5) or to keep the overall function $f(X)$ constant and modify (simplifying in our case) the resultant core function $f'(X)$ (see Fig. 4.10).

The full development of this synthesis technique may be found in Edwards[23,24] and elsewhere.[22,25] As an example, consider the given function

$$f(X) = x_1x_2x_3 + x_1x_3\bar{x}_4 + \bar{x}_1x_2x_4 + \bar{x}_2\bar{x}_3x_4$$

whose spectrum **S** is

s_0	s_1	s_2	s_3	s_4	s_{12}	s_{13}	s_{14}	s_{23}	s_{24}	s_{34}	s_{123}	s_{124}	s_{134}	s_{234}	s_{1234}
2;	2	2	2	6;	2	−6	6	−6	−2	6;	2	−2	−2	6;	−2

Replacing x_1 by $\{x_1 \oplus x_4\}$, we obtain the core function $f(\tilde{x}_1, x_2, x_3, x_4)$:

$$2;\quad 6\,2\,2\,6;\quad -2\,-2\,2\,-6\,-2\,6;\quad -2\,2\,-6\,6;\quad 2$$

Replacing x_3 by $\{x_3 \oplus x_4\}$, we obtain the core function $f(\tilde{x}_1, x_2, \tilde{x}_3, x_4)$:

$$2;\quad 6\,2\,6\,6;\quad -2\,-6\,2\,6\,-2\,2;\quad 2\,2\,-2\,-6;\quad -2$$

Replacing x_2 by $\{x_2 \oplus \tilde{x}_3\}$, we obtain the core function $f(\tilde{x}_1, \tilde{x}_2, \tilde{x}_3, x_4)$:

$$2;\quad 6\,6\,6\,6;\quad 2\,-6\,2\,2\,-6\,2;\quad -2\,-2\,-2\,-2;\quad 2$$

This has maximised all the first-order coefficients as far as possible; s_0, however, has not been maximised by these three operations. If we calculate the complexity factor $C(f)$ as illustrated in Fig. 4.9, we obtain

$$C(f) \text{ for } f(X)\text{: } 32$$

$$C(f) \text{ for } f(\tilde{x}_1, x_2, x_3, x_4)\text{: } 32$$

$$C(f) \text{ for } f(\tilde{x}_1, x_2, \tilde{x}_3, x_4)\text{: } 36$$

$$C(f) \text{ for } f(\tilde{x}_1, \tilde{x}_2, \tilde{x}_3, x_4)\text{: } 40$$

Accepting this as a realisation for $f(X)$, the synthesis stages and the final realisation are shown in Fig. 4.11.

However, if we proceed so as to maximise all the zero- and first-order coefficients, then we may finally obtain the spectrum

$$6;\quad 6\,6\,6\,-6;\quad -2\,-2\,2\,-2\,2\,2;\quad -2\,2\,2\,2;\quad -6$$

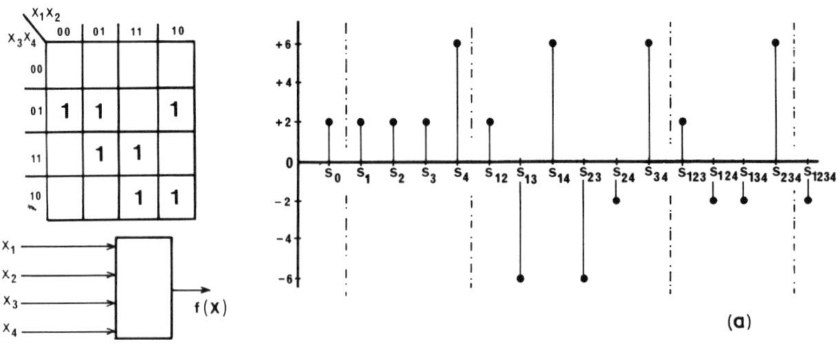

Fig. 4.11. Spectral translation synthesis of $f(X) = x_1x_2x_3 + x_1x_3\bar{x}_4 + \bar{x}_1x_2x_4 + \bar{x}_2\bar{x}_3x_4$. (a) Given function; (b) linear operation $\tilde{x}_1 = \{x_1 \oplus x_4\}$; (c) linear operation $\tilde{x}_3 = \{x_3 \oplus x_4\}$; (d) linear operation $\tilde{x}_2 = \{x_2 \oplus \tilde{x}_3\}$; (e) Final realisation.

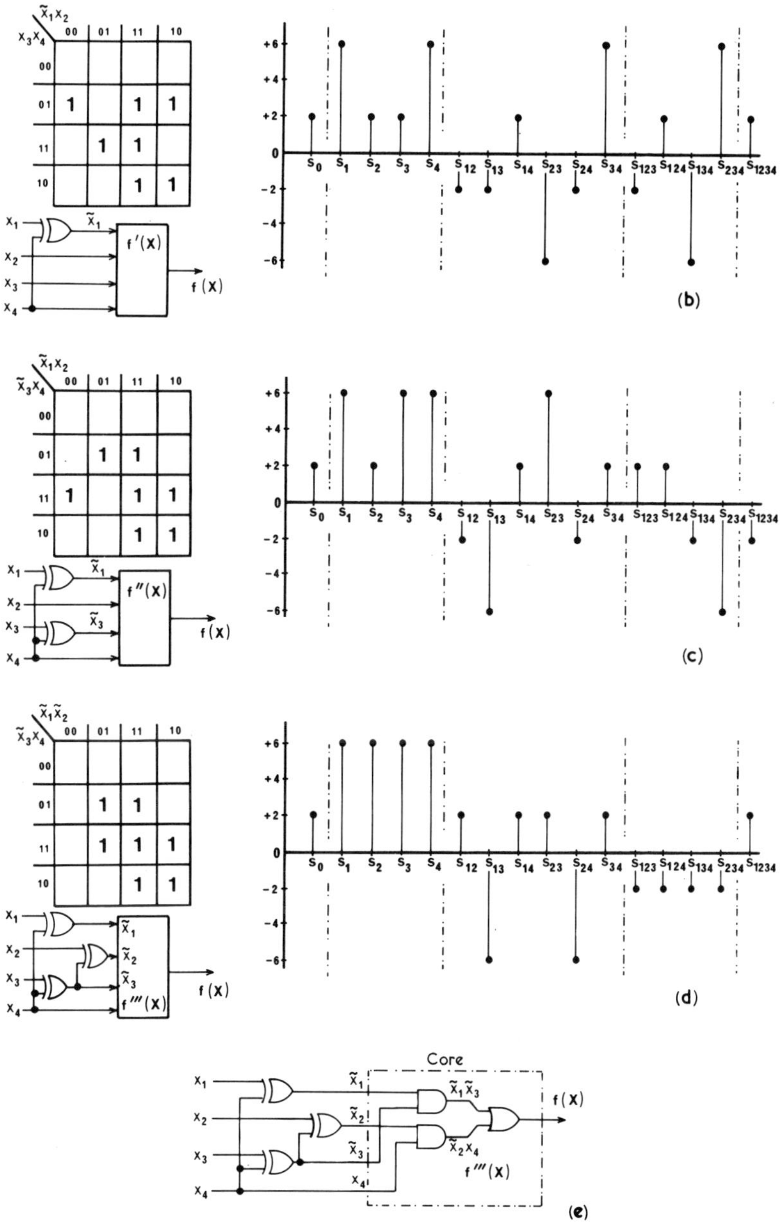

Fig. 4.11. (Continued)

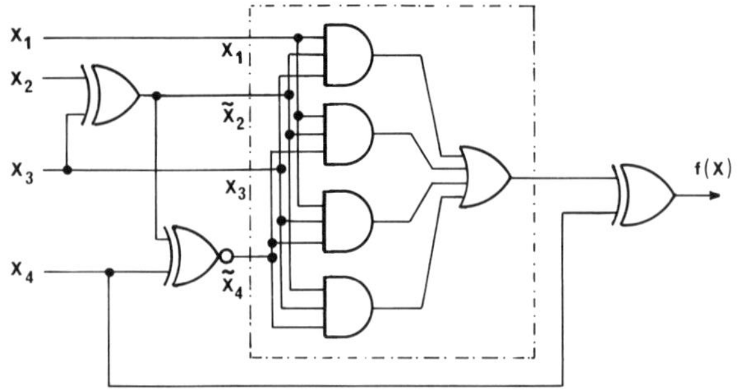

Fig. 4.12. Alternative spectral translation synthesis of Fig. 4.11.

This spectrum is obtained by

(a) performing the spectral operation $f(X) = \{f'(X) \oplus x_4\}$;
(b) replacing x_2 by $\{x_2 \oplus x_3\}$ to give $f'(x_1, \tilde{x}_2, x_3, x_4)$; and
(c) replacing x_4 by $\{x_4 \oplus \tilde{x}_2\}$ to give $f'(x_1, \tilde{x}_2, x_3, \tilde{x}_4)$.

Note that in maximising both the zero- and the first-order coefficients, the nth order coefficient s_{1234} has been left as a high-magnitude coefficient; $C(f)$ for this core function will be found to be 42. The core function, however, is $x_1\tilde{x}_2x_3 + x_1\tilde{x}_2\bar{\tilde{x}}_4 + x_1x_3\bar{\tilde{x}}_4 + \tilde{x}_2x_3\bar{\tilde{x}}_4$, which is a simple linearly separable 3-out-of-4 majority gate but not a very attractive AND OR core function. This alternative realisation is shown in Fig. 4.12.

4.4.2 Further Considerations

The preceding example illustrates the fundamental difficulty, as in all synthesis methods, of defining a "best" solution, and the criteria to give such a solution short of exhaustive synthesis. The best solution is clearly technology-dependent; if Exclusive functions are as freely available as AND OR functions, then the cost of the pre- and postlinear filter networks in this synthesis technique and syntheses such as Fig. 4.11 may be acceptable. On the other hand, should they be at a high premium, then other syntheses should be considered.

Further study is required to define more selective complexity factors than $C(f)$ in order to give guidance on which and how many spectral translation operations are necessary for an optimum synthesis under given technology constraints. In an automated CAD program, it has so far been found necessary to maintain interactive working, so that the designer can retain control of the acceptance or rejection of the design at each stage of the synthesis program, particularly if spectral translation and other synthesis methods are being combined so as to produce an ideal final solution under given constraints.

4.5 MULTIPLEXER SYNTHESIS

Multiplexer elements are frequently used as general-purpose combinatorial logic elements.[1,3,25] The basic multiplexer cell is a three-input element capable of realising all two-variable ($n = 2$) functions from the input set $\{0, 1, x_1, \bar{x}_1, x_2\}$, as shown in Fig. 4.13a. It may also be recognised as one of the six possible basic ULG.2 configurations developed in Chapter 3 (see Fig. 3.10e), being an NP-complete ULG-2 candidate.

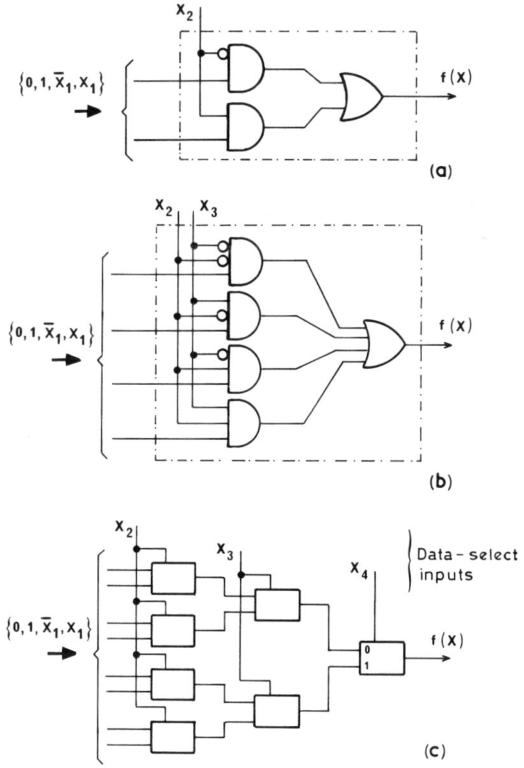

Fig. 4.13. Multiplexer topologies; (a) basic multiplexer module, $n \leq 2$ capability; (b) next size multiplexer module, $n \leq 3$ capability; (c) assemblies of (a) to provide increased capability.

For $n > 2$, either larger multiplexer elements or assemblies of elements become necessary, as in Figs. 4.13b and c. However, it should be noted that above $n = 2$, multiplexing structures do not produce ULG assemblies with the minimum possible number of input terminals; for example, for $n = 3$, six input terminals are required on a multiplexer assembly compared with five on an optimum ULG.3 element. This difference arises because the multiplexer assembly maintains a fixed subdivision of the x_i input variables between the data-select and the data-input terminals, whereas an optimum ULG.3 cell allows free permutation of input signals between all five input terminals.[26,27] This discrepancy in input terminal count becomes wider as n increases.

However, if one employs multiplexer elements for synthesis purposes, the question of which variables to allocate as the data-select inputs arises,

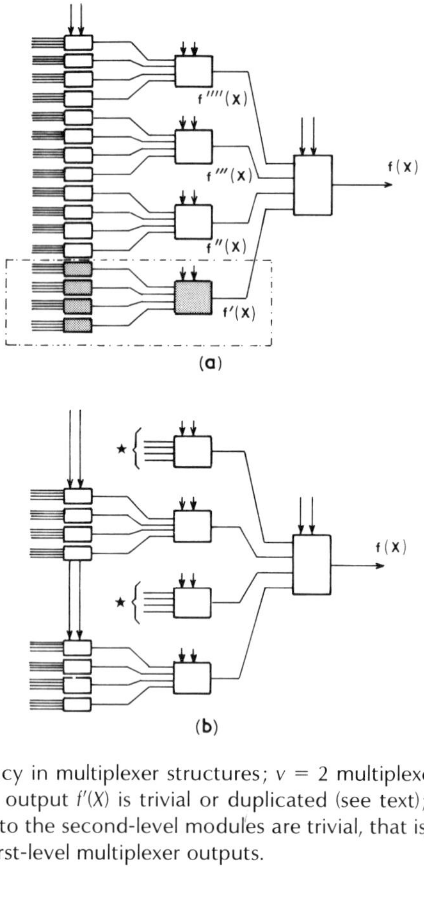

Fig. 4.14. Redundancy in multiplexer structures; $v = 2$ multiplexer cells shown. (a) Five modules redundant if output $f'(X)$ is trivial or duplicated (see text); (b) eight modules redundant if all inputs * to the second-level modules are trivial, that is, $\{0, 1, \bar{x}_1, x_1, \bar{x}_2, x_2\}$, or available from other first-level multiplexer outputs.

since an optimum choice can minimise the number of multiplexer elements (modules) required in a multiple-cell realisation, or optimise the data inputs so that, for example, x_i and $\bar{x}_i$ are not both required. This question of the selection of the data-select variables has received considerable attention using algebraic approaches[27–31]; here we shall consider the spectral approach to this problem.

A multiplexer synthesis is the direct realisation of a Shannon decomposition of the given function $f(X)$. Our earlier considerations in Section 4.2.1 and Eqs. (4.2)–(4.8) are therefore relevant. However, to highlight the effects of variable selection, consider the general structure shown in Fig. 4.14. If each multiplexer module has v data-select terminals, there will be one module at the final output level, 2^v modules at the next level, and so on, giving a total of $\{1 + 2^v + 2^{2v} + \cdots\}$ modules in all. It is also clear that finding a module redundant near the final level is generally most advantageous in reducing the total number of modules necessary to realise a given function, since all lower-level modules leading to the former module themselves become unnecessary. A module becomes redundant

(i) when its output is the trivial function 0 or 1 or x_i or $\bar{x}_i$, $i = 1$ to n; or
(ii) when its output is identical to that of another module, so that the one or the other becomes redundant.

These module output conditions may equally be considered as individual data input functions to a next-level module, and therefore module redundancy is indicated by determining the data input functions of all modules under the chosen pattern of data-select variables.

Consider a five-variable function realised using (at most) five multiplexer cells of the type shown in Fig. 4.13b. Let x_1, x_2 be the two data-select inputs at the final-level module. Then

$$f(X) = \bar{x}_1\bar{x}_2 f_0(X) + \bar{x}_1 x_2 f_1(X) + x_1\bar{x}_2 f_2(X) + x_1 x_2 f_3(X)$$

where $f_0(X), \ldots, f_3(X)$ are all functions of at most x_3, x_4, x_5. From Eq. (4.8), we have

$$[\mathbf{S}_0\,\mathbf{S}_1\,\mathbf{S}_2\,\mathbf{S}_3] = \tfrac{1}{4}\{[\mathbf{S}^0\,\mathbf{S}^1\,\mathbf{S}^2\,\mathbf{S}^3]\mathbf{T}^2\} \tag{4.8}$$

where $\mathbf{S}_0, \ldots, \mathbf{S}_3$ are the spectra of the subfunctions $f_0(X), \ldots, f_3(X)$ respectively, and where $\mathbf{S}^0$, $\mathbf{S}^1$, $\mathbf{S}^2$, $\mathbf{S}^3$ is the ordered partition of the spectrum $\mathbf{S}$ of $f(X)$. This may be rewritten in the form of Eq. (4.8a), if required.

Hence it is straightforward to determine the spectrum of each subfunction, which is the output from each preceding-level multiplexer module. The two redundancy conditions just stated will be

(i) when any subfunction spectrum $\mathbf{S}_0, \ldots$ is maximally valued ($\pm 2^n$) in a zero- or first-order coefficient; and

(ii) when any subfunction spectrum is identical to any other subfunction spectrum.

However, it still remains theoretically necessary to apply Eq. (4.8) to all possible combinations of the data-select input variables, that is, all possible reordered partitions of the spectrum **S**, in order to ensure a best choice of the data-select control signals.

An alternative approach is based upon the spectral addition techniques of Section 4.3. It will be recalled that the arithmetic summation of an orthogonal subset of coefficients of **S** can define a sub-area of $f(X)$, the fewer the coefficients in the subset then the larger the area of $f(X)$ that may be defined. If we continue with our five-variable example with a Shannon decomposition about x_1, x_2, then consideration of the subset of coefficients

$$s_0 \quad s_1 \quad s_2 \quad s_{12}$$

will provide information about the areas $\bar{x}_1\bar{x}_2, \bar{x}_1x_2, x_1\bar{x}_2$ and x_1x_2 of $f(X)$. Suppose we had

$$s_0 = +8, \quad s_1 = +8, \quad s_2 = -8, \quad s_{12} = +24$$

We may now apply the transform operation to find the result of all allowable $\pm$ summations of these coefficients:

$$\begin{bmatrix} 1 & 1 & 1 & 1 \\ 1 & -1 & 1 & -1 \\ 1 & 1 & -1 & -1 \\ 1 & -1 & -1 & 1 \end{bmatrix} \begin{bmatrix} 8 \\ 8 \\ -8 \\ 24 \end{bmatrix} \begin{matrix} s_0 \\ s_1 \\ s_2 \\ s_{12} \end{matrix} = \begin{bmatrix} +32 \\ -32 \\ 0 \\ +32 \end{bmatrix} \begin{matrix} (\bar{x}_1\bar{x}_2) \\ (x_1\bar{x}_2) \\ (\bar{x}_1x_2) \\ (x_1x_2) \end{matrix}$$

the results of which indicate

(a) area $\bar{x}_1\bar{x}_2$ of $f(X)$ = all logic 0, indicated by the summation to $+2^5$;
(b) area $x_1\bar{x}_2$ = all logic 1, indicated by the summation to -2^5;
(c) area $\bar{x}_1x_2$ has an equal number of logic 0 and logic 1 minterms, indicated by the summation to zero; and
(d) area x_1x_2 = all logic 0, indicated by the summation to $+2^5$.

Maximum summations to $\pm 2^n$ therefore are immediately useful results; summations to zero are possible useful results, since the subfunction may be merely $\bar{x}_i$ or x_i, $i \neq$ the data-select variables; summations to other than 0 or $\pm 2^n$ are immediately useful results, since they indicate that the subfunction *cannot* be a trivial function 0, 1, $\bar{x}_i$ or x_i.

The advantage of this approach is that only a small number of coefficients are involved at each stage; if single-variable data-select modules as in

Fig. 4.13c are involved, then the effect of the data-select input variables is rapidly investigated by consideration of the coefficient-pair summations

$$s_0 \pm s_1, \quad s_0 \pm s_2, \quad \ldots, \quad s_0 + s_n$$

corresponding to all possible choices for the data-select control input. The disadvantage of this approach is that a summation to zero is not explicit and requires further checking to define whether the subfunction is trivial or not; for example, an Exclusive-OR subfunction will give a zero summation value, but clearly does not eliminate the need for a previous-level multiplexer module.

Table 4.2. The flow chart for optimal multiplexer data-select variable selection, based upon spectral subset summation.

Given function $f(X)$

↓

(1) Take spectrum **S**.

↓

(2) Determine all final-level subset spectral coefficient summations, corresponding to all available choices of data-select variables $x_1, \ldots, x_n$.

↓ (branches to (3) and (4))

(3) Discard all bad choices.

(4) If no one "best" choice, investigate all potentially good choices, particularly to confirm when $\sum s_i = 0$ cases are trivial (acceptable).

↓

(5) Choose output-level decomposition. Determine spectra **S** of all non-trivial input subfunctions (see Eq. (4.6)) = preceding level multiplexer outputs.

↓

(6) Repeat (2), (3), (4), (5) for each preceding-level multiplexer module.

↓

(7) Continue (2) to (6) until realisation complete.

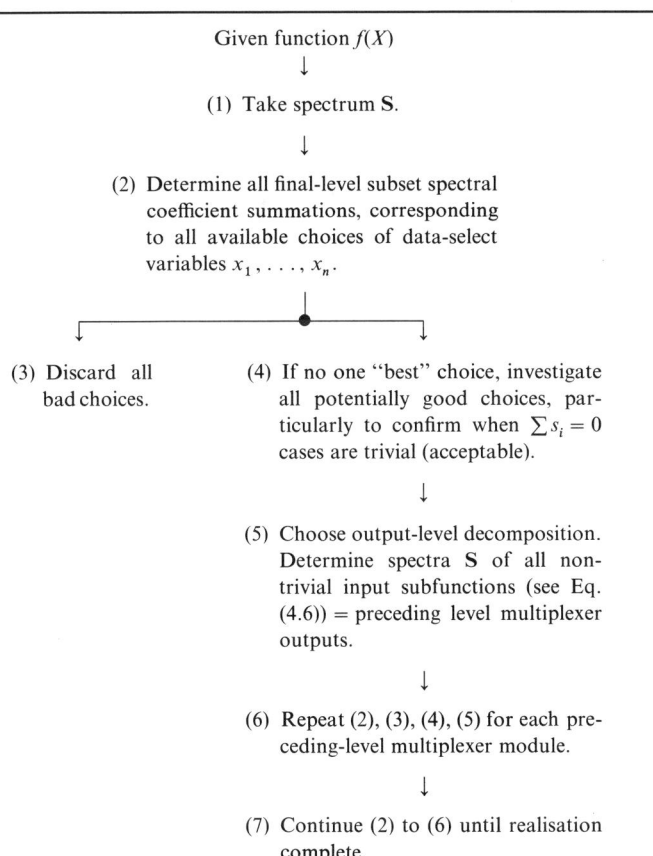

Lloyd has investigated this problem in detail.[32] Among his suggestions is the maintenance of a running record of the maximum and minimum number of multiplexer modules necessary for each choice of data-select variables at each level of synthesis, which is readily obtained from the occurrence of summation values 0, $\pm 2^n$, or otherwise, from which "bad" choices may be discarded from further consideration. The flow chart of this procedure is shown in Table 4.2.

Two features should be noted in this procedure. First, the choice of the data-select control variable(s) for any module at any level can be independent of all other modules, although in general a common choice at any given level, such as indicated in Fig. 4.13c, may be preferred. Second, this procedure, in common with other published procedures, does not cater for the possibility of a better overall realisation if optimisation at one particular level is relaxed; Fig. 4.14b illustrates how it may be possible to reduce the total number of multiplexer elements by not optimising from the final-level module backwards. Equally, certain algebraic techniques that optimise from the input-level forwards may not provide an optimum overall realisation.

In summary, it should be emphasized that there is no way of guaranteeing the minimum number of multiplexer modules for the realisation of a given function $f(X)$ short of exhaustive consideration of all decompositions of $f(X)$. The spectral methods maintain the advantages of the spectral domain in comparison with algebraic techniques, particularly as n increases. As a

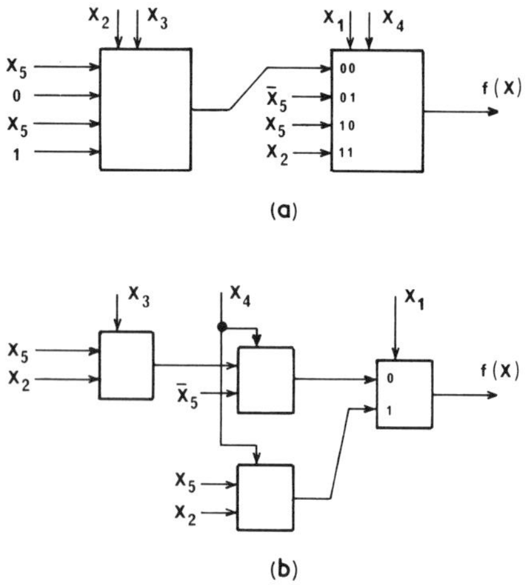

Fig. 4.15. Multiplexer realisations of $f(X) = \bar{x}_1 x_4 \bar{x}_5 + x_1 x_2 x_4 + \bar{x}_1 x_2 x_3 \bar{x}_4 + \bar{x}_3 \bar{x}_4 x_5 + x_1 \bar{x}_4 x_5$. (a) Using 4-to-1-line multiplexer modules; (b) using 2-to-1-line multiplexer modules (alternative minimal realisations possible).

final example, Fig. 4.15 illustrates minimal realisations for the function

$$f(X) = \bar{x}_1 x_4 \bar{x}_5 + x_1 x_2 x_4 + \bar{x}_1 x_2 x_3 \bar{x}_4 + \bar{x}_3 \bar{x}_4 x_5 + x_1 \bar{x}_4 x_5$$

Realisation (a) is discussed in Lloyd[33]; it is left as an exercise for the reader to confirm realisation (b), given that the spectrum for $f(X)$ is

$$\begin{aligned}\mathbf{S} = 0;\ \ 0\ 12\ 0\ 0\ 4;\ \ -4\ 0\ 0\ -12\ -4\ -4\ 0\ 0\ 4\ 20;\\ -4\ 12\ 0\ 0\ 4\ 4\ -4\ 0\ 0\ 4;\ \ -4\ 0\ 0\ 4\ 0;\ \ 0\end{aligned}$$

Both syntheses are developed from the output-level module backwards. Note that in realisation (b), x_2 and x_5 may immediately be discounted as profitable final-level data-select input variables, since examination of $\mathbf{S}$ shows that neither $s_0 \pm s_2$ nor $s_0 \pm s_5$ sums to 0 or $\pm 2^n$. This will also be found to be so at the next level of realisation.

4.6 SEQUENTIAL NETWORK SYNTHESIS

The subdivision of sequential networks into (a) asynchronous (unclocked) networks, and (b) synchronous (clocked) networks, is detailed in all standard logic texts.[1–5] The former is desirable when ultimate speed of response to changing input data is required, operating speed being dependent only upon the response time of the logic elements and their interconnections, but the problem of race hazards in other than simple networks is critical. Most texts consider race hazard aspects on the basis that only one data input changes at a time, which may not be acceptable in practice. Hence for this and other reasons such as testability, clocked operation is usually more acceptable in practice.†

Furthermore, many texts concentrate upon algorithms for state minimisation, in order to reduce the number of storage elements per system realisation. This historical emphasis is based largely upon the assumed cost of using storage elements in comparison with combinatorial logic gates, without regard to the cost of the latter. Also, complexity of interconnection between storage elements and gates, which in i.c. fabrication involves on-chip silicon area, is not a parameter of concern. Both of these factors are debatable with current i.c. technology; indeed, increasing emphasis may be observed in the use of shift register configurations for sequential applications, which is prolific in the use of storage elements, but has great practical advantages in ease of silicon layout and testability.[33,34]

† For a more detailed consideration of the uses of the terminology asynchronous and synchronous, see Muroga.[1]

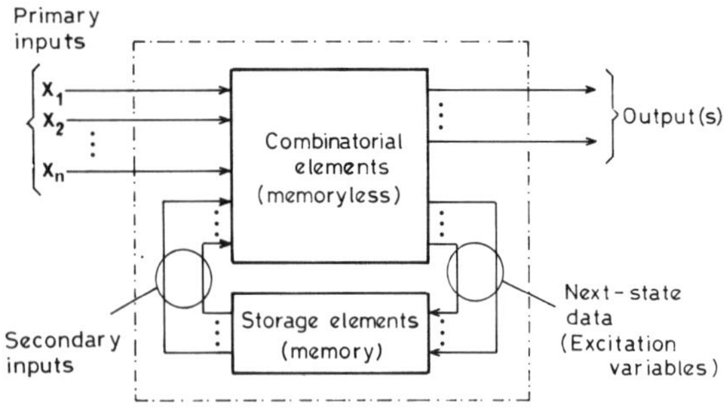

Fig. 4.16. General schematic of a sequential network with memory. Note that the storage may be clocked (synchronous) or unclocked (asynchronous).

However, neglecting for the present such considerations, the design of any sequential network may be classically subdivided into

(i) the state assignment procedure, to define the states through which the network progresses in realising the required system; and
(ii) realisation of the combinatorial logic network(s) required to execute the chosen state assignment.

Figure 4.16 illustrates this classic subdivision of the realising network. The combinatorial logic block may be subdivided into a next-state (or excitation) function block and a present-state output function block, if desired.

Optimal state assignment remains one of the outstanding theoretical problems in digital logic theory. Classical methods based upon flow tables and finite state machine theory are comprehensively cited in, for example, references (1)–(5), (35) and (36). The problem is compounded by the presence of incompletely specified states in many network requirements.

Karpovsky[6] currently is the only known publication that pursues the application of spectral techniques to this problem of state assignment. His realisation topology, however, remains the arithmetic structure of Fig. 4.1, consisting of a function generator, a coefficient block (memory) and an arithmetic summation unit. The output from the latter provides the changing

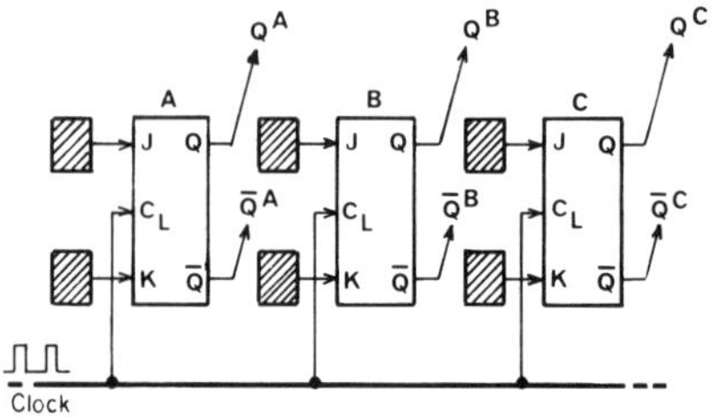

Fig. 4.17. Synchronously operating assembly, type JK bistable storage elements shown; clock steering functions to each J,K to be designed. (Similarly, for type RS, T and D circuit elements.)

coefficient data stored in the memory. Recall from Section 4.1 that the arithmetic output from such an assembly is not constrained to a single binary digit. Optimisation of the overall design, which includes the state assignment problem, is treated in terms of a search for the minimum number of non-vanishing (non-zero) coefficients in the coefficient memory block, appropriate procedures for this search being suggested. Again, the Harr orthogonal functions and spectral coefficients rather than the Hadamard/Rademacher–Walsh variants may be preferable in this arithmetic situation.

Outside this specialised topology, the use of the spectral design concepts discussed in preceding sections and in Chapter 5 may be applied to the combinatorial blocks of sequential networks, once their logical requirements have been defined by appropriate (conventional) means. There is, however, one particular application of spectral methods to sequential network design that may be mentioned here, namely the design of the clock-steering functions of conventional synchronously operating assemblies such as counters and accumulators.

Figure 4.17 shows the general arrangement of a synchronously operating assembly. The design of the clock-steering input functions by algebraic or mapping techniques may be found in most standard texts.[1–5] Here we shall consider the use of spectral data to perform this synthesis.

Let us first recall the algebraic equations that may be employed in this synthesis. We have two sets of equations that must be satisfied in the final realisation,† namely,

(i) the defining next-state *characteristic equation*, which defines the type of clocked bistable storage element being used; and

(ii) the next-state *system equation* for each of the bistable elements, in terms of the present-state outputs of the whole assembly.

The characteristic equation for each of the four main types of clocked bistable storage elements is

$$
\begin{aligned}
\text{Type } JK\text{:}\quad Q_{n+1} &= \bar{Q}_n J + Q_n \bar{K} \\
\text{Type } RS\text{:}\quad Q_{N+1} &= \bar{R}S + Q_n \bar{R} \\
&\quad \text{where } R \text{ and } S \text{ are disjoint} \qquad (4.17) \\
\text{Type } T\text{:}\quad Q_{n+1} &= \bar{Q}_n T + Q_n \bar{T} \\
&= T \oplus Q_n \\
\text{Type } D\text{:}\quad Q_{n+1} &= D
\end{aligned}
$$

The system equation for a given requirement is derived for each bistable element by compiling the truthtable of the present-state outputs $Q_n^A, Q_n^B, \ldots,$

† Unless otherwise specified, the output at Q of a bistable circuit is always considered; Q_n = output Q at time n, Q_{n+1} = output Q at time $n+1$, one clock pulse later.

and the required next-state output Q_{n+1} of the element in question, from which the next-state system output equation

$$Q_{n+1} = f(Q_n^A, Q_n^B, \ldots) \tag{4.18}$$

is derived for the element. The design procedure now involves the generation of the J, K (or R, S, T, or D) clock-steering logic functions for each element, such that when these functions are included in the appropriate characteristic equation (4.17), then the required next-state system equation (4.18) results.

Algebraic methods of solving these simultaneous Boolean equations may be found in Mano[2] and elsewhere. However, consider the spectral method following. Consider first the use of type JK circuits.

The JK characteristic equation will be seen to be a Shannon decomposition about Q_n, the two reduced functions being J and $\bar{K}$, respectively. Hence the following spectral synthesis technique readily follows:

(i) take the spectrum $\mathbf{S}$ of the required next-state system output equation (4.18);

(ii) decompose into two disjoint spectra $\mathbf{S}_0$, $\mathbf{S}_1$ about the appropriate variable, using the normal spectral decomposition

$$\mathbf{S}_0 = \tfrac{1}{2}(\mathbf{S}^0 + \mathbf{S}^1) \quad \text{and} \quad \mathbf{S}_1 = \tfrac{1}{2}(\mathbf{S}^0 - \mathbf{S}^1) \tag{4.6}$$

(iii) spectrum $\mathbf{S}_0$ now represents the required clock-steering input function J (see Eq. (4.17)), and $\mathbf{S}_1$ represents the negation of the required clock-steering input function K.

The inverse transform on $\mathbf{S}_0, \mathbf{S}_1$ will therefore lead to the required clock-steering functions for J and $\bar{K}$. Note that in the original disclosure of this method,[38] the negation required on K was incorporated by interchanging the spectral coefficients in the summation $\mathbf{S}_1 = \tfrac{1}{2}(\mathbf{S}^0 - \mathbf{S}^1)$, which is equivalent to multiplying the coefficients by -1, which in turn is equivalent to the required negation of K.

Syntheses using type RS elements are complicated by the need to ensure that the R and S clock-steering functions are disjoint. Assuming this is present, then we may revise the RS characteristic equation to

$$\bar{Q}_n S + Q_n \bar{R}$$

which will be seen to be identical in form to the JK characteristic equation. Hence the above spectral procedure may be identically followed, where $S \equiv J$ and $R \equiv K$, but a final check must be undertaken to ensure that our resulting functions for R and S are disjoint. If they *are* disjoint, the realisation may be immediately accepted; if they are *not* disjoint, then they may always be made disjoint by further multiplying the S function by $\bar{Q}_n$ and the R function by Q_n.

Syntheses using type T elements are very straightforward, since the characteristic equation $T \oplus Q_n$ will be recognised as a spectral translation operation (see Eqs. (3.17 and (4.16)). Hence the interchange of spectral coefficients

$$\begin{gathered} s_i \leftrightarrow s_0 \\ s_{ij} \leftrightarrow s_j \\ \vdots \end{gathered} \tag{4.16}$$

of the spectrum $\mathbf{S}$ of the next-state system equation immediately gives the spectrum of the required clock-steering input function T.

Syntheses using type D circuits require no comment; the clock-steering input function is directly given by the next-state system equation. As a short working example of these spectral design procedures, consider the following three-digit continuous coding sequence:

Present-state output Q_n			Next-state output Q_{n+1}		
Q^A	Q^B	Q^C	Q^A	Q^B	Q^C
0	0	0	0	0	1
0	0	1	0	1	0
0	1	0	0	1	1
0	1	1	1	1	1
1	1	1	1	1	0
1	1	0	1	0	1
1	0	1	1	0	0
1	0	0	0	0	0

Considering, say, storage element A, the next-state system equation for A is

$$Q_{n+1}^A = Q_n^A Q_n^B + Q_n^A Q_n^C + Q_n^B Q_n^C$$

which has the spectrum

$$\begin{array}{lcccc} & s_0; & s_A\ s_B\ s_C; & s_{AB}\ s_{AC}\ s_{BC}; & s_{ABC} \\ \mathbf{S} = & 0 & 4\ 4\ 4 & 0\ \ 0\ \ 0 & -4 \end{array}$$

Decomposing about $\bar{Q}^A$, Q^A, we have

$$\begin{aligned} \mathbf{S}_0 &= \tfrac{1}{2}\{\ \ 4; 4\,4; -4\} \\ &= \quad\ 2; 2\,2; -2, = \text{function } Q^B Q^C \\ \mathbf{S}_1 &= \tfrac{1}{2}\{-4; 4\,4;\ \ 4\} \\ &= \ -2; 2\,2;\ \ 2, = \text{function } Q^B + Q^C \end{aligned}$$

Hence a solution for the clock-steering functions for A is

$$J^A = Q^B Q^C$$
$$K^A = \overline{Q^B + Q^C} = \bar{Q}^B \bar{Q}^C$$

It will be noted that these two clock-steering functions are disjoint. Hence we may immediately accept this solution for a type RS realisation, giving us

$$S^A = Q^B Q^C, \qquad R^A = \bar{Q}^B \bar{Q}^C$$

Type T synthesis for A requires the disjoint spectral translation operations $s_0 \leftrightarrow s_A, \ldots,$ on $\mathbf{S}$, giving us the spectrum for T^A of

s_0;	s_A	s_B	s_C;	s_{AB}	s_{AC}	s_{BC};	s_{ABC}
4	0	0	0	4	4	−4	0

whence the inverse transform immediately gives

$$T^A = \bar{Q}^A Q^B Q^C + Q^A \bar{Q}^B \bar{Q}^C$$

From the above, the simplicity of the JK and T synthesis procedures will be evident; their power becomes increasingly attractive compared with algebraic means as the number of state variables increases. However, the type RS procedure is not so attractive, should the initial decomposition functions not be disjoint. This problem is discussed in the original disclosure,[36] but further work on type RS synthesis may be useful in order to define a more direct and optimal design procedure using the spectral domain data.

4.7 CHAPTER SUMMARY

In this chapter we have considered a number of possible techniques for the synthesis of combinatorial logic networks, principally using conventional vertex and Exclusive-OR gates. By choice we have not concentrated upon direct synthesis using the spectral coefficient values themselves in an arithmetic form of realisation, but rather have used the spectral data to provide information about decompositions of the function being synthesised, or other structural information.

The ease with which Shannon and other decompositions of a given Boolean function may be processed in the spectral domain, and the method by which prime implicant terms may be detected by consideration of a small subset of coefficients, are particularly powerful features of several of these methods. The advantages of numerical computations in the spectral domain compared with algebraic working in the functional domain should be apparent from our discussions. These advantages will be reinforced by our considerations in Chapter 5, when we consider the detection of symmetries in Boolean functions and the use of symmetries in function realisation.

As in Chapter 3, we have confined all our detailed discussions herewith on the spectral domain data obtained using Hadamard transformations from the functional domain. This represents the published state of the art of the procedures we have here considered. It remains an outstanding area of work to reconsider these and further concepts using the Haar spectral data,[38] particularly since the row functions of the Haar transform do not involve Exclusive-OR operators and hence may have practical advantages.

References

1. Muroga, S., "Logic Design and Switching Theory." Wiley, New York, 1979.
2. Mano, Morris M., "Digital Logic and Computer Design." Prentice-Hall, New Jersey, 1979.
3. Lewin, D. "Computer-Aided Design of Digital Systems." Crane-Russak, New York and Edward Arnold, London, 1977.
4. Friedman, A. D., and Menon, P. R., "Theory and Design of Switching Circuits." Computer Science Press, Woodland Hills, California, 1975.
5. Lee, S. C., "Digital Circuits and Logic Design." Prentice-Hall, New Jersey, 1976.
6. Karpovsky, M. G., "Finite Orthogonal Series in the Design of Digital Devices." Wiley, New York, 1976.
7. Shannon, C., "The synthesis of two-terminal switching circuits, *Bell System Tech. J.* **28,** 59–98 (1949).
8. Curtiss, H. A., "A New Approach to the Design of Switching Circuits." D. Van Nostrand, London, 1962.
9. Karp, R., Functional decomposition and switching circuit design, *SIAM J.* **11,** 291–335 (1963).
10. Davio, M., Deschamps, J. P., and Thayse, A., "Discrete and Switching Functions." McGraw-Hill, New York, 1978.
11. Tokmen, V. H., Disjoint decomposibility of multi-valued functions by spectral means, *Proc. IEEE 10th Internat. Symp. Multiple-Valued Logic*, 88–93 (1980).
12. Muzio, J. C., The decomposition of Rademacher–Walsh spectra, Technical Report CS7707-R, Virginia Polytechnic Institute and State University Blacksburg, February 1977.
13. Tokmen, V. H., An Investigation into the Properties of Multi-Valued Spectral Logic, Ph.D. Thesis, University of Bath, United Kingdom, 1980.
14. Lloyd, A. M., A consideration of orthogonal matrices, other than the Rademacher–Walsh types, for the synthesis of digital networks, *Internat. J. Electron.* **47,** 205–212 (1979).
15. Lloyd, A. M., Spectral addition techniques for the synthesis of multivariable logic networks, *IEE Comp. Digital Tech.* **1,** 152–164 (1978).
16. Lechner, R. J., Harmonic analysis of switching functions, *in* "Recent Developments in Switching Theory" (A. Mukhopadhyay, ed.). Academic Press, New York, 1971.
17. Besslich, P. W., On the Walsh–Hadamard transform and prime implicant extraction, *Trans. IEEE,* **EMC-20,** 516–519 (1978).
18. Besslich, P. W., Computer-aided design of logic circuits using transform methods, *Proc. IEE Internat. Conf. Comp. Aided Design Manufacture*, 75–79 (July 1979).

19. Besslich, P. W., Determination of the irredundant forms of a Boolean function using Walsh–Hadamard analysis and dyadic groups, *IEE Comp. Digital Tech.* **1,** 143–151 (1978).
20. Besslich, P. W., Fast transform procedure for the generation of near-minimal covers of Boolean functions, *Proc. IEE* **128,** 250–254 (1981).
21. Yablonski, S. W., On algorithmic obstacles to the synthesis of minimal contact networks, *Problemy Kibernet. No. 2,* 75–121 (1955) (in Russian).
22. Miller, D. M., A spectral estimate of function complexity, *in* NATO Final Report 1623, "An Integrated Theory of Digital Logic Employing Rademacher–Walsh Transforms and Two-Place Decompositions," University of Bath, United Kingdom, 1980.
23. Edwards, C. R., The application of the Rademacher–Walsh transform to Boolean function classification and threshold logic synthesis, *Trans. IEEE* **C-24,** 48–62 (1975).
24. Edwards, C. R., The design of easily tested circuits using mapping and spectral techniques, *Radio Electron. Eng.* **47,** 321–342 (1977).
25. Hurst, S. L., "The Logical Processing of Digital Signals." Crane-Russak, New York and Edward Arnold, London, 1978.
26. Chen, X., and Hurst, S. L., A consideration of the number of input terminals on universal logic gates and their realization, *Internat. J. Electron.* **50,** 1–13 (1981).
27. Yau, S. S., and Tang, C. K., Universal logic modules and their application, *Trans. IEEE* **C-19,** 141–149 (1970).
28. Tabloski, T. F., and Moule, F. J., A numerical expansion technique and its application to minimal multiplexer circuits, *Trans. IEEE* **C-25,** 684–702 (1976).
29. Almaini, A. E. A., and Woodward, M. E., An approach to the control variable selection problem for universal logic modules, *Digital Process.* **3,** 189–206 (1977).
30. Lotfi, Z. M., and Tosser, A. F., Graphical exhaustive analysis of minimum MUX synthesis of switching functions, *Comput. Electr. Engrg.* **7,** 235–242 (1980).
31. Voith, R. P., ULM implicants for minimisation of universal logic module circuits, *Trans. IEEE* **C-26,** 417–424 (1977).
32. Lloyd, A. M., Design of multiplexer universal-logic-module networks using spectral techniques, *Proc. IEE* **127E,** 31–36 (1980).
33. Bennetts, R. G., "Introduction to Digital Board Testing." Crane-Russak, New York and Edward Arnold, London, 1981.
34. Pradhan, D. K., Hsiao, M. Y., Patel, A. M., and Su, S. Y., Shift registers designed for on-line fault detection, *Proc. IEEE Eighth Ann. Internat. Conf. Fault-Tolerant Computing,* 173–184 (1978).
35. Clare, C. R., "Designing Logic Systems Using State Machines." McGraw-Hill, New York, 1973.
36. Haring, D. R., "Sequential Circuit Synthesis: State Assignment Aspects." MIT Press, Cambridge, Massachusetts, 1966.
37. Picton, P. D., Clock-steering synthesis using spectral techniques, *IEE Electron. Lett.* **16,** 409–411 (1980).
38. Hurst, S. L., The Haar transform in digital network synthesis, *Proc. IEEE 11th Internat. Symp. Multiple-Valued Logic,* 10–18 (1981).
39. Miller, D. M., Muzio, J. C., and Hurst, S. L., Spectral method of Boolean function complexity, *IEE Electron. Lett.* **18,** 572–574 (1982).

5

Symmetry Conditions in Boolean Functions

5.1 INTRODUCTION

Every Boolean function possesses some simple or complex form of symmetry, recognition of which may lead to a more efficient synthesis of the function. The most powerful form of symmetry is that of the totally symmetric function. A totally symmetric function $f(X)$ is unchanged by any permutation of its n input variables. Examples of totally symmetric functions are $f(x_1, x_2, x_3) = x_1 + x_2 + x_3$, the majority function $f(x_1, x_2, x_3) = x_1x_2 + x_1x_3 + x_2x_3$ and other similar functions such as $f(x_1, x_2, x_3) = \bar{x}_1x_2 + \bar{x}_1\bar{x}_3 + x_2\bar{x}_3$. The variables of symmetry in the first two cases are $\{x_1, x_2, x_3)$, whilst in the third case they are $\{\bar{x}_1, x_2, \bar{x}_3\}$. This means, for example, that interchanging $\bar{x}_1$ and x_2 in the third function does not change the function.

Clearly only a very small proportion of functions will be totally symmetric. However, we can define a rather more restricted form of symmetry involving smaller subsets of the variables. A function $f(X)$ is symmetric in some non-empty subset of k input variables if any interchanges within this subset leave $f(X)$ invariant. For example, $f\{x_1, x_2, x_3, x_4\} = x_1\bar{x}_3 + x_2\bar{x}_3 + x_1x_2x_4$ is symmetric in $\{x_1, x_2\}$ since interchanging the variables x_1, x_2 in the function leaves it unchanged.

Let us consider this function in a little more detail. The function is illustrated in the map shown in Fig. 5.1.

Since this function is symmetric in $\{x_1, x_2\}$, we can interchange x_1 and x_2 without changing this function, that is, $f(0, 1, x_3, x_4) = f(1, 0, x_3, x_4)$. This is indicated in Fig. 5.1 by the two columns marked =, which have to be identical. This identification of equal portions of a function is what will concern us for the remainder of this chapter.

Classical methods for the detection of simple symmetries can be found in the literature[1–5], but the reader is warned to be careful of some confusion in the terminology among various authors.

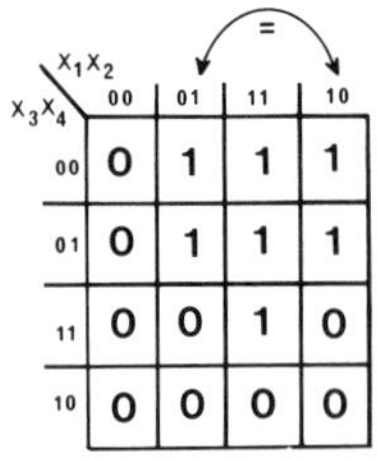

Fig. 5.1. $f(X) = x_1\bar{x}_3 + x_2\bar{x}_3 + x_1x_2x_4$, symmetric in $\{x_1, x_2\}$.

5.2 SYMMETRIES OF DEGREE 2

We have seen in Section 5.1 an example where a symmetry in a pair of variables can be identified with two equal columns of the Karnaugh map. This property of equal subfunctions within a function is what we now address. Throughout this section we shall be examining functions for properties of the type $f(x_1, \ldots, a, \ldots, b, \ldots, x_n) = f(x_1, \ldots, c, \ldots, d, \ldots, x_n)$ for some $a, b, c, d \in \{0, 1\}$ and these will be all the possible symmetries of degree 2 for the function. The degree 2 refers to the number of variables that are being fixed—in our case two. There are a number of different possibilities, which are considered below. The terminology for some of these symmetries, together with their description, was introduced by Edwards and Hurst.[6,7] Symmetries of degree 2 are discussed in detail by Hurst.[8]

5.2.1 Equivalence Symmetry in $\{x_i, x_j\}$

A function of $f(x_1, \ldots, x_i, \ldots, x_j, \ldots, x_n)$ possesses an equivalence symmetry in $\{x_i, x_j\}$, written $E\{x_i, x_j\}$, if $f(x_i, \ldots, 0, \ldots, 0, \ldots, x_n) = f(x_1, \ldots, 1, \ldots, 1, \ldots, x_n)$.

An equivalence symmetry implies equal rows, columns or other groupings on the Karnaugh map of the function. The six possible equivalence symmetries for functions of four variables are illustrated in Fig. 5.2.

It is possible for more than one of these symmetries to be present in any given function. For example, the function $f(x_1, x_2, x_3, x_4) = \bar{x}_1\bar{x}_2\bar{x}_3 + \bar{x}_1\bar{x}_2x_4 + \bar{x}_1\bar{x}_3x_4 + \bar{x}_2\bar{x}_3x_4 + x_1x_2x_3 + x_1x_3\bar{x}_4 + x_2x_3\bar{x}_4$, which is shown in Fig. 5.3, has equivalence symmetries in both x_2, x_4 and x_1, x_4. Notice that the relationship is not transitive in that $E\{x_1, x_4\}$ and $E\{x_2, x_4\}$ do not imply $E\{x_1, x_2\}$, although all three can exist simultaneously in a function.

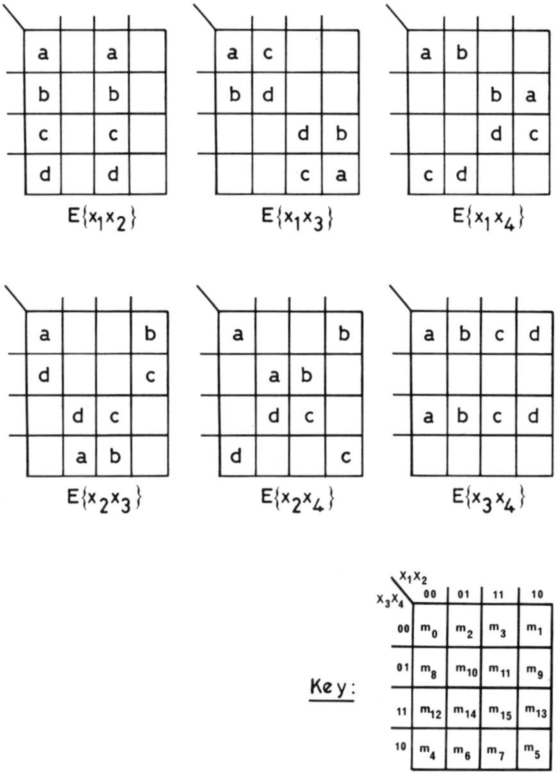

Fig. 5.2. The six equivalence symmetries, any of which may exist in a four-variable function $f(x_1, x_2, x_3, x_4)$. The specified entries indicate like-valued minterms; outside these specified pairs any pattern of 0 and 1 may exist.

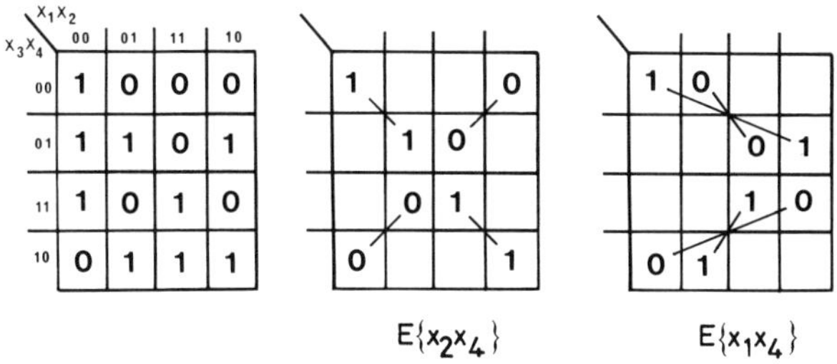

Fig. 5.3. The function $f(x_1, x_2, x_3, x_4) = \bar{x}_1\bar{x}_2\bar{x}_3 + \bar{x}_1\bar{x}_2x_4 + \bar{x}_1\bar{x}_3x_4 + \bar{x}_2\bar{x}_3x_4 + x_1x_2x_3 + x_1x_3\bar{x}_4 + x_2x_3\bar{x}_4$, which possesses $E\{x_2, x_4\}$ and $E\{x_1, x_4\}$.

5.2.2 Non-Equivalence Symmetry in $\{x_i\, x_j\}$

A function $f(x_1, \ldots, x_i, \ldots, x_j, \ldots, x_n)$ possesses a non-equivalence symmetry in $\{x_i, x_j\}$ written $N\{x_i, x_j\}$, if $f(x_1, \ldots, 0, \ldots, 1, \ldots, x_n) = f(x_1, \ldots, 1, \ldots, 0, \ldots, x_n)$. All the six possible non-equivalence symmetries for functions of four variables are illustrated in Fig. 5.4.

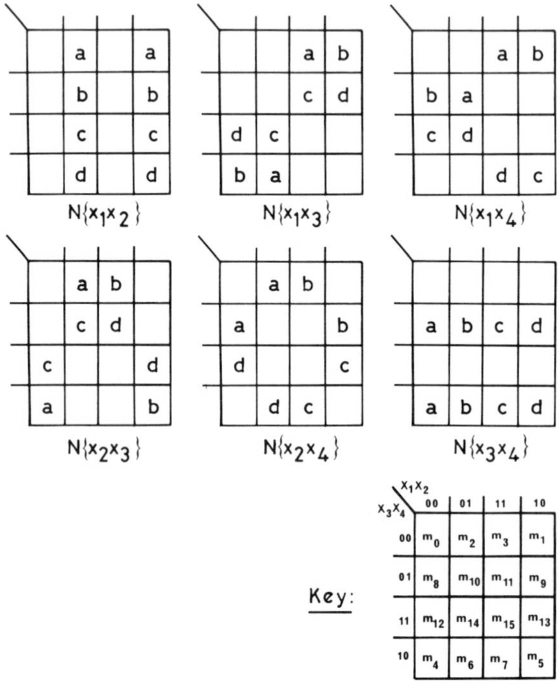

Fig. 5.4. The six non-equivalence symmetries, any of which may exist in a four-variable function $f(x_1, x_2, x_3, x_4)$. The specified entries indicate like-valued minterms; outside these specified pairs any pattern of 0 and 1 may exist.

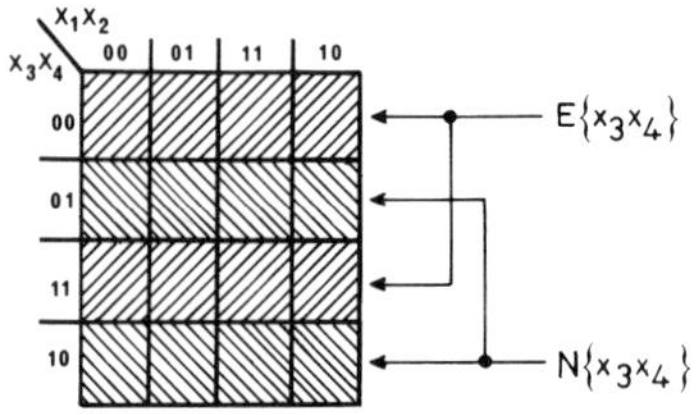

Fig. 5.5. Symmetries $E\{x_3, x_4\}$ and $N\{x_3, x_4\}$.

If we consider $f(x_1, x_2, x_3, x_4)$ and look at $E\{x_3, x_4\}$ and $N\{x_3, x_4\}$, we see that we are looking for identical pairs of non-adjacent rows on the Karnaugh map, as illustrated in Fig. 5.5. However, it is also possible to have symmetries involving adjacent rows on the map, and it is these that we discuss now.

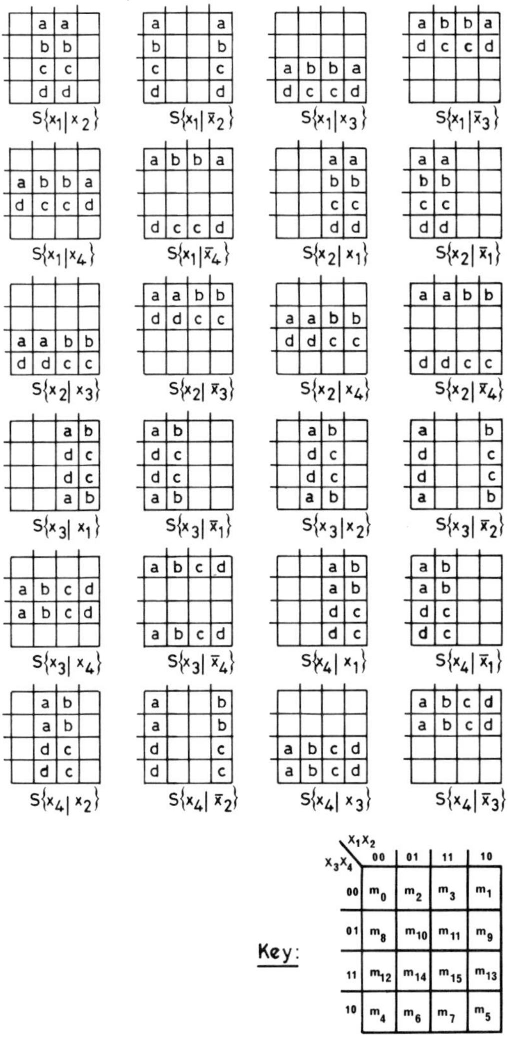

Fig. 5.6. The 24 single-variable symmetries, any of which may exist in a four-variable function $f(x_1, x_2, x_3, x_4)$. The specified entries indicate like-valued minterms; outside these specified pairs any pattern of 0 and 1 may exist.

5.2.3 Single-Variable Symmetries

There are two possible single-variable symmetries we can define as follows:

A function $f(x_1, \ldots, x_i, \ldots, x_j, \ldots, x_n)$ possesses a single-variable symmetry in x_j over x_i, written $S\{x_j|x_i\}$, if $f(x_1, \ldots, 1, \ldots, 0, \ldots, x_n) = f(x_1, \ldots, 1, \ldots, 1, \ldots, x_n)$.

A function $f(x_1, \ldots, x_i, \ldots, x_j, \ldots, x_n)$ possesses a single-variable symmetry in x_j over $\bar{x}_i$, written $S\{x_j|\bar{x}_i\}$, if $f(x_1, \ldots, 0, \ldots, 0, \ldots, x_n) = f(x_1, \ldots, 0, \ldots, 1, \ldots, x_n)$.

There are 24 possible single-variable symmetries for a four-variable function $f(x_1, x_2, x_3, x_4)$. These are illustrated in Fig. 5.6.

For each pair of variables, there are six possible symmetries that can be listed as follows for x_3, x_4, using the row labelling shown in Fig. 5.7:

if row 0 = row 1, the function possesses $S\{x_4|\bar{x}_3\}$;
if row 0 = row 2, the function possesses $E\{x_3, x_4\}$;
if row 0 = row 3, the function possesses $S\{x_3|\bar{x}_4\}$;
if row 1 = row 2, the function possesses $S\{x_3|x_4\}$;
if row 1 = row 3, the function possesses $N\{x_3, x_4\}$;
if row 2 = row 3, the function possesses $S\{x_4|x_3\}$.

Recalling our notation that

$$f_0(x_1, \ldots, x_{n-2}) = f(x_1, \ldots, x_{n-2}, 0, 0)$$

$$f_1(x_1, \ldots, x_{n-2}) = f(x_1, \ldots, x_{n-2}, 1, 0)$$

$$f_2(x_1, \ldots, x_{n-2}) = f(x_1, \ldots, x_{n-2}, 0, 1)$$

$$f_3(x_1, \ldots, x_{n-2}) = f(x_1, \ldots, x_{n-2}, 1, 1)$$

then we see that for $f(x_1, x_2, x_3, x_4)$

$$f_0 = f_1 \text{ is } S\{x_3|\bar{x}_4\}$$

$$f_0 = f_2 \text{ is } S\{x_4|\bar{x}_3\}$$

$$f_0 = f_3 \text{ is } E\{x_3, x_4\}$$

$$f_1 = f_2 \text{ is } N\{x_3, x_4\}$$

$$f_1 = f_3 \text{ is } S\{x_4|x_3\}$$

$$f_2 = f_3 \text{ is } S\{x_3|x_4\}$$

For an n-variable function, there are $2n(n-1)$ possible single-variable symmetries, and $\frac{1}{2}n(n-1)$ possible equivalence and the same number of non-equivalence symmetries, giving $3n(n-1)$ possible symmetries.

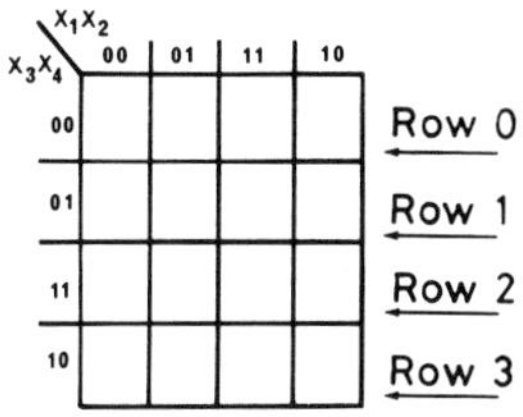

Fig. 5.7. Row labelling for definition of x_3, x_4 symmetries.

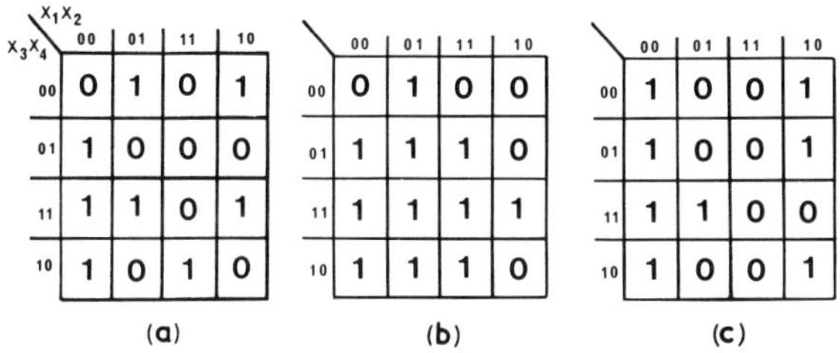

Fig. 5.8. Functions with multiple symmetries: (a) $f(X) = \bar{x}_1\bar{x}_2(x_3 + x_4) + x_3x_4(\bar{x}_1 + \bar{x}_2) + \bar{x}_3\bar{x}_4(\bar{x}_1x_2 + x_1\bar{x}_2) + x_1x_2x_3\bar{x}_4$; (b) $f(X) = \bar{x}_1x_2 + \bar{x}_1x_3 + \bar{x}_1x_4 + x_2x_3 + x_2x_4 + x_3x_4$; (c) $f(X) = \bar{x}_1\bar{x}_2 + x_1\bar{x}_2(\bar{x}_3 + \bar{x}_4) + \bar{x}_1x_3x_4$.

Consider the functions shown in Fig. 5.8 and all the symmetries they possess. The function in Fig. 5.8a does not possess any single-variable symmetries, but it does have $N\{x_1, x_2\}$, $E(x_1, x_3\}$ and $E(x_2, x_3\}$. For the function in Fig. 5.8b, we have $E\{x_1, x_2\}$, $E\{x_1, x_3\}$, $E\{x_1, x_4\}$, $N\{x_2, x_3\}$, $N\{x_2, x_4\}$ and $N\{x_3, x_4\}$. Finally, the function in Fig. 5.8c possesses $N\{x_3, x_4\}$, $S\{x_3|\bar{x}_4\}$, $S\{x_4|\bar{x}_3\}$, $S\{x_1|\bar{x}_3\}$ and $S\{x_1|\bar{x}_4\}$. Note that for this latter function any two of $N\{x_3, x_4\}$, $S\{x_3|\bar{x}_4\}$, $S\{x_4|\bar{x}_3\}$ will imply the third, since all that is being demonstrated is that the three rows in the Karnaugh map are identical. This connection between the symmetries will always exist for completely specified functions.

5.3 SPECTRAL CONDITIONS FOR SYMMETRIES OF DEGREE 2

In Section 5.2, we saw the conditions on the minterms of a function in order for it to possess particular symmetries. Here we derive conditions on the spectral coefficients for the determination of these properties. The results described herein are primarily due to Hurst and Edwards.[8–10]

Consider a function $f(x_1, \ldots, x_n)$ and the four subfunctions $f_0(x_1, \ldots, x_{n-2})$, $f_1(x_1, \ldots, x_{n-2})$, $f_2(x_1, \ldots, x_{n-2})$ and $f_3(x_1, \ldots, x_{n-2})$. As noted before, symmetries in x_{n-1}, x_n are concerned with the equality of pairs of these subfunctions. The function $f(X)$ will have a truthtable column vector **Y**, where 0, 1 entries for $f(X)$ are coded by $+1$, -1, respectively, in **Y** (see Section 2.2).

We have $\mathbf{S} = \mathbf{T}^n\mathbf{Y}$ and $\mathbf{S}_u = \mathbf{T}^{n-2}\mathbf{Y}_u$, each u $0 \leq u \leq 3$, to determine the spectra of $f(X)$ and the four subfunctions.

Our symmetry conditions are concerned with the equality between pairs of the $f_u(x_1, \ldots, x_{n-2})$ or, equivalently, with equality between the corresponding $\mathbf{S}_u$. However, we do not wish to evaluate $\mathbf{S}_u$ for each subfunction, but rather to use the coefficients from **S**. This connection exists directly using Tokmen's theorem (see Eq. (4.9)), which gives

$$[\mathbf{S}_0\mathbf{S}_1\mathbf{S}_2\mathbf{S}_3] = \tfrac{1}{4}[\mathbf{S}^0\mathbf{S}^1\mathbf{S}^2\mathbf{S}^3]\mathbf{T}^2$$

where

$$\mathbf{S} = \begin{matrix} \mathbf{S}^0 \\ \mathbf{S}^1 \\ \mathbf{S}^2 \\ \mathbf{S}^3 \end{matrix}\Bigg]$$

i.e.,

$$\begin{aligned} 4\mathbf{S}_0 &= \mathbf{S}^0 + \mathbf{S}^1 + \mathbf{S}^2 + \mathbf{S}^3 \\ 4\mathbf{S}_1 &= \mathbf{S}^0 - \mathbf{S}^1 + \mathbf{S}^2 - \mathbf{S}^3 \\ 4\mathbf{S}_2 &= \mathbf{S}^0 + \mathbf{S}^1 - \mathbf{S}^2 - \mathbf{S}^3 \\ 4\mathbf{S}_3 &= \mathbf{S}^0 - \mathbf{S}^1 - \mathbf{S}^2 + \mathbf{S}^3 \end{aligned} \tag{5.1}$$

From (5.1) it is easy to write down conditions to determine the equality of pairs of the $\mathbf{S}_u$, and the six possible symmetries in x_{n-1}, x_n are given in Table 5.1. The conditions come directly from (5.1) and are the result of the differences between pairs of rows of $\mathbf{T}^2$. Note that these conditions have been derived for symmetries involving x_{n-1}, x_n. For symmetries in other pairs of variables, we shall have to be careful to select the correct coefficients in $\mathbf{S}^0, \mathbf{S}^1, \mathbf{S}^2, \mathbf{S}^3$. This can be summarized as follows for a pair of variables x_i, x_j $(i < j)$:

$\mathbf{S}^0$ includes all coefficients that involve neither of x_i, x_j;
$\mathbf{S}^1$ includes all coefficients that involve x_i but not x_j;
$\mathbf{S}^2$ includes all coefficients that involve x_j but not x_i;
$\mathbf{S}^3$ includes all coefficients that involve both of x_i, x_j.

Table 5.1. Spectral symmetry tests for symmetries in $\{x_{n-1}, x_n\}$ for completely specified functions.

Symmetry	Condition	Test
$\mathbf{Y}_0 = \mathbf{Y}_1$ Single variable, x_{n-1} over $\bar{x}_n$ $f(x_1, \ldots, x_{n-2}, 0, 0) = f(x_1, \ldots, x_{n-2}, 1, 0)$	$\mathbf{S}_0 = \mathbf{S}_1$	$\mathbf{S}^1 + \mathbf{S}^3 = \mathbf{0}$
$\mathbf{Y}_0 = \mathbf{Y}_2$ Single variable, x_n over $\bar{x}_{n-1}$ $f(x_1, \ldots, x_{n-2}, 0, 0) = f(x_1, \ldots, x_{n-2}, 0, 1)$	$\mathbf{S}_0 = \mathbf{S}_2$	$\mathbf{S}^2 + \mathbf{S}^3 = \mathbf{0}$
$\mathbf{Y}_0 = \mathbf{Y}_3$ Equivalence $\{x_n, x_{n-1}\}$ $f(x_1, \ldots, x_{n-2}, 0, 0) = f(x_1, \ldots, x_{n-2}, 1, 1)$	$\mathbf{S}_0 = \mathbf{S}_3$	$\mathbf{S}^1 + \mathbf{S}^2 = \mathbf{0}$
$\mathbf{Y}_1 = \mathbf{Y}_2$ Nonequivalence, $\{x_n, x_{n-1}\}$ $f(x_1, \ldots, x_{n-2}, 1, 0) = f(x_1, \ldots, x_{n-2}, 0, 1)$	$\mathbf{S}_1 = \mathbf{S}_2$	$\mathbf{S}^1 - \mathbf{S}^2 = \mathbf{0}$
$\mathbf{Y}_1 = \mathbf{Y}_3$ Single variable, x_n over x_{n-1} $f(x_1, \ldots, x_{n-2}, 1, 0) = f(x_1, \ldots, x_{n-2}, 1, 1)$	$\mathbf{S}_1 = \mathbf{S}_3$	$\mathbf{S}^2 - \mathbf{S}^3 = \mathbf{0}$
$\mathbf{Y}_2 = \mathbf{Y}_3$ Single variable, x_{n-1} over x_n $f(x_1, \ldots, x_{n-2}, 0, 1) = f(x_1, \ldots, x_{n-2}, 1, 1)$	$\mathbf{S}_2 = \mathbf{S}_3$	$\mathbf{S}^1 - \mathbf{S}^3 = \mathbf{0}$

It is easy to tell which coefficients involve particular variables since the subscripts contain this information. For example, s_{134} involves x_1, x_3, x_4 but not x_2, etc. Notice that each of the tests involves two of the $\mathbf{S}^u$ and so uses 2^{n-1}, or exactly half, of the coefficients since each $\mathbf{S}^u$ contains 2^{n-2} coefficients.

Consider the function $f(X) = \bar{x}_1\bar{x}_2(x_3 + x_4) + x_3x_4(\bar{x}_1 + \bar{x}_2) + \bar{x}_3\bar{x}_4(\bar{x}_1x_2 + x_1\bar{x}_2) + x_1x_2x_3\bar{x}_4$ shown in Fig. 5.9. This function has the

$x_3x_4 \backslash x_1x_2$	00	01	11	10
00	0	1	0	1
01	1	0	0	0
11	1	1	0	1
10	1	0	1	0

Fig. 5.9. Function $f(X) = \bar{x}_1\bar{x}_2(x_3 + x_4) + x_3x_4(\bar{x}_1 + \bar{x}_2) + \bar{x}_3\bar{x}_4(\bar{x}_1x_2 + x_1\bar{x}_2) + x_1x_2x_3\bar{x}_4$.

following set of spectral coefficients:

s_0	s_1	s_2	s_3	s_4	s_{12}	s_{13}	s_{14}	s_{23}	s_{24}	s_{34}	s_{123}	s_{124}	s_{134}	s_{234}	s_{1234}
0;	−4	−4	4	0;	0	0	4	0	4	−4;	4	0	0	0;	12

For $\{x_1, x_2\}$, we have

$$\mathbf{S}^0 = \begin{bmatrix} s_0 \\ s_3 \\ s_4 \\ s_{34} \end{bmatrix} = \begin{bmatrix} 0 \\ 4 \\ 0 \\ -4 \end{bmatrix}, \quad \mathbf{S}^1 = \begin{bmatrix} s_1 \\ s_{13} \\ s_{14} \\ s_{134} \end{bmatrix} = \begin{bmatrix} -4 \\ 0 \\ 4 \\ 0 \end{bmatrix},$$

$$\mathbf{S}^2 = \begin{bmatrix} s_2 \\ s_{23} \\ s_{24} \\ s_{234} \end{bmatrix} = \begin{bmatrix} -4 \\ 0 \\ 4 \\ 0 \end{bmatrix}, \quad \mathbf{S}^3 = \begin{bmatrix} s_{12} \\ s_{123} \\ s_{124} \\ s_{1234} \end{bmatrix} = \begin{bmatrix} 0 \\ 4 \\ 0 \\ 12 \end{bmatrix}$$

and the only test that is satisfied is $\mathbf{S}^1 - \mathbf{S}^2 = \mathbf{0}$, so that $f(X)$ possesses $N\{x_1, x_2\}$.

For $\{x_1, x_3\}$, we have

$$\mathbf{S}^0 = \begin{bmatrix} s_0 \\ s_1 \\ s_4 \\ s_{24} \end{bmatrix} = \begin{bmatrix} 0 \\ -4 \\ 0 \\ 4 \end{bmatrix}, \quad \mathbf{S}^1 = \begin{bmatrix} s_1 \\ s_{12} \\ s_{14} \\ s_{124} \end{bmatrix} = \begin{bmatrix} -4 \\ 0 \\ 4 \\ 4 \end{bmatrix},$$

$$\mathbf{S}^2 = \begin{bmatrix} s_3 \\ s_{23} \\ s_{34} \\ s_{234} \end{bmatrix} = \begin{bmatrix} 4 \\ 0 \\ -4 \\ 0 \end{bmatrix}, \quad \mathbf{S}^3 = \begin{bmatrix} s_{13} \\ s_{123} \\ s_{134} \\ s_{1234} \end{bmatrix} = \begin{bmatrix} 0 \\ 4 \\ 0 \\ 12 \end{bmatrix}$$

and $\mathbf{S}^1 + \mathbf{S}^2 = \mathbf{0}$, so $f(X)$ possesses $E\{x_1, x_3\}$.

The easiest way to check the conditions is to write down the matrix $[\mathbf{S}^0\,\mathbf{S}^1\,\mathbf{S}^2\,\mathbf{S}^3]$ and check it. The pattern of the coefficients is easily established, and we can even put explicit labels on the rows and columns. The four columns are labelled with the combinations of the variables in the symmetry, and the rows with the combination of the remaining variables. The coefficient in any position just combines the row and column headings. In practice, the x's may be omitted and just the variable number retained. For the first column and first row, which do not involve any variables, we use the labelling ϕ. Hence for $\{x_1, x_4\}$ for a four variable function $f(x_1, x_2, x_3, x_4)$, we have

$$[\mathbf{S}^0\,\mathbf{S}^1\,\mathbf{S}^2\,\mathbf{S}^3] = \begin{array}{c} \begin{matrix} \phi & x_1 & x_4 & x_1, x_4 \end{matrix} \\ \left[\begin{matrix} s_0 & s_1 & s_4 & s_{14} \\ s_2 & s_{12} & s_{24} & s_{124} \\ s_3 & s_{13} & s_{34} & s_{134} \\ s_{23} & s_{123} & s_{234} & s_{1234} \end{matrix}\right] \end{array} \begin{matrix} \phi \\ x_2 \\ x_3 \\ x_2, x_3 \end{matrix}$$

For the example $f(X)$, this gives

$$[\mathbf{S}^0\,\mathbf{S}^1\,\mathbf{S}^2\,\mathbf{S}^3] = \begin{array}{c} \begin{matrix} \phi & 1 & 4 & 1,4 \end{matrix} \\ \left[\begin{matrix} 0 & -4 & 0 & 4 \\ -4 & 0 & 4 & 0 \\ 4 & 0 & -4 & 0 \\ 0 & 4 & 0 & 12 \end{matrix}\right] \end{array} \begin{matrix} \phi \\ 2 \\ 3 \\ 2,3 \end{matrix}$$

and the function does not possess any symmetries in $\{x_1, x_4\}$.
For the remaining three pairs of variables, we have

For $\{x_2, x_3\}$,

$$[\mathbf{S}^0\,\mathbf{S}^1\,\mathbf{S}^2\,\mathbf{S}^3] = \begin{array}{c} \begin{matrix} \phi & 2 & 3 & 2,3 \end{matrix} \\ \left[\begin{matrix} 0 & -4 & 4 & 0 \\ -4 & 0 & 0 & 4 \\ 0 & 4 & -4 & 0 \\ 4 & 0 & 0 & 12 \end{matrix}\right] \end{array} \begin{matrix} \phi \\ 1 \\ 4 \\ 1,4 \end{matrix}$$

so $f(X)$ possesses $E\{x_2, x_3\}$.

For $\{x_2, x_4\}$,

$$[\mathbf{S}^0\,\mathbf{S}^1\,\mathbf{S}^2\,\mathbf{S}^3] = \begin{array}{c} \begin{matrix} \phi & 2 & 4 & 2,4 \end{matrix} \\ \left[\begin{matrix} 0 & -4 & 0 & 4 \\ -4 & 0 & 4 & 0 \\ 4 & 0 & -4 & 0 \\ 0 & 4 & 0 & 12 \end{matrix}\right] \end{array} \begin{matrix} \phi \\ 1 \\ 3 \\ 1,3 \end{matrix}$$

so $f(X)$ possesses no symmetries in $\{x_2, x_4\}$.

Table 5.2. Spectral coefficient tests for symmetry in $\{x_i, x_j\}$ for $f(x_1, \ldots, x_i, \ldots, x_j, \ldots, x_n)$.

Symmetry	First necessary test	Set of necessary and sufficient conditions[a]
Equivalence symmetry $E\{x_i, x_j\}$ $f(x_1, \ldots, 0, \ldots, 0, \ldots, x_n) = f(x_1, \ldots, 1, \ldots, 1, \ldots, x_n)$	$s_i + s_j = 0$	$s_{i\alpha} + s_{j\alpha} = 0$
Non-equivalence symmetry $N\{x_i, x_j\}$ $f(x_1, \ldots, 0, \ldots, 1, \ldots, x_n) = f(x_1, \ldots, 1, \ldots, 0, \ldots, x_n)$	$s_i - s_j = 0$	$s_{i\alpha} - s_{j\alpha} = 0$
Single-variable symmetry $S\{x_i \| \bar{x}_j\}$ $f(x_1, \ldots, 0, \ldots, 0, \ldots, x_n) = f(x_1, \ldots, 1, \ldots, 0, \ldots, x_n\}$	$s_i + s_{ij} = 0$	$s_{i\alpha} + s_{ij\alpha} = 0$
Single-variable symmetry $S\{x_i \| x_j\}$ $f(x_1, \ldots, 0, \ldots, 1, \ldots, x_n) = f(x_1, \ldots, 1, \ldots, 1, \ldots, x_n)$	$s_i - s_{ij} = 0$	$s_{i\alpha} - s_{ij\alpha} = 0$

[a] The necessary and sufficient conditions consist of the set of 2^{n-2} equations deduced by substituting for α all possible different combinations of subscripts, excluding i and j.

For $\{x_3, x_4\}$,

$$
\begin{array}{c c}
 & \begin{array}{cccc} \phi & 3 & 4 & 3,4 \end{array} \\
[\mathbf{S}^0\, \mathbf{S}^1\, \mathbf{S}^2\, \mathbf{S}^3] = & \left[\begin{array}{cccc} 0 & 4 & 0 & -4 \\ -4 & 0 & 4 & 0 \\ -4 & 0 & 4 & 0 \\ 0 & 4 & 0 & 12 \end{array}\right] \begin{array}{l} \phi \\ 1 \\ 2 \\ 1,2 \end{array}
\end{array}
$$

so $f(X)$ possesses no symmetries in $\{x_3, x_4\}$.

Note that this matrix is not necessarily square but will always possess

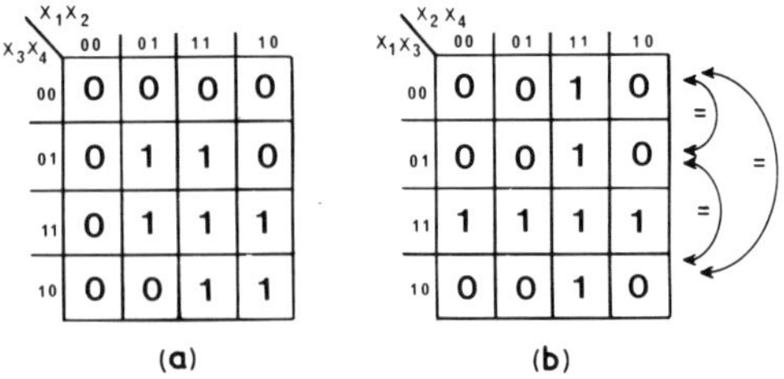

Fig. 5.10. Function with multiple symmetries in x_1, x_3, $f(X) = x_1x_3 + x_2x_4$; (a) conventional map axes identification; (b) reordered axes identification to show x_1, x_3 symmetries more clearly.

four columns and 2^{n-2} rows. For example, $\{x_1, x_4\}$ for the five-variable function $f(x_1, x_2, x_3, x_4, x_5)$, we have

$$
\begin{array}{cccccc}
 & \phi & 1 & 4 & 1,4 & \\
[\mathbf{S}^0\,\mathbf{S}^1\,\mathbf{S}^2\,\mathbf{S}^3] = & \left[\begin{array}{cccc}
s_0 & s_1 & s_4 & s_{14} \\
s_2 & s_{12} & s_{24} & s_{124} \\
s_3 & s_{13} & s_{34} & s_{134} \\
s_{23} & s_{123} & s_{234} & s_{1234} \\
s_5 & s_{15} & s_{45} & s_{145} \\
s_{25} & s_{125} & s_{245} & s_{1245} \\
s_{35} & s_{135} & s_{345} & s_{1345} \\
s_{235} & s_{1235} & s_{2345} & s_{12345}
\end{array}\right] & \begin{array}{l} \phi \\ 2 \\ 3 \\ 2,3 \\ 5 \\ 2,5 \\ 3,5 \\ 2,3,5 \end{array}
\end{array}
$$

Instead of listing the conditions in terms of the $\mathbf{S}^u$ as we did in Table 5.1, we can explicitly list the coefficients that have to be inspected. For each symmetry there are 2^{n-2} equations to be verified and in Table 5.2 we list a single initial condition that can be used as a necessary initial check for a symmetry, and then the general form of all the remaining conditions that must be verified.

Of course it is possible for a function to possess several different symmetries in the same pair of variables. Consider the function $f(X) = x_1x_3 + x_2x_4$, which is shown in Fig. 5.10. $f(X)$ has the spectrum

s_0	s_1	s_2	s_3	s_4	s_{12}	s_{13}	s_{14}	s_{23}	s_{24}	s_{34}	s_{123}	s_{124}	s_{134}	s_{234}	s_{1234}
2;	6	6	6	6;	2	-6	2	2	-6	2;	-2	-2	-2	-2	2

For $\{x_1, x_3\}$ we have

$$
\begin{array}{cccccc}
 & \phi & 1 & 3 & 1,3 & \\
[\mathbf{S}^0\,\mathbf{S}^1\,\mathbf{S}^2\,\mathbf{S}^3] = & \left[\begin{array}{rrrr}
2 & 6 & 6 & -6 \\
6 & 2 & 2 & -2 \\
6 & 2 & 2 & -2 \\
-6 & -2 & -2 & 2
\end{array}\right] & \begin{array}{l} \phi \\ 2 \\ 4 \\ 2,4 \end{array}
\end{array}
$$

and

$$\mathbf{S}^1 = \mathbf{S}^2, \text{ so } f(X) \text{ possesses } N\{x_1, x_3\};$$

$$\mathbf{S}^1 + \mathbf{S}^3 = \mathbf{0}, \quad \text{so } f(X) \text{ possesses } S\{x_3 | \bar{x}_1\};$$

$$\mathbf{S}^2 + \mathbf{S}^3 = \mathbf{0}, \quad \text{so } f(X) \text{ possesses } S\{x_1 | \bar{x}_3\}.$$

As was noted earlier, two of these symmetries will always imply the third for a completely specified function.

The absence of any single-variable symmetries can often be detected very quickly for a function. When considering the tests for single-variable symmetries, it becomes apparent that the highest-order coefficient is always used. The final equation to check any single-variable symmetry is always the sum or difference of the highest-order coefficient and a coefficient of order $n - 1$. Consequently, when the highest-order coefficient has a magnitude distinct from that of all the coefficients of order $n - 1$, it is clear that the function cannot possess any single-variable symmetries. This is illustrated by the function shown in Fig. 5.9 for which $s_{1234} = 12$, while all the third-order coefficients are either 0 or -4. Consequently, it is immediate that this function cannot possess any single-variable symmetries.

5.4 SYMMETRIES IN THE SYNTHESIS OF DIGITAL CIRCUITS

Having considered how to detect symmetries, it is time to consider how they might assist us in deriving a circuit realization for a function. As soon as a symmetry is detected, this information can be used to start the design of a circuit. The aim of our procedure is illustrated in Fig. 5.11. Given some function $f(X)$ with a symmetry, it will be used to define the required function g and hence to determine the image function h. If the repeated application of this procedure is to lead eventually to a circuit realization for f, we need to be sure that either

(a) one of $\tilde{x}_{n-1}$ or $\tilde{x}_n$ is unnecessary—consequently, h is a function of only $(n - 1)$ variables; or

(b) if both $\tilde{x}_{n-1}$ and $\tilde{x}_n$ are necessary, then h must, in some sense, be simpler

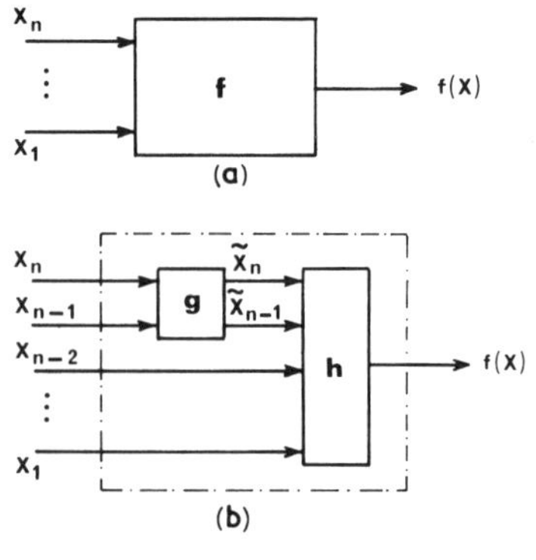

Fig. 5.11. Circuit realizations of $f(X)$: (a) original given function; (b) the design structure.

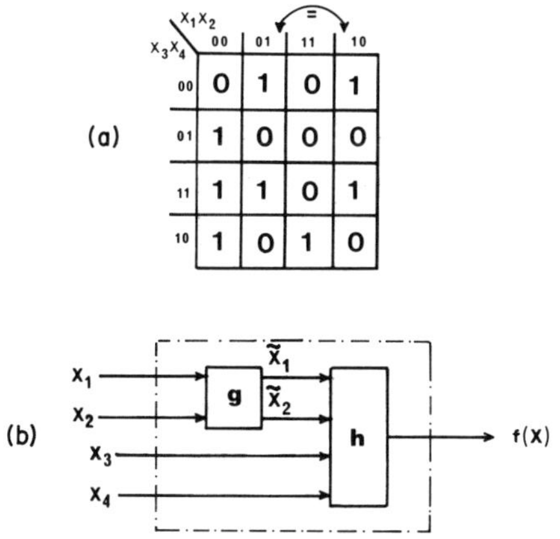

Fig. 5.12. Realization of $f(X) = \bar{x}_1\bar{x}_2(x_3 + x_4) + x_3x_4(\bar{x}_1 + \bar{x}_2) + \bar{x}_3\bar{x}_4(\bar{x}_1x_2 + x_1\bar{x}_2) + x_1x_2x_3\bar{x}_4$: (a) given function; (b) realisation in the form $f(X) = h(\tilde{x}_1, \tilde{x}_2, x_3, x_4)$.

than f. For our purposes, we shall take this to mean that h has more unspecified minterms (don't cares) than f.

For example, consider the function $f(X) = \bar{x}_1\bar{x}_2(x_3 + x_4) + x_3x_4(\bar{x}_1 + \bar{x}_2) + \bar{x}_3\bar{x}_4(\bar{x}_1x_2 + x_1\bar{x}_2) + x_1x_2x_3\bar{x}_4$, which is illustrated in Fig. 5.12a. This is the same function that we were considering in the last section (see Fig. 5.9), and it possesses $N\{x_1, x_2\}$. The question that is now addressed is how this knowledge can assist us toward a realization of the type shown in Fig. 5.12b.

Since $f(X)$ possesses $N\{x_1, x_2\}$, we have that $f(1,0,x_3,x_4) = f(0,1,x_3,x_4)$, i.e., two rows in Table 5.3 are identical.

We want to remap the variables x_1, x_2 to two new variables $\tilde{x}_1, \tilde{x}_2$ that will take advantage of the fact that the second and third rows of Table 5.3 are identical, that is, there are only three distinct cases to be distinguished, not four. Suppose we remap x_1, x_2 as shown in Table 5.4. It is clear that the two equal quarters of $f(X)$, namely, $f(1, 0, x_3, x_4)$ and $f(0, 1, x_3, x_4)$, are both going to be mapped onto the case $(\tilde{x}_2, \tilde{x}_1) = (0, 1)$, while the other two cases are left unchanged.

Table 5.3. Function with symmetry $N\{x_1, x_2\}$.

x_2	x_1	$f(X)$
0	0	$f(0, 0, x_3, x_4)$
0	1	$f(1, 0, x_3, x_4)$
1	0	$f(0, 1, x_3, x_4)$
1	1	$f(1, 1, x_3, x_4)$

Table 5.4. Function with symmetry $N\{x_1, x_2\}$ with input 1, 0 remapped to 0, 1.

x_2	x_1	$\tilde{x}_2$	$\tilde{x}_1$
0	0	0	0
0	1	0	1
1	0	0	1
1	1	1	1

We now have

$$\tilde{x}_1 = x_1 + x_2, \qquad \tilde{x}_2 = x_1 x_2$$

and $f(x_1, x_2, x_3, x_4) = h(\tilde{x}_1, \tilde{x}_2, x_3, x_4)$.

We notice that h is identical to f except that one quarter of its map has been replaced by don't cares. This is illustrated in Fig. 5.13 and we can see the beginning of a realization for $f(X)$ in Fig. 5.13a. We regard h as being simpler than f in that its specification contains more don't cares, and con-

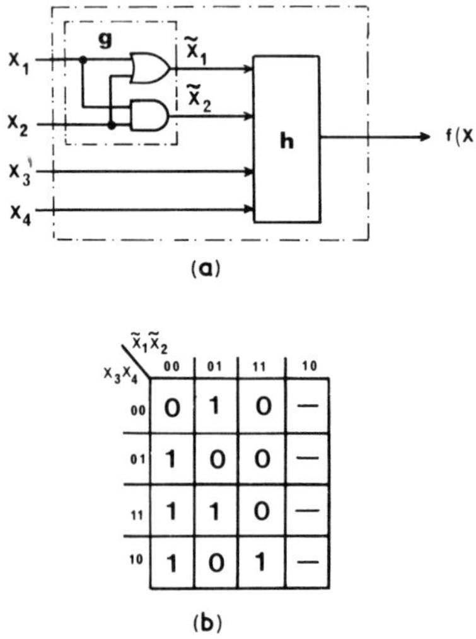

Fig. 5.13. Continued realization of Fig. 5.12: (a) the beginning of the realization; (b) the resulting map for $h(\tilde{x}_1, \tilde{x}_2, x_3, x_4)$.

sequently is easier to realize than the original function $f(X)$. We can now repeat the above procedure on $h(\tilde{x}_1, \tilde{x}_2, x_3, x_4)$.

We shall explicitly consider all the possible symmetries that can simultaneously exist for a function in a particular pair of variables and, for each, describe the resulting remapping functions that are required and the spectral conditions that have to be satisfied.

Recall the notation for subfunctions of $f(X)$ that

$$f_0(x_1, \ldots, x_{n-2}) = f(x_1, \ldots, x_{n-2}, 0, 0)$$
$$f_1(x_1, \ldots, x_{n-2}) = f(x_1, \ldots, x_{n-2}, 1, 0)$$
$$f_2(x_1, \ldots, x_{n-2}) = f(x_1, \ldots, x_{n-2}, 0, 1)$$
$$f_3(x_1, \ldots, x_{n-2}) = f(x_1, \ldots, x_{n-2}, 1, 1)$$

and that the symmetries are concerned with the equality of pairs of these subfunctions.

Initially we consider the four cases of single symmetries in x_n, x_{n-1}, proceeding from these to the possible multiple symmetries, that is, when the function simultaneously possesses more than one symmetry. For each symmetry we shall give the simplest remapping functions, following the general structure of Fig. 5.12b. The general conditions for the symmetries were listed in Table 5.1 and we use these results here.

5.4.1 $S\{x_n | \bar{x}_{n-1}\}$, $f_0 = f_2$

Condition: $\mathbf{S}^1 + \mathbf{S}^3 = \mathbf{0}$.

Requirement

	x_n	x_{n-1}	$\tilde{x}_n$	$\tilde{x}_{n-1}$
f_0	0	0	a	0
f_1	0	1	b	1
f_2	1	0	a	0
f_3	1	1	c	1

$a, b, c \in \{0, 1\}$, $b \neq c$. The simplest remapping function is (with $a = 1$, $b = 1$, $c = 0$):

$$\tilde{x}_{n-1} = x_{n-1}, \qquad \tilde{x}_n = \overline{x_n x_{n-1}}$$

The circuit is illustrated in Fig. 5.14a. Clearly whenever we have a realization for a remapping function, we can also use the complement function.

5.4.2 $E\{x_n, x_{n-1}\}$, $f_0 = f_3$

Condition: $\mathbf{S}^1 + \mathbf{S}^2 = \mathbf{0}$.

Requirement

	x_n	x_{n-1}	$\tilde{x}_n$	$\tilde{x}_{n-1}$
f_0	0	0	a	a_1
f_1	0	1	b	b_1
f_2	1	0	c	c_1
f_3	1	1	a	a_1

where (a, a_1), (b, b_1), (c, c_1) are three distinct triples from $\{(0, 0), (0, 1), (1, 0), (1, 1)\}$. One possibility is shown below.

x_n	x_{n-1}	$\tilde{x}_n$	$\tilde{x}_{n-1}$
0	0	0	0
0	1	0	1
1	0	1	0
1	1	0	0

$\tilde{x}_{n-1} = \bar{x}_n x_{n-1}$, $\tilde{x}_n = x_n \bar{x}_{n-1}$. This solution is illustrated in Fig. 5.14b.

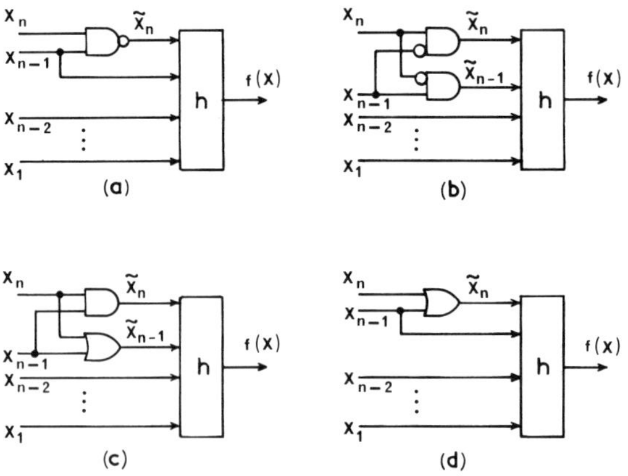

Fig. 5.14. Realizations using a single symmetry: (a) $S\{x_n|\bar{x}_{n-1}\}$, $f_0 = f_2$; (b) $E\{x_n, x_{n-1}\}$, $f_0 = f_3$; (c) $N\{x_n, x_{n-1}\}$, $f_1 = f_2$; (d) $S\{x_n|x_{n-1}\}$, $f_1 = f_3$.

5.4.3 $N\{x_n, x_{n-1}\}, f_1 = f_2$

Condition: $\mathbf{S}^1 - \mathbf{S}^2 = \mathbf{0}$.

Requirement

	x_n	x_{n-1}	$\tilde{x}_n$	$\tilde{x}_{n-1}$
f_0	0	0	b	b_1
f_1	0	1	a	a_1
f_2	1	0	a	a_1
f_3	1	1	c	c_1

where (a, a_1), (b, b_1), (c, c_1) are three distinct triples from $\{(0, 0), (0, 1), (1, 0), (1, 1)\}$. This case is similar to that for $E\{x_n, x_{n-1}\}$.

Possible solution

x_n	x_{n-1}	$\tilde{x}_n$	$\tilde{x}_{n-1}$
0	0	0	0
0	1	0	1
1	0	0	1
1	1	1	1

$\tilde{x}_{n-1} = x_{n-1} + x_n$, $\tilde{x}_n = x_{n-1}x_n$. This solution is illustrated in Fig. 5.14c.

5.4.4 $S\{x_n | x_{n-1}\}, f_1 = f_3$.

Condition: $\mathbf{S}^1 - \mathbf{S}^3 = \mathbf{0}$.

Requirement

	x_n	x_{n-1}	$\tilde{x}_n$	$\tilde{x}_{n-1}$
f_0	0	0	b	0
f_1	0	1	a	1
f_2	1	0	c	0
f_3	1	1	a	1

$a, b, c \in \{0, 1\}$, $b \neq c$. Possible solution: $\tilde{x}_{n-1} = x_{n-1}$, $\tilde{x}_n = x_{n-1} + x_n$. This solution is illustrated in Fig. 5.14d.

When multiple symmetries in the same pair of variables are present, simpler circuit structures result and these are considered now.

5.4.5 $E\{x_n, x_{n-1}\}$ and $N\{x_n, x_{n-1}\}$, $f_0 = f_3$ and $f_1 = f_2$

Condition: $\mathbf{S}^1 = \mathbf{S}^2 = \mathbf{0}$.

Requirement

	x_n	x_{n-1}	$\tilde{x}_{n-1}$
f_0	0	0	a
f_1	0	1	b
f_2	1	0	b
f_3	1	1	a

$a, b \in \{0, 1\}$, $a \neq b$. Since $f_0 = f_3$ and $f_1 = f_2$, there are only two cases to distinguish, hence a single variable will suffice. Possible solution: $\tilde{x}_{n-1} = x_{n-1} \oplus x_n$. This solution is illustrated in Fig. 5.15a.

5.4.6 $S\{x_n|\bar{x}_{n-1}\}$ and $S\{x_n|x_{n-1}\}$, $f_0 = f_2$ and $f_1 = f_3$

Condition: $\mathbf{S}^1 = \mathbf{S}^3 = \mathbf{0}$.

The two conditions $f_0 = f_2$, $f_1 = f_3$ imply that $f(x_1, \ldots, x_{n-1}, 1) = f(x_1, \ldots, x_{n-1}, 0)$ so that x_n is redundant as an input.

Conditions 5.4.5 and 5.4.6 are the only two in which a completely specified

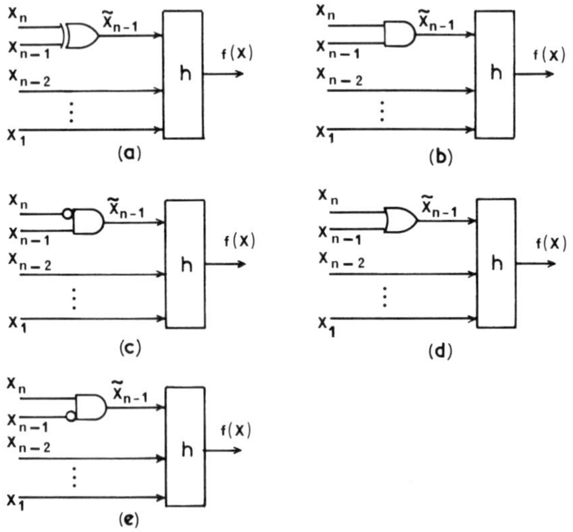

Fig. 5.15. Realizations using multiple symmetries: (a) $E\{x_n, x_{n-1}\}$, $N\{x_n, x_{n-1}\}$ $f_0 = f_3$ and $f_1 = f_2$; (b) $S\{x_n|\bar{x}_{n-1}\}$, $S\{x_{n-1}|\bar{x}_n\}$, $N\{x_n, x_{n-1}\}$, $f_0 = f_1 = f_2$; (c) $S\{x_n|\bar{x}_{n-1}\}$, $S\{x_{n-1}|x_n\}$, $E\{x_n, x_{n-1}\}$ $f_0 = f_2 = f_3$; (d) $S\{x_n|x_{n-1}\}$, $S\{x_{n-1}|x_n\}$, $N\{x_n, x_{n-1}\}$, $f_1 = f_2 = f_3$; (e) $S\{x_{n-1}|\bar{x}_n\}$, $S\{x_n|x_{n-1}\}$, $E\{x_n, x_{n-1}\}$, $f_0 = f_1 = f_3$.

function can possess exactly two symmetries. There are, however, four cases where the function can possess exactly three symmetries.

5.4.7 $S\{x_n|\bar{x}_{n-1}\}, S\{x_{n-1}|\bar{x}_n\}$ and $N\{x_n, x_{n-1}\}, f_0 = f_1 = f_2$

Condition: $\mathbf{S}^1 = \mathbf{S}^2 = -\mathbf{S}^3$.

Requirement

	x_n	x_{n-1}	$\tilde{x}_{n-1}$
f_0	0	0	a
f_1	0	1	a
f_2	1	0	a
f_3	1	1	b

$a, b \in \{0, 1\}$, $a \neq b$. Possible solution: $\tilde{x}_{n-1} = x_{n-1}x_n$. This solution is illustrated in Fig. 5.15b.

5.4.8 $S\{x_n|\bar{x}_{n-1}\}, S\{x_{n-1}|x_n\}$ and $E\{x_n, x_{n-1}\}, f_0 = f_2 = f_3$

Condition: $-\mathbf{S}^1 = \mathbf{S}^2 = \mathbf{S}^3$.

Requirement

	x_n	x_{n-1}	$\tilde{x}_{n-1}$
f_0	0	0	a
f_1	0	1	b
f_2	1	0	a
f_3	1	1	a

$a, b \in \{0, 1\}$, $a \neq b$. Possible solution: $\tilde{x}_{n-1} = x_{n-1}\bar{x}_n$. This solution is illustrated in Fig. 5.15c.

5.4.9 $S\{x_n|x_{n-1}\}, S\{x_{n-1}|x_n\}$ and $N\{x_n, x_{n-1}\}, f_1 = f_2 = f_3$

Condition: $\mathbf{S}^1 = \mathbf{S}^2 = \mathbf{S}^3$.

Requirement

	x_n	x_{n-1}	$\tilde{x}_{n-1}$
f_0	0	0	b
f_1	0	1	a
f_2	1	0	a
f_3	1	1	a

$a, b \in \{0, 1\}$, $a \neq b$. Possible solution: $\tilde{x}_{n-1} = x_{n-1} + x_n$. This solution is illustrated in Fig. 5.15d.

5.4.10 $S\{x_{n-1}|\bar{x}_n\}$, $S\{x_n|x_{n-1}\}$ and $E\{x_n, x_{n-1}\}$, $f_0 = f_1 = f_3$

Condition: $\mathbf{S}^1 = -\mathbf{S}^2 = \mathbf{S}^3$.

Requirement

	x_n	x_{n-1}	$\tilde{x}_{n-1}$
f_0	0	0	a
f_1	0	1	a
f_2	1	0	b
f_3	1	1	a

$a, b \in \{0, 1\}$, $a \neq b$. Possible solution: $\tilde{x}_{n-1} = \bar{x}_{n-1}x_n$. This solution is illustrated in Fig. 5.15e.

5.4.11 Examples

1. Consider the function $f(X) = \bar{x}_1x_2 + \bar{x}_1x_3 + \bar{x}_1x_4 + x_2(x_3 + x_4) + x_3x_4$, which is illustrated in Fig. 5.16a. $f(X)$ has the spectrum

s_0	s_1	s_2	s_3	s_4	s_{12}	s_{13}	s_{14}	s_{23}	s_{24}	s_{34}	s_{123}	s_{124}	s_{134}	s_{234}	s_{1234}
−6;	−6	6	6	6;	−2	−2	−2	2	2	2;	2	2	2	−2;	6

Initially this function cannot possess any single-variable symmetries since s_{1234} has a different magnitude from any of the third-order coefficients.

Consider symmetries in $\{x_3, x_4\}$.

$$
\begin{array}{c} \\ [\mathbf{S}^0\,\mathbf{S}^1\,\mathbf{S}^2\,\mathbf{S}^3] = \end{array}
\begin{array}{c} \begin{array}{cccc} \phi & 3 & 4 & 3,4 \end{array} \\ \begin{bmatrix} -6 & 6 & 6 & 2 \\ -6 & -2 & -2 & 2 \\ 6 & 2 & 2 & -2 \\ 2 & 2 & 2 & 6 \end{bmatrix} \end{array}
\begin{array}{l} \\ \phi \\ 1 \\ 2 \\ 1,2 \end{array}
$$

Since $\mathbf{S}^1 = \mathbf{S}^2$, this function possesses $N\{x_3, x_4\}$ so we may use the substitution

$$\tilde{x}_4 = x_3x_4, \qquad \tilde{x}_3 = x_3 + x_4$$

leading to the circuit shown in Fig. 5.16b with $h(x_1, x_2, \tilde{x}_3, \tilde{x}_4)$ specified as shown in Fig. 5.16c.

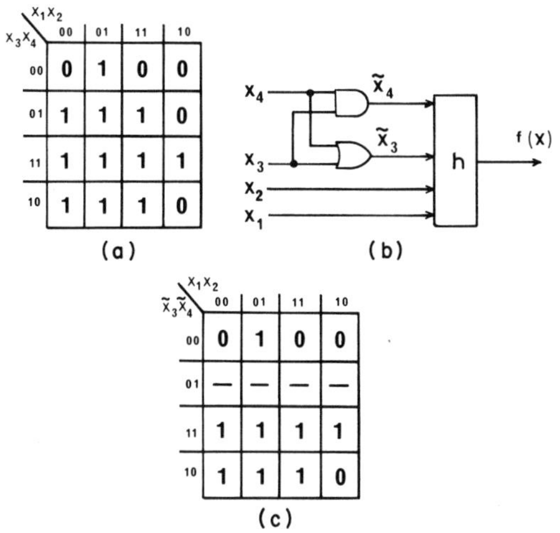

Fig. 5.16. Example synthesis: (a) given function $f(X) = \bar{x}_1(x_2 + x_3 + x_4) + x_2(x_3 + x_4) + x_3x_4$; (b) initial synthesis; (c) specification for $h(x_1, x_2, \tilde{x}_3, \tilde{x}_4)$.

2. Consider the function $f(X) = x_1\bar{x}_3x_4 + x_3(\bar{x}_1 + x_2 + \bar{x}_4)$, which is illustrated in Fig. 5.17a. This function has the spectrum

s_0	s_1	s_2	s_3	s_4	s_{12}	s_{13}	s_{14}	s_{23}	s_{24}	s_{34}	s_{123}	s_{124}	s_{134}	s_{234}	s_{1234}
−2;	2	2	10	2;	−2	6	−2	−2	−2	6;	2	2	−6	2;	−2

Consider the symmetries in $\{x_1, x_4\}$:

$$
\begin{array}{c}
\qquad\qquad\quad \phi \quad\ \ 1 \quad\ \ 4 \quad 1,4 \\
[\mathbf{S}^0\,\mathbf{S}^1\,\mathbf{S}^2\,\mathbf{S}^3] = \left[\begin{array}{rrrr} -2 & 2 & 2 & -2 \\ 2 & -2 & -2 & 2 \\ 10 & 6 & 6 & -6 \\ -2 & 2 & 2 & -2 \end{array}\right]\begin{array}{l} \phi \\ 2 \\ 3 \\ 2,3 \end{array}
\end{array}
$$

and we have $\mathbf{S}^1 = \mathbf{S}^2 = -\mathbf{S}^3$.

Hence $f(X)$ possesses $S\{x_1|\bar{x}_4\}$, $S\{x_4|\bar{x}_1\}$ and $N\{x_4, x_1\}$. Both these variables can be replaced by the single variable $\tilde{x}_1$ as shown in the remapping

x_4	x_1	$\tilde{x}_1$
0	0	0
0	1	0
1	0	0
1	1	1

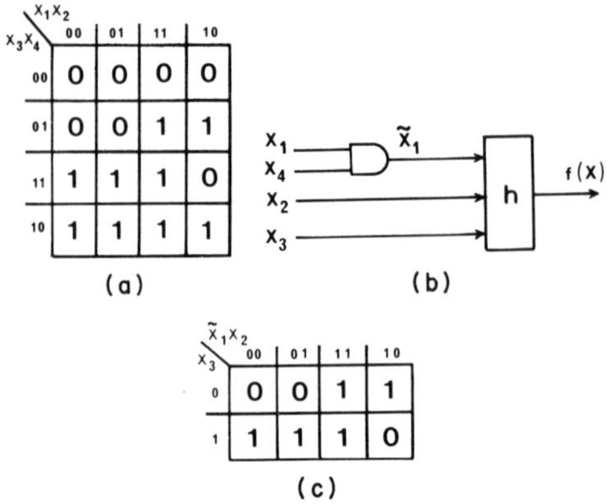

Fig. 5.17. Example synthesis: (a) given function $f(X) = x_1\bar{x}_3x_4 + x_3(\bar{x}_1 + x_2 + \bar{x}_4)$; (b) initial synthesis; (c) specification for $h(\tilde{x}_1, x_2, x_3)$.

$\tilde{x}_1 = x_1x_4$, $f(x_1, x_2, x_3, x_4) = h(\tilde{x}_1, x_2, x_3)$. The circuit is illustrated in Fig. 5.17b with $h(\tilde{x}_1, x_2, x_3) = \tilde{x}_1\bar{x}_3 + \tilde{x}_1x_2 + \bar{\tilde{x}}_1x_3$ illustrated in Fig. 5.17c. We can continue by analyzing the spectrum of h, namely,

$$\begin{matrix} s_0 & s_1 & s_2 & s_3 & s_{12} & s_{13} & s_{23} & s_{123} \\ -2; & 2 & 2 & 2; & -2 & 6 & -2; & 2 \end{matrix}$$

and for symmetries in $\{x_2, x_3\}$ we have

$$\begin{matrix} & \phi & 2 & 3 & 2,3 & \\ [\mathbf{S}^0\,\mathbf{S}^1\,\mathbf{S}^2\,\mathbf{S}^3] = & \begin{bmatrix} -2 & 2 & 2 & -2 \\ 2 & -2 & 6 & 2 \end{bmatrix} & & & & \begin{matrix} \phi \\ 1 \end{matrix} \end{matrix}$$

so that $h(\tilde{x}_1, x_2, x_3)$ possesses the symmetry $S\{x_2|\bar{x}_3\}$ leading to the next

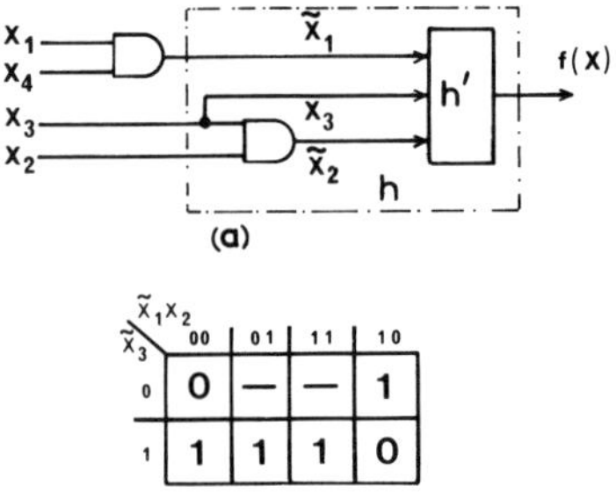

Fig. 5.18. Continued synthesis of Fig. 5.17: (a) second stage of realization for $f(X)$; (b) specification for $h'(\tilde{x}_1, x_2, \tilde{x}_3)$.

stage of a realization shown in Fig. 5.18a with

$$\tilde{x}_3 = x_3, \qquad \tilde{x}_2 = x_2 x_3$$

To complete both of these examples, we require knowledge of the symmetries for incompletely specified functions and this is discussed in the following section.

5.5 SYMMETRIES OF DEGREE 2 FOR INCOMPLETELY SPECIFIED FUNCTIONS

A serious problem with the synthesis procedures described in the last section may have occurred to the reader. When symmetries are used to help design circuits, it is inevitable that incompletely specified functions arise during the procedure, even if the original function is completely specified. However, the development of symmetry conditions in Section 5.3 only applied to completely specified functions. That omission will now be remedied by studying the symmetries of incompletely specified functions. These results are due to Muzio *et al.*[11]

The representation so far adopted for the function $f(X)$ has used the vector **Y**, where

$$y_i = \begin{cases} +1 & \text{if} \quad f(X) = 0 \\ -1 & \text{if} \quad f(X) = 1 \end{cases}$$

To incorporate the don't cares requires a more complex coding scheme and two different ones will be considered.

5.5.1 A Coding Scheme for Incompletely Specified Functions

Here we use two vectors $\hat{\mathbf{Y}}$ and $\hat{\hat{\mathbf{Y}}}$ to represent the function. $\hat{\mathbf{Y}}$ has all the don't cares coded as logic zeroes, while $\hat{\hat{\mathbf{Y}}}$ has all the don't cares coded as logic ones, that is,

$$\hat{y}_i = \begin{cases} +1 & \text{if} \quad f(X) = 0 \quad \text{or} \quad f(X) \text{ not specified} \\ -1 & \text{if} \quad f(X) = 1 \end{cases}$$

$$\hat{\hat{y}}_i = \begin{cases} +1 & \text{if} \quad f(X) = 0 \\ -1 & \text{if} \quad f(X) = 1 \quad \text{or} \quad f(X) \text{ not specified} \end{cases}$$

Previously, to detect symmetries, we were looking for equal subfunctions. However, the presence of don't cares implies that our search must be for subfunctions that can be made equal with the appropriate assignment of some or all of the don't cares.

Two functions $f(x_1, \ldots, x_n)$ and $g(x_1, \ldots, x_n)$ are "compatible" if there exists an assignment to some or all of their don't cares such that they are equal.

The problem of identifying potential symmetries in $\{x_n, \text{not}.\, x_{n-1}\}$ reduces to that of determining when two of the four subfunctions f_0, f_1, f_2 or f_3 are compatible. Each of these $f_u(x_1, \ldots, x_{n-2})$ has its two corresponding vectors $\hat{\mathbf{Y}}_u$ and $\hat{\hat{\mathbf{Y}}}_u$. Consider two of these subfunctions, f_u and f_v $(0 \leq u, v \leq 3, u \neq v)$. For them to be compatible requires that, for all possible values of i, the ith element of either $(\hat{\mathbf{Y}}_u - \hat{\mathbf{Y}}_v)$ or $(\hat{\hat{\mathbf{Y}}}_u - \hat{\hat{\mathbf{Y}}}_v)$ is zero. A convenient necessary condition for this to occur is that the inner product of $(\hat{\mathbf{Y}}_u - \hat{\mathbf{Y}}_v)$ and $(\hat{\hat{\mathbf{Y}}}_u - \hat{\hat{\mathbf{Y}}}_u)$ is zero, that is,

$$(\hat{\mathbf{Y}}_u - \hat{\mathbf{Y}}_v)^t(\hat{\hat{\mathbf{Y}}}_u - \hat{\hat{\mathbf{Y}}}_v) = 0$$

Since the product of the corresponding entries of $(\hat{\mathbf{Y}}_u - \hat{\mathbf{Y}}_v)$ and $(\hat{\hat{\mathbf{Y}}}_u - \hat{\hat{\mathbf{Y}}}_v)$ cannot be negative, it follows that this condition is also sufficient. To convert this condition to the spectral domain we have

$$\hat{\mathbf{S}}_u = \mathbf{T}^{n-2}\hat{\mathbf{Y}}_u, \qquad \hat{\hat{\mathbf{S}}}_u = \mathbf{T}^{n-2}\hat{\hat{\mathbf{Y}}}_u$$

for each u, $0 \leq u \leq 3$. Alternatively,

$$\hat{\mathbf{Y}}_u = \frac{1}{2^{n-2}}\mathbf{T}^{n-2}\hat{\mathbf{S}}_u, \qquad \hat{\hat{\mathbf{Y}}}_u = \frac{1}{2^{n-2}}\mathbf{T}^{n-2}\hat{\hat{\mathbf{S}}}_u$$

and similarly for $\hat{\mathbf{Y}}_v$, $\hat{\hat{\mathbf{Y}}}_v$. Hence,

$$(\hat{\mathbf{Y}}_u - \hat{\mathbf{Y}}_v)^t(\hat{\hat{\mathbf{Y}}}_u - \hat{\hat{\mathbf{Y}}}_v) = \frac{1}{2^{n-2}}(\hat{\mathbf{S}}_u - \hat{\mathbf{S}}_v)^t\mathbf{T}^{n-2}\frac{1}{2^{n-2}}\mathbf{T}^{n-2}(\hat{\hat{\mathbf{S}}}_u - \hat{\hat{\mathbf{S}}}_v)$$

$$= \frac{1}{2^{n-2}}(\hat{\mathbf{S}}_u - \hat{\mathbf{S}}_v)^t(\hat{\hat{\mathbf{S}}}_u - \hat{\hat{\mathbf{S}}}_v)$$

Consequently, $(\hat{\mathbf{Y}}_u - \hat{\mathbf{Y}}_v)^t(\hat{\hat{\mathbf{Y}}}_u - \hat{\hat{\mathbf{Y}}}_v) = 0$ if and only if

$$(\hat{\mathbf{S}}_u - \hat{\mathbf{S}}_v)^t(\hat{\hat{\mathbf{S}}}_u - \hat{\hat{\mathbf{S}}}_v) = 0 \tag{5.2}$$

We can link these $\hat{\mathbf{S}}_u$, $\hat{\hat{\mathbf{S}}}_u$ back to the spectra of the whole function exactly as we did in Section 5.3.

If

$$\hat{\mathbf{S}} = \mathbf{T}^n\hat{\mathbf{Y}}, \qquad \hat{\hat{\mathbf{S}}} = \mathbf{T}^n\hat{\hat{\mathbf{Y}}}$$

and

$$\hat{\mathbf{S}} = \begin{bmatrix} \hat{\mathbf{S}}^0 \\ \hat{\mathbf{S}}^1 \\ \hat{\mathbf{S}}^2 \\ \hat{\mathbf{S}}^3 \end{bmatrix}, \qquad \hat{\hat{\mathbf{S}}} = \begin{bmatrix} \hat{\hat{\mathbf{S}}}^0 \\ \hat{\hat{\mathbf{S}}}^1 \\ \hat{\hat{\mathbf{S}}}^2 \\ \hat{\hat{\mathbf{S}}}^3 \end{bmatrix}$$

then, by Tokmen's theorem (see Eq. (4.9)).

$$[\hat{\mathbf{S}}_0\, \hat{\mathbf{S}}_1\, \hat{\mathbf{S}}_2\, \hat{\mathbf{S}}_3] = \tfrac{1}{4}[\hat{\mathbf{S}}^0\, \hat{\mathbf{S}}^1\, \hat{\mathbf{S}}^2\, \hat{\mathbf{S}}^3]\mathbf{T}^2$$

and

$$[\hat{\hat{\mathbf{S}}}_0\, \hat{\hat{\mathbf{S}}}_1\, \hat{\hat{\mathbf{S}}}_2\, \hat{\hat{\mathbf{S}}}_3] = \tfrac{1}{4}[\hat{\hat{\mathbf{S}}}^0\, \hat{\hat{\mathbf{S}}}^1\, \hat{\hat{\mathbf{S}}}^2\, \hat{\hat{\mathbf{S}}}^3]\mathbf{T}^2$$

giving

$$\begin{aligned} 4\hat{\mathbf{S}}_0 &= \hat{\mathbf{S}}^0 + \hat{\mathbf{S}}^1 + \hat{\mathbf{S}}^2 + \hat{\mathbf{S}}^3 \\ 4\hat{\mathbf{S}}_1 &= \hat{\mathbf{S}}^0 - \hat{\mathbf{S}}^1 + \hat{\mathbf{S}}^2 - \hat{\mathbf{S}}^3 \\ 4\hat{\mathbf{S}}_2 &= \hat{\mathbf{S}}^0 + \hat{\mathbf{S}}^1 - \hat{\mathbf{S}}^2 - \hat{\mathbf{S}}^3 \\ 4\hat{\mathbf{S}}_3 &= \hat{\mathbf{S}}^0 - \hat{\mathbf{S}}^1 - \hat{\mathbf{S}}^2 + \hat{\mathbf{S}}^3 \end{aligned} \tag{5.3}$$

together with the equivalent set of results for $\hat{\hat{\mathbf{S}}}_u$.

From these it is easy to write down the conditions that must be satisfied for each of the symmetries. This is done explicitly in Table 5.5, which parallels Table 5.1 for completely specified functions. These results show that to test for the compatibility of two subfunctions, we apply the appropriate test to the function with all the don't cares coded as logic zeroes and then a second time with all the don't cares coded as logic ones. If the inner product of the resulting vectors is zero, then the compatibility exists.

Much more care is needed when checking for multiple symmetries in incompletely specified functions than was the case for completely specified functions. The problem arises because the compatibility relation is not transitive. For example, for completely specified functions we saw that any two of $N\{x_3, x_4\}$, $S\{x_3|\bar{x}_4\}$, $S\{x_4|\bar{x}_3\}$ will imply the third. This is not necessarily true for incompletely specified functions, because the symmetries concerned may imply different assignments to the don't cares.

Consider the function illustrated in Fig. 5.19a. It is clear that this function possesses $S\{x_4|\bar{x}_3\}$ and $N\{x_4, x_3\}$ but not $S\{x_3|\bar{x}_4\}$. The presence of $S\{x_4|\bar{x}_3\}$ will imply the assignment of some of the don't cares as illustrated in Fig. 5.19b, while for $N\{x_3, x_4\}$ we require the assignment shown in Fig. 5.19c. It is clear that the exact choice of symmetry used at any stage in a synthesis procedure will have an impact on the complexity of the final circuit.

Table 5.5. Spectral symmetry tests for symmetries in $\{x_{n-1}, x_n\}$ for incompletely specified functions.

Symmetry	Test
$\mathbf{Y}_0 = \mathbf{Y}_1$ Single variable, x_{n-1} over $\bar{x}_n$ $f(x_1, \ldots, x_{n-2}, 0, 0) = f(x_1, \ldots, x_{n-2}, 1, 0)$	$(\hat{\mathbf{S}}^1 + \hat{\mathbf{S}}^3)^t(\hat{\hat{\mathbf{S}}}^1 + \hat{\hat{\mathbf{S}}}^3) = 0$
$\mathbf{Y}_0 = \mathbf{Y}_2$ Single variable, x_n over $\bar{x}_{n-1}$ $f(x_1, \ldots, x_{n-2}, 0, 0) = f(x_1, \ldots, x_{n-2}, 0, 1)$	$(\hat{\mathbf{S}}^2 + \hat{\mathbf{S}}^3)^t(\hat{\hat{\mathbf{S}}}^2 + \hat{\hat{\mathbf{S}}}^3) = 0$
$\mathbf{Y}_0 = \mathbf{Y}_3$ Equivalence, $\{x_n, x_{n-1}\}$ $f(x_1, \ldots, x_{n-2}, 0, 0) = f(x_1, \ldots, x_{n-2}, 1, 1)$	$(\hat{\mathbf{S}}^1 + \hat{\mathbf{S}}^2)^t(\hat{\hat{\mathbf{S}}}^1 + \hat{\hat{\mathbf{S}}}^2) = 0$
$\mathbf{Y}_1 = \mathbf{Y}_2$ Nonequivalence, $\{x_n, x_{n-1}\}$ $f(x_1, \ldots, x_{n-2}, 1, 0) = f(x_1, \ldots, x_{n-2}, 0, 1)$	$(\hat{\mathbf{S}}^1 - \hat{\mathbf{S}}^2)^t(\hat{\hat{\mathbf{S}}}^1 - \hat{\hat{\mathbf{S}}}^2) = 0$
$\mathbf{Y}_1 = \mathbf{Y}_3$ Single variable, x_n over x_{n-1} $f(x_1, \ldots, x_{n-2}, 1, 0) = f(x_1, \ldots, x_{n-2}, 1, 1)$	$(\hat{\mathbf{S}}^2 - \hat{\mathbf{S}}^3)^t(\hat{\hat{\mathbf{S}}}^2 - \hat{\hat{\mathbf{S}}}^3) = 0$
$\mathbf{Y}_2 = \mathbf{Y}_3$ Single variable, x_{n-1} over x_n $f(x_1, \ldots, x_{n-2}, 0, 1) = f(x_1, \ldots, x_{n-2}, 1, 1)$	$(\hat{\mathbf{S}}^1 - \hat{\mathbf{S}}^3)^t(\hat{\hat{\mathbf{S}}}^1 - \hat{\hat{\mathbf{S}}}^3) = 0$

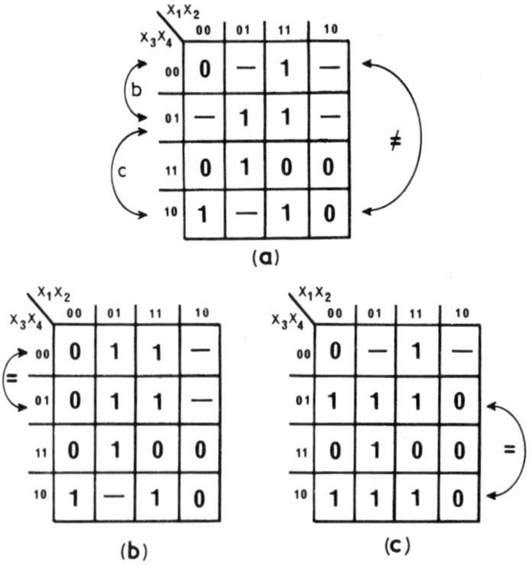

Fig. 5.19. Function symmetries with don't-care minterms: (a) example function, two possible x_3x_4 symmetries; (b) the function with $S\{x_4|\bar{x}_3\}$; (c) the function with $N\{x_3, x_4\}$.

Consider the function illustrated in Fig. 5.20. For this function we shall derive all the symmetries in $\{x_3, x_4\}$.

Initially we have

$$\hat{\mathbf{Y}} = 1 \; -1 \quad 1 \; -1 \quad -1 \quad 1 \; 1 \; -1 \quad 1 \quad 1 \; 1 \; -1 \quad 1 \; -1 \; 1 \quad 1]^t$$

$$\hat{\hat{\mathbf{Y}}} = 1 \; -1 \; -1 \; -1 \quad -1 \; -1 \; 1 \; -1 \quad -1 \; -1 \; 1 \; -1 \quad 1 \; -1 \; 1 \; -1]^t$$

giving two sets of spectral coefficients:

$\hat{s}_0$	$\hat{s}_1$	$\hat{s}_2$	$\hat{s}_3$	$\hat{s}_4$	$\hat{s}_{12}$	$\hat{s}_{13}$	$\hat{s}_{14}$	$\hat{s}_{23}$	$\hat{s}_{24}$	$\hat{s}_{34}$	$\hat{s}_{123}$	$\hat{s}_{124}$	$\hat{s}_{134}$	$\hat{s}_{234}$	$\hat{s}_{1234}$
4;	8	0	0	−4;	−4	4	0	4	0	0;	0	−4	4	−4;	8

$\hat{\hat{s}}_0$	$\hat{\hat{s}}_1$	$\hat{\hat{s}}_2$	$\hat{\hat{s}}_3$	$\hat{\hat{s}}_4$	$\hat{\hat{s}}_{12}$	$\hat{\hat{s}}_{13}$	$\hat{\hat{s}}_{14}$	$\hat{\hat{s}}_{23}$	$\hat{\hat{s}}_{24}$	$\hat{\hat{s}}_{34}$	$\hat{\hat{s}}_{123}$	$\hat{\hat{s}}_{124}$	$\hat{\hat{s}}_{134}$	$\hat{\hat{s}}_{234}$	$\hat{\hat{s}}_{1234}$
−6;	10	−2	−2	−2;	−2	−2	−2	2	2	2;	2	2	2	6;	6

Hence

$$[\hat{\mathbf{S}}^0\, \hat{\mathbf{S}}^1\, \hat{\mathbf{S}}^2\, \hat{\mathbf{S}}^3] = \begin{array}{c} \begin{matrix} \phi & 3 & 4 & 34 \end{matrix} \\ \left[\begin{matrix} 4 & 0 & -4 & 0 \\ 8 & 4 & 0 & 4 \\ 0 & 4 & 0 & -4 \\ -4 & 0 & -4 & 8 \end{matrix}\right] \end{array} \begin{matrix} \phi \\ 1 \\ 2 \\ 12 \end{matrix}$$

$$[\hat{\hat{\mathbf{S}}}^0\, \hat{\hat{\mathbf{S}}}^1\, \hat{\hat{\mathbf{S}}}^2\, \hat{\hat{\mathbf{S}}}^3] = \begin{array}{c} \begin{matrix} \phi & 3 & 4 & 34 \end{matrix} \\ \left[\begin{matrix} -6 & -2 & -2 & 2 \\ 10 & -2 & -2 & 2 \\ -2 & 2 & 2 & 6 \\ -2 & 2 & 2 & 6 \end{matrix}\right] \end{array} \begin{matrix} \phi \\ 1 \\ 2 \\ 12 \end{matrix}$$

There are six symmetries to check for, and for each we evaluate the test shown in Table 5.5. If the result of the inner product is zero, then there is an assignment to the don't cares for which the symmetry exists.

$x_3x_4 \backslash x_1x_2$	00	01	11	10
00	0	—	1	1
01	—	0	1	—
11	0	0	—	1
10	1	0	1	—

Fig. 5.20. Further example function with don't-care conditions.

(a) $S\{x_3|\bar{x}_4\}$

$$(\hat{\mathbf{S}}^1 + \hat{\mathbf{S}}^3)^t(\hat{\hat{\mathbf{S}}}^1 + \hat{\hat{\mathbf{S}}}^3) = [0 \quad 8 \quad 0 \quad 8]\begin{bmatrix}0\\0\\8\\8\end{bmatrix} = 64$$

Consequently, there is no assignment to the don't cares to ensure that this function will possess $S\{x_3|\bar{x}_4\}$. This is easily seen from the map.

(b) $S\{x_4|\bar{x}_3\}$

$$(\hat{\mathbf{S}}^2 + \hat{\mathbf{S}}^3)^t(\hat{\hat{\mathbf{S}}}^2 + \hat{\hat{\mathbf{S}}}^3) = [-4 \quad 4 \quad -4 \quad 4]\begin{bmatrix}0\\0\\8\\8\end{bmatrix} = 0$$

It follows that there is an assignment such that the resulting function will possess $S\{x_4|\bar{x}_3\}$.

(c) $E\{x_3, x_4\}$

$$(\hat{\mathbf{S}}^1 + \hat{\mathbf{S}}^2)^t(\hat{\hat{\mathbf{S}}}^1 + \hat{\hat{\mathbf{S}}}^2) = 0$$

so $E\{x_3, x_4\}$ exists for the correct choice of the don't cares.

(d) $N\{x_3, x_4\}$

$$(\hat{\mathbf{S}}^1 - \hat{\mathbf{S}}^2)^t(\hat{\hat{\mathbf{S}}}^1 - \hat{\hat{\mathbf{S}}}^2) = 0$$

so $N\{x_3, x_4\}$ exists for the correct choice of the don't cares.

(e) $S\{x_4|x_3\}$

$$(\hat{\mathbf{S}}^2 - \hat{\mathbf{S}}^3)^t(\hat{\hat{\mathbf{S}}}^2 - \hat{\hat{\mathbf{S}}}^3) = 64$$

so there is no assignment to the don't cares for which the function possesses $S\{x_4|x_3\}$.

(f) $S\{x_3|x_4\}$

$$(\hat{\mathbf{S}}^1 - \hat{\mathbf{S}}^3)^t(\hat{\hat{\mathbf{S}}}^1 - \hat{\hat{\mathbf{S}}}^3) = 0$$

so $S\{x_3|x_4\}$ exists for the correct choice of the don't cares.

5.5.2 An Alternative Coding Scheme

The whole of the above development has been in terms of the two vectors $\hat{\mathbf{Y}}$, $\hat{\hat{\mathbf{Y}}}$ to represent the function. An alternative is to use the single vector $\mathbf{Y}^*$

where

$$y_i^* = \begin{cases} +1 & \text{if } f(X) = 0 \\ 0 & \text{if } f(X) \text{ is not specified} \\ -1 & \text{if } f(X) = 1 \end{cases}$$

The transformation between $\hat{\mathbf{Y}}$, $\hat{\hat{\mathbf{Y}}}$ and $\mathbf{Y}^*$ is straightforward since

$$y_i^* = \tfrac{1}{2}(\hat{y}_i + \hat{\hat{y}}_i)$$

while

$$\hat{y}_i = y_i^*(1 - y_i^*) + 1$$

$$\hat{\hat{y}}_i = y_i^*(1 + y_i^*) - 1$$

In the following derivation, it is convenient to have explicitly available the location of all the don't cares, which can easily be done by using $\hat{\mathbf{Y}}^*$, defined by

$$\hat{y}_i^* = 1 - (y_i^*)^2$$

Corresponding to $\mathbf{Y}^*$ and $\hat{\mathbf{Y}}^*$, we shall have $\mathbf{S}^*$ and $\hat{\mathbf{S}}^*$ given by $\mathbf{S}^* = \mathbf{T}^n\mathbf{Y}^*$ and $\hat{\mathbf{S}}^* = \mathbf{T}^n\hat{\mathbf{Y}}^*$.

Equations identical in form to those in (5.3) will follow for the $\mathbf{S}_u^*$ and $\hat{\mathbf{S}}_u^*$. It is easily shown that

$$\mathbf{S}^* = \tfrac{1}{2}(\hat{\mathbf{S}} + \hat{\hat{\mathbf{S}}}), \qquad \hat{\mathbf{S}}^* = \tfrac{1}{2}(\hat{\mathbf{S}} - \hat{\hat{\mathbf{S}}})$$

or equivalently,

$$\hat{\mathbf{S}} = \mathbf{S}^* + \mathbf{S}^*, \qquad \hat{\hat{\mathbf{S}}} = \mathbf{S}^* - \hat{\mathbf{S}}^*$$

Recall the criterion given in (5.2) that $(\hat{\mathbf{Y}}_u - \hat{\mathbf{Y}}_v)^t(\hat{\hat{\mathbf{Y}}}_u - \hat{\hat{\mathbf{Y}}}_v) = 0$ if and only if $(\hat{\mathbf{S}}_u - \hat{\mathbf{S}}_v)^t(\hat{\hat{\mathbf{S}}}_u - \hat{\hat{\mathbf{S}}}_v) = 0$. This condition can be reexpressed in the form

$$(\mathbf{S}_u^* - \mathbf{S}_v^* + \hat{\mathbf{S}}_u^* - \hat{\mathbf{S}}_v^*)^t(\mathbf{S}_u^* - \mathbf{S}_v^* - \hat{\mathbf{S}}_u^* + \hat{\mathbf{S}}_v^*) = 0$$

which reduces to the condition

$$(\mathbf{S}_u^* - \mathbf{S}_v^*)^t(\mathbf{S}_u^* - \mathbf{S}_v^*) = (\hat{\mathbf{S}}_u^* - \hat{\mathbf{S}}_v^*)^t(\hat{\mathbf{S}}_u^* - \hat{\mathbf{S}}_v^*)$$

Although this criterion looks rather complex in appearance, it is in fact a reasonably simple sum-of-squares condition, which in practice is straightforward to check. The procedure involved is to evaluate the symmetry conditions as usual using $\mathbf{Y}^*$, and repeat the procedure using $\hat{\mathbf{Y}}^*$. The symmetry exists in the incompletely specified function if the sum of the squares of the results of the symmetry checks for $\mathbf{Y}^*$ is equal to that for $\hat{\mathbf{Y}}^*$. The method is clarified by repeating the example used above for $\mathbf{Y}^*$ and $\hat{\mathbf{Y}}^*$. The function concerned is illustrated in Fig. 5.20.

We have

$$\mathbf{Y}^* = 1\ \ -1\ \ 0\ \ -1\quad -1\ \ 0\ \ 1\ \ -1\quad 0\ \ 0\ \ 1\ \ -1\quad 1\ \ -1\ \ 1\ \ 0]^t$$

$$\hat{\mathbf{Y}}^* = 0\quad 0\ \ 1\quad 0\quad 0\ \ 1\ \ 0\quad 0\ \ 1\ \ 1\ \ 0\quad 0\ \ 0\quad 0\ \ 0\ \ 1]^t$$

with the corresponding spectral coefficients

s_0^*	s_1^*	s_2^*	s_3^*	s_4^*	s_{12}^*	s_{13}^*	s_{14}^*	s_{23}^*	s_{24}^*	s_{34}^*	s_{123}^*	s_{124}^*	s_{134}^*	s_{234}^*	s_{1234}^*
-1;	9	-1	-1	-3;	-3	1	-1	3	1	1;	1	-1	3	1;	7

$\hat{s}_0^*$	$\hat{s}_1^*$	$\hat{s}_2^*$	$\hat{s}_3^*$	$\hat{s}_4^*$	$\hat{s}_{12}^*$	$\hat{s}_{13}^*$	$\hat{s}_{14}^*$	$\hat{s}_{23}^*$	$\hat{s}_{24}^*$	$\hat{s}_{34}^*$	$\hat{s}_{123}^*$	$\hat{s}_{124}^*$	$\hat{s}_{134}^*$	$\hat{s}_{234}^*$	$\hat{s}_{1234}^*$
5;	-1	1	1	-1;	-1	3	1	1	-1	-1;	-1	-3	1	-5;	1

For symmetries in $\{x_3, x_4\}$ we have

$$[\mathbf{S}^{*0}\,\mathbf{S}^{*1}\,\mathbf{S}^{*2}\,\mathbf{S}^{*3}] = \begin{array}{c} \begin{array}{cccc} \phi & 3 & 4 & 34 \end{array} \\ \left[\begin{array}{rrrr} -1 & -1 & -3 & 1 \\ 9 & 1 & -1 & 3 \\ -1 & 3 & 1 & 1 \\ -3 & 1 & -1 & 7 \end{array}\right] \end{array} \begin{array}{l} \phi \\ 1 \\ 2 \\ 12 \end{array}$$

and

$$[\hat{\mathbf{S}}^{*0}\,\hat{\mathbf{S}}^{*1}\,\hat{\mathbf{S}}^{*2}\,\hat{\mathbf{S}}^{*3}] = \begin{array}{c} \begin{array}{cccc} \phi & 3 & 4 & 34 \end{array} \\ \left[\begin{array}{rrrr} 5 & 1 & -1 & -1 \\ -1 & 3 & 1 & 1 \\ 1 & 1 & -1 & -5 \\ -1 & -1 & -3 & 1 \end{array}\right] \end{array} \begin{array}{l} \phi \\ 1 \\ 2 \\ 12 \end{array}$$

(a) $\mathbf{S}\{x_3|\bar{x}_4\}$

$$(\mathbf{S}^{*1} + \mathbf{S}^{*3})^t(\mathbf{S}^{*1} + \mathbf{S}^{*3}) = [0\quad 4\quad 4\quad 8] \left.\begin{array}{r} 0 \\ 4 \\ 4 \\ 8 \end{array}\right] = 96$$

$$(\hat{\mathbf{S}}^{*1} + \hat{\mathbf{S}}^{*3})^t(\hat{\mathbf{S}}^{*1} + \hat{\mathbf{S}}^{*3}) = [0\quad 4\quad -4\quad 0] \left.\begin{array}{r} 0 \\ 4 \\ -4 \\ 0 \end{array}\right] = 32$$

Since these values are distinct, the symmetry does not exist.

(b) $S\{x_4|\bar{x}_3\}$

$$(\mathbf{S}^{*2} + \mathbf{S}^{*3})^t = (-2 \quad 2 \quad 2 \quad 6): \qquad \text{sum of squares} = 48$$

$$(\hat{\mathbf{S}}^{*2} + \hat{\mathbf{S}}^{*2})^t = (-2 \quad 2 \quad -6 \quad -2): \qquad \text{sum of squares} = 48$$

Since the two values are equal, the symmetry exists for the function.

The other four symmetries can be investigated in a similar fashion.

5.5.3 An Example

We are now in a position to complete the example that was commenced in Section 5.4.11. In the first example considered there,

$$f(X) = \bar{x}_1(x_2 + x_3 + x_4) + x_2(x_3 + x_4) + x_3x_4$$

was analyzed and, using an initial symmetry, the first portion of a circuit was derived. However, the resulting remainder function h was incompletely specified, with the map shown in Fig. 5.21a. The original synthesis was illustrated in Fig. 5.16.

An analysis of the function illustrated in Fig. 5.21a reveals that there are eight symmetries that it possesses, namely, $E\{x_1, x_2\}$, $N\{x_2, \tilde{x}_4\}$, $S\{x_2|\tilde{x}_4\}$, $S\{\tilde{x}_4|x_2\}$, $E\{x_1, \tilde{x}_4\}$, $S\{\tilde{x}_4|\bar{x}_1\}$, $S\{x_1|\tilde{x}_4\}$, $N\{x_2, \tilde{x}_3\}$. In addition to these, there are three symmetries in $\{\tilde{x}_3, \tilde{x}_4\}$ which involve assigning all the don't care entries to be identical to one of the other rows in the map. These are not useful to us and are omitted.

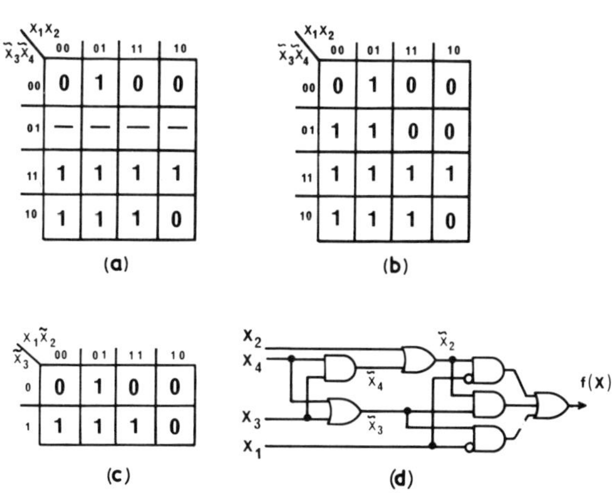

Fig. 5.21. Completion of the synthesis of Fig. 5.16: (a) the don't-care minterms from Fig. 5.16c; (b) the don't-care assignment; (c) the new core function; (d) the final realization for $f(X)$.

We notice that it is possible for the function to possess all of $N\{x_2, \tilde{x}_4\}$, $S\{x_2|\tilde{x}_4\}$ and $S\{\tilde{x}_4|x_2\}$ simultaneously, using the assignment to the don't cares that is illustrated in Fig. 5.21b. Using the substitution $\tilde{x}_2 = x_2 + \tilde{x}_4$ results in a reduction of one variable, leaving us with the new residual function shown in Fig. 5.21c, which leads to the complete circuit in Fig. 5.21d.

5.6 SYMMETRIES OF AN ARBITRARY DEGREE

In the earlier sections we confined our discussions entirely to symmetries of degree 2, that is, to symmetries in which two variables are fixed. For example, the function $f(x_1, x_2, x_3, x_4)$, which is illustrated in Fig. 5.22, possesses an equivalence symmetry $E\{x_3, x_4\}$. This, of course, means that $f(x_1, x_2, 0, 0) = f(x_1, x_2, 1, 1)$ or $\mathbf{Y}_0 = \mathbf{Y}_3$. What we have is that two particular quarters of the function are equal. Unfortunately, the proportion of functions that possess symmetries of degree 2 declines as n, the number of variables, increases. Any three-variable function will possess a symmetry of degree 2, but there exist a small number of four-variable functions that do not possess any symmetries of degree 2. From this point on, the proportion of functions that do not possess any symmetries of degree 2 increases rapidly.

For example, the function $f(X) = \bar{x}_1x_3\bar{x}_4 + x_1\bar{x}_2x_3x_4 + \bar{x}_1\bar{x}_2\bar{x}_4 + x_1x_2\bar{x}_3 + x_1\bar{x}_3\bar{x}_4$, which is illustrated in Fig. 5.23, does not possess any symmetries of degree 2. In practice, this means that having failed to find any pairs of equal blocks of minterms, each of which contains one quarter of the minterms, then we shall look for smaller blocks that are equal. For example, in Fig. 5.23 we know that $f(0, 0, x_3, x_4) \neq f(0, 1, x_3, x_4)$, but we do have that $f(0, 0, 1, x_4) = f(0, 1, 1, x_4)$ and $f(0, 0, x_3, 1) = f(0, 1, x_3, 1)$, which are two particular symmetries of degree 3. In this section we derive criteria for the detection of these higher-degree symmetries and show how they can be useful in the synthesis of a function. These higher-degree symmetries were first described by Muzio *et al.*[11]

Let us start by recalling our definition of the subfunction $f_u(x_1, \ldots, x_m)$

$x_3x_4 \backslash x_1x_2$	00	01	11	10
00	1	0	0	1
01	0	1	0	0
11	1	0	0	1
10	1	0	0	0

Fig. 5.22. Equivalence symmetry in function $f(X) = \bar{x}_1\bar{x}_2(x_3 + \bar{x}_4) + \bar{x}_2(\bar{x}_3\bar{x}_4 + x_3x_4) + \bar{x}_1x_2\bar{x}_3x_4$.

x_3x_4 \ x_1x_2	00	01	11	10
00	1	0	1	1
01	0	0	1	0
11	0	0	0	1
10	1	1	0	0

Fig. 5.23. Function $f(X) = \bar{x}_1 x_3 \bar{x}_4 + x_1 \bar{x}_2 x_3 x_4 + \bar{x}_1 \bar{x}_2 \bar{x}_4 + x_1 x_2 \bar{x}_3 + x_1 \bar{x}_3 \bar{x}_4$, which does not possess any degree-2 symmetries.

$= f(x_1, \ldots, x_m, u_1, \ldots, u_{n-m})$, where u is a constant with the binary expansion $u = \sum_{i=1}^{n-m} u_i 2^{i-1}$, with its corresponding column vector $\mathbf{Y}_u$ and spectrum $\mathbf{S}_u$.

Our interest is in the equality of pairs of these subfunctions. Suppose $(u_1, \ldots, u_{n-m})$ and $(v_1, \ldots, v_{n-m})$ are the binary expansions of u and v, respectively. A function $f(x_1, \ldots, x_n)$ possesses a multivariable symmetry in $x_{m+1}, \ldots, x_n$ if there exist u, v such that $f_u(x_1, \ldots, x_m) = f_v(x_1, \ldots, x_m)$ $(u \neq v, 0 \leq u, v < 2^{n-m})$ for all 2^m sets of assignments to the variables $x_1, \ldots, x_m$. The symmetry is said to be of degree k, $k = n - m$.

The number of symmetries of degree k that are possible increases as k rises. As we have already seen, there are six distinct symmetries of degree 2 in a particular pair of variables. There will be 28 distinct symmetries of degree 3 for a particular set of three variables and, in general, for a set of k variables there will be $\binom{2^k}{2}$ distinct symmetries of degree k.

For example, the function which was illustrated in Fig. 5.23, possesses all the following symmetries of degree 3:

(a) In $\{x_1, x_2, x_3\}$:

$$f(0, 0, 0, x_4) = f(0, 1, 1, x_4) = f(0, 0, 1, x_4) = f(1, 0, 0, x_4)$$

$$f(0, 1, 0, x_4) = f(1, 1, 1, x_4)$$

(b) In $\{x_1, x_2, x_4\}$:

$$f(0, 0, x_3, 1) = f(0, 1, x_3, 1)$$

$$f(0, 1, x_3, 0) = f(1, 0, x_3, 1)$$

$$f(1, 0, x_3, 0) = f(1, 1, x_3, 0) = f(1, 1, x_3, 1)$$

(c) In $\{x_1, x_3, x_4\}$:

$$f(0, x_2, 0, 0) = f(1, x_2, 1, 1)$$

$$f(0, x_2, 0, 1) = f(0, x_2, 1, 1) = f(1, x_2, 1, 0)$$

$$f(0, x_2, 1, 0) = f(1, x_2, 0, 0)$$

(d) In $\{x_2, x_3, x_4\}$:

$$f(x_1, 0, 0, 1) = f(x_1, 1, 1, 1)$$
$$f(x_1, 0, 1, 0) = f(x_1, 1, 1, 0)$$
$$f(x_1, 0, 1, 1) = f(x_1, 1, 0, 0) = f(x_1, 1, 0, 1)$$

Because of the increasingly large number of possible symmetries, we do not give them specific names as we did for symmetries of degree 2, but simply identify them by the particular equal subfunctions.

The condition which must be satisfied for symmetry to be present is that

$$\mathbf{Y}_u = \mathbf{Y}_v, \qquad u \neq v, \quad 0 \leq u, v < 2^k$$

or equivalently, the spectral condition that

$$\mathbf{S}_u = \mathbf{S}_v, \qquad u \neq v, \quad 0 \leq u, v < 2^k$$

This last condition is not directly useful, since we wish to avoid calculating all possible $\mathbf{S}_u$, the spectra of all the subfunctions, and to use conditions on $\mathbf{S}$, the spectrum of $f(X)$. We shall now derive such a condition and, although the resulting criterion may look a little complex, it will turn out to be quite straightforward to use in practice.

5.6.1 Conditions for Multivariable Symmetries for Completely Specified Functions

As was done earlier, our initial consideration is limited to completely specified functions, and the results will be widened to incompletely specified functions in the next section.

Consider a function $f(x_1, \ldots, x_n)$ and the 2^k subfunctions that result from fixing the variables $x_{n-k+1}, \ldots, x_n$. For each value of w, $0 \leq w \leq 2^k - 1$, we have

$$\mathbf{S}_w = \mathbf{T}^m \mathbf{Y}_w \tag{5.4}$$

where $m = n - k$ (since each $\mathbf{Y}_w$ is the truthtable column vector for a function of m variables, namely, $f_w(x_1, \ldots, x_m)$).

For convenience, let $\beta = 2^k - 1 = 2^{n-m} - 1$. Writing all the resulting (5.4) together, we have

$$\begin{bmatrix} \mathbf{S}_0 \\ \mathbf{S}_1 \\ \vdots \\ \mathbf{S}_\beta \end{bmatrix} = [\mathbf{I}^k \circledast \mathbf{T}^m] \begin{bmatrix} \mathbf{Y}_0 \\ \mathbf{Y}_1 \\ \vdots \\ \mathbf{Y}_\beta \end{bmatrix}$$

where ⊛ is the Kronecker product operator. Now

$$\begin{bmatrix} \mathbf{Y}_0 \\ \mathbf{Y}_1 \\ \vdots \\ \mathbf{Y}_\beta \end{bmatrix} = \mathbf{Y} = \frac{1}{2^n}\mathbf{T}^n\mathbf{S}$$

which gives

$$\begin{bmatrix} \mathbf{S}_0 \\ \mathbf{S}_1 \\ \vdots \\ \mathbf{S}_\beta \end{bmatrix} = \frac{1}{2^n}[\mathbf{I}^k \circledast \mathbf{T}^m]\mathbf{T}^n\mathbf{S}$$

However, as was seen in Section 2.2.1, we can use the Kronecker product ⊛ to express $\mathbf{T}^n$ in the form

$$\mathbf{T}^n = \mathbf{T}^k \circledast \mathbf{T}^m$$

Consequently,

$$\mathbf{T}^n = \mathbf{T}^k \circledast \mathbf{T}^m = \begin{bmatrix} t_{00}\mathbf{T}^m & t_{01}\mathbf{T}^m & \cdots & t_{0\beta}\mathbf{T}^m \\ t_{10}\mathbf{T}^m & t_{11}\mathbf{T}^m & \cdots & t_{1\beta}\mathbf{T}^m \\ \vdots & & & \vdots \\ t_{\beta 0}\mathbf{T}^m & t_{\beta 1}\mathbf{T}^m & \cdots & t_{\beta\beta}\mathbf{T}^m \end{bmatrix}$$

Since $\mathbf{I}^k \circledast \mathbf{T}^m$ is just a block diagonal matrix, *viz*

$$\mathbf{I}^k \circledast \mathbf{T}^m = \begin{bmatrix} \mathbf{T}^m & & & 0 \\ & \mathbf{T}^m & & \\ & & \ddots & \\ 0 & & & \mathbf{T}^m \end{bmatrix}$$

and $\mathbf{T}^m\mathbf{T}^m = 2^m\mathbf{I}^m$.

We have

$$(\mathbf{I}^k \circledast \mathbf{T}^m)\mathbf{T}^n = (\mathbf{I}^k \circledast \mathbf{T}^m)(\mathbf{T}^k \circledast \mathbf{T}^m)$$

$$= 2^m \begin{bmatrix} t_{00}\mathbf{I}^m & t_{01}\mathbf{I}^m & \cdots & t_{0\beta}\mathbf{I}^m \\ t_{01}\mathbf{I}^m & t_{11}\mathbf{I}^m & \cdots & t_{1\beta}\mathbf{I}^m \\ \vdots & & & \vdots \\ t_{\beta 0}\mathbf{I}^m & t_{\beta 1}\mathbf{I}^m & \cdots & t_{\beta\beta}\mathbf{I}^m \end{bmatrix}$$

$$= 2^m(\mathbf{T}^k \circledast \mathbf{I}^m)$$

Hence,

$$\begin{bmatrix} \mathbf{S}_0 \\ \mathbf{S}_1 \\ \vdots \\ \mathbf{S}_\beta \end{bmatrix} = \frac{1}{2^{n-m}}(\mathbf{T}^k \circledast \mathbf{I}^m)\mathbf{S} = \frac{1}{2^k}(\mathbf{T}^k \circledast \mathbf{I}^m) \begin{bmatrix} \mathbf{S}^0 \\ \mathbf{S}^1 \\ \vdots \\ \mathbf{S}^\beta \end{bmatrix} \tag{5.5}$$

This result enables us to write down a condition for the equality of two of the $\mathbf{S}_u$ in terms of parts of $\mathbf{S}$ and, as was pointed out earlier, we have $\mathbf{S}$ available but do not want to have to calculate all the $\mathbf{S}_u$ for each possible symmetry.

To test for $f_u(x_1, \ldots, x_m) = f_v(x_1, \ldots, x_m)$ for some u, v $(u \neq v, 0 \leq u, v \leq \beta)$, we have to check whether $\mathbf{S}_u = \mathbf{S}_v$. Now from (5.5),

$$\mathbf{S}_u = \frac{1}{2^k}\sum_{i=0}^{\beta} t_{ui}\mathbf{S}^i, \qquad \mathbf{S}_v = \frac{1}{2^k}\sum_{i=0}^{\beta} t_{vi}\mathbf{S}^i \tag{5.6}$$

which leads immediately to the following condition.

Criterion 5.1 A necessary and sufficient condition for $\mathbf{Y}_u = \mathbf{Y}_v$ $(u \neq v, 0 \leq u, v \leq \beta)$ is that

$$\sum_{i=0}^{\beta} (t_{ui} - t_{vi})\mathbf{S}^i = \mathbf{0}$$

An alternative statement of this same condition is that

$$[(\mathbf{T}^k_{u*} - \mathbf{T}^k_{v*}) \circledast \mathbf{I}^m]\mathbf{S} = \mathbf{0}$$

In the first statement of the condition, the multipliers that have to be applied to the $\mathbf{S}^i$ are just the difference of the entries of two rows of $\mathbf{T}^k$. Exactly half of these will always be zero, so the summation given in the criterion will involve 2^{k-1} terms, each term being a column vector containing 2^m entries. The symmetry test consequently always involves exactly half the coefficients and is independent of the value of k.

Although the result looks a little complex, it is quite straightforward to apply as illustrated in the following two examples.

Consider a check for a particular symmetry of degree 3 for an n-variable

x_3x_4 \ x_1x_2	00	01	11	10
00	1	0	1	1
01	0	0	1	0
11	0	0	0	1
10	1	1	0	0

Fig. 5.24. The function of Fig. 5.23 (repeated).

function. For example, suppose we wish to determine whether $f(x_1, \ldots, x_{n-3}, 0, 1, 0) = f(x_1, \ldots, x_{n-3}, 1, 0, 1)$ or, equivalently, whether $f_2(x_1, \ldots, x_{n-3}) = f_5(x_1, \ldots, x_{n-3})$.

According to our criterion (we have $k = 3, u = 2, v = 5$), the check will involve rows 2 and 5 of $\mathbf{T}^3$ (these rows are noted with asterisks; note that the rows are numbered 0 to 7).

$$\mathbf{T}^3 = \begin{bmatrix} 1 & 1 & 1 & 1 & 1 & 1 & 1 & 1 \\ 1 & -1 & 1 & -1 & 1 & -1 & 1 & -1 \\ 1 & 1 & -1 & -1 & 1 & 1 & -1 & -1 \\ 1 & -1 & -1 & 1 & 1 & -1 & -1 & 1 \\ 1 & 1 & 1 & 1 & -1 & -1 & -1 & -1 \\ 1 & -1 & 1 & -1 & -1 & 1 & -1 & 1 \\ 1 & 1 & -1 & -1 & -1 & -1 & 1 & 1 \\ 1 & -1 & -1 & 1 & -1 & 1 & 1 & -1 \end{bmatrix} \begin{matrix} \\ \\ * \\ \\ \\ * \\ \\ \\ \end{matrix}$$

The criterion involves the difference of these two rows and, since they are equal in columns 0, 3, 5 and 6, the check will not involve $\mathbf{S}^0$, $\mathbf{S}^3$, $\mathbf{S}^5$ or $\mathbf{S}^6$. We have the condition that $\mathbf{Y}_2 = \mathbf{Y}_5$, if and only if $\mathbf{S}^1 - \mathbf{S}^2 + \mathbf{S}^4 - \mathbf{S}^7 = \mathbf{0}$. Note that each of the $\mathbf{S}^i$ will contain 2^{n-3} coefficients.

Consider the function illustrated in Fig. 5.24, which is the same one that we considered earlier in this section. This function has the spectrum

s_0	s_1	s_2	s_3	s_4	s_{12}	s_{13}	s_{14}	s_{23}	s_{24}	s_{34}	s_{123}	s_{124}	s_{134}	s_{234}	s_{1234}
2;	2	−2	−2	−6;	−2	6	−6	2	−2	−2;	−6	−2	6	−6;	2

We shall consider some of the symmetries of degree 3 in $\{x_2, x_3, x_4\}$. Since we are considering symmetries of degree 3, the multipliers used in criterion 1 are the differences in rows of $\mathbf{T}^3$.

To check for $\mathbf{Y}_2 = \mathbf{Y}_3$, we use the difference between rows 2 and 3 of $\mathbf{T}^3$, which leads to the condition

$$\mathbf{S}^1 - \mathbf{S}^3 + \mathbf{S}^5 - \mathbf{S}^7 = \begin{bmatrix} s_2 \\ s_{12} \end{bmatrix} - \begin{bmatrix} s_{23} \\ s_{123} \end{bmatrix} + \begin{bmatrix} s_{24} \\ s_{124} \end{bmatrix} - \begin{bmatrix} s_{234} \\ s_{1234} \end{bmatrix}$$

$$\begin{bmatrix} -2 \\ -2 \end{bmatrix} - \begin{bmatrix} 2 \\ -6 \end{bmatrix} + \begin{bmatrix} -2 \\ -2 \end{bmatrix} - \begin{bmatrix} -6 \\ 2 \end{bmatrix} = \begin{bmatrix} 0 \\ 0 \end{bmatrix}$$

so it follows that $\mathbf{Y}_2 = \mathbf{Y}_3$.

However, for $\mathbf{Y}_2 = \mathbf{Y}_5$ the check will be

$$\mathbf{S}^1 - \mathbf{S}^2 + \mathbf{S}^4 - \mathbf{S}^7 = \begin{bmatrix} -2 \\ -2 \end{bmatrix} - \begin{bmatrix} -2 \\ 6 \end{bmatrix} + \begin{bmatrix} -6 \\ -6 \end{bmatrix} - \begin{bmatrix} -6 \\ 2 \end{bmatrix} = \begin{bmatrix} 0 \\ -16 \end{bmatrix}$$

and consequently $\mathbf{Y}_2 \neq \mathbf{Y}_5$, since all values in the resulting column vector have to be 0 for the condition to be satisfied.

Since the statement of the test is independent of n and only depends on k, the degree of the symmetry, it is possible to list all the tests for symmetries of degree 3 and these are all given in Table 5.6. Of course, the number of entries in each of the $\mathbf{S}^i$ is 2^{n-3}. The table is divided into three parts for convenience according to the distance of the symmetry. This distance is the Hamming distance between the u and v that define the subfunctions in the symmetry. More formally, we say that $\mathbf{Y}_u = \mathbf{Y}_v$, where $(u_1, \ldots, u_k)$ and

Table 5.6. Tests for degree-3 symmetries in $\{x_n, x_{n-1}, x_{n-2}\}$

		Symmetries	
		Symmetry	Test
Distance 1	1	$\mathbf{Y}_0 = \mathbf{Y}_1$	$\mathbf{S}^1 + \mathbf{S}^3 + \mathbf{S}^5 + \mathbf{S}^7$
	2	$\mathbf{Y}_0 = \mathbf{Y}_2$	$\mathbf{S}^2 + \mathbf{S}^3 + \mathbf{S}^6 + \mathbf{S}^7$
	3	$\mathbf{Y}_0 = \mathbf{Y}_4$	$\mathbf{S}^4 + \mathbf{S}^5 + \mathbf{S}^6 + \mathbf{S}^7$
	4	$\mathbf{Y}_1 = \mathbf{Y}_3$	$\mathbf{S}^2 - \mathbf{S}^3 + \mathbf{S}^6 - \mathbf{S}^7$
	5	$\mathbf{Y}_1 = \mathbf{Y}_5$	$\mathbf{S}^4 - \mathbf{S}^5 + \mathbf{S}^6 - \mathbf{S}^7$
	6	$\mathbf{Y}_2 = \mathbf{Y}_3$	$\mathbf{S}^1 - \mathbf{S}^3 + \mathbf{S}^5 - \mathbf{S}^7$
	7	$\mathbf{Y}_2 = \mathbf{Y}_6$	$\mathbf{S}^4 + \mathbf{S}^5 - \mathbf{S}^6 - \mathbf{S}^7$
	8	$\mathbf{Y}_3 = \mathbf{Y}_7$	$\mathbf{S}^4 - \mathbf{S}^5 - \mathbf{S}^6 + \mathbf{S}^7$
	9	$\mathbf{Y}_4 = \mathbf{Y}_5$	$\mathbf{S}^1 + \mathbf{S}^3 - \mathbf{S}^5 - \mathbf{S}^7$
	10	$\mathbf{Y}_4 = \mathbf{Y}_6$	$\mathbf{S}^2 + \mathbf{S}^3 - \mathbf{S}^6 - \mathbf{S}^7$
	11	$\mathbf{Y}_5 = \mathbf{Y}_7$	$\mathbf{S}^2 - \mathbf{S}^3 - \mathbf{S}^6 + \mathbf{S}^7$
	12	$\mathbf{Y}_6 = \mathbf{Y}_7$	$\mathbf{S}^1 - \mathbf{S}^3 - \mathbf{S}^5 + \mathbf{S}^7$
Distance 2	13	$\mathbf{Y}_0 = \mathbf{Y}_3$	$\mathbf{S}^1 + \mathbf{S}^2 + \mathbf{S}^5 + \mathbf{S}^6$
	14	$\mathbf{Y}_0 = \mathbf{Y}_5$	$\mathbf{S}^1 + \mathbf{S}^3 + \mathbf{S}^4 + \mathbf{S}^6$
	15	$\mathbf{Y}_0 = \mathbf{Y}_6$	$\mathbf{S}^2 + \mathbf{S}^3 + \mathbf{S}^4 + \mathbf{S}^5$
	16	$\mathbf{Y}_1 = \mathbf{Y}_2$	$\mathbf{S}^1 - \mathbf{S}^2 + \mathbf{S}^5 - \mathbf{S}^6$
	17	$\mathbf{Y}_1 = \mathbf{Y}_4$	$\mathbf{S}^1 + \mathbf{S}^3 - \mathbf{S}^4 - \mathbf{S}^6$
	18	$\mathbf{Y}_1 = \mathbf{Y}_7$	$\mathbf{S}^2 - \mathbf{S}^3 + \mathbf{S}^4 - \mathbf{S}^5$
	19	$\mathbf{Y}_2 = \mathbf{Y}_4$	$\mathbf{S}^2 + \mathbf{S}^3 - \mathbf{S}^4 - \mathbf{S}^5$
	20	$\mathbf{Y}_2 = \mathbf{Y}_7$	$\mathbf{S}^1 - \mathbf{S}^3 + \mathbf{S}^4 - \mathbf{S}^6$
	21	$\mathbf{Y}_3 = \mathbf{Y}_5$	$\mathbf{S}^2 - \mathbf{S}^3 - \mathbf{S}^4 + \mathbf{S}^5$
	22	$\mathbf{Y}_3 = \mathbf{Y}_6$	$\mathbf{S}^1 - \mathbf{S}^3 - \mathbf{S}^4 + \mathbf{S}^6$
	23	$\mathbf{Y}_4 = \mathbf{Y}_7$	$\mathbf{S}^1 + \mathbf{S}^2 - \mathbf{S}^5 - \mathbf{S}^6$
	24	$\mathbf{Y}_5 = \mathbf{Y}_6$	$\mathbf{S}^1 - \mathbf{S}^2 - \mathbf{S}^5 + \mathbf{S}^6$
Distance 3	25	$\mathbf{Y}_0 = \mathbf{Y}_7$	$\mathbf{S}^1 + \mathbf{S}^2 + \mathbf{S}^4 + \mathbf{S}^7$
	26	$\mathbf{Y}_1 = \mathbf{Y}_6$	$\mathbf{S}^1 - \mathbf{S}^2 - \mathbf{S}^4 + \mathbf{S}^7$
	27	$\mathbf{Y}_2 = \mathbf{Y}_5$	$\mathbf{S}^1 - \mathbf{S}^2 + \mathbf{S}^4 - \mathbf{S}^7$
	28	$\mathbf{Y}_3 = \mathbf{Y}_4$	$\mathbf{S}^1 + \mathbf{S}^2 - \mathbf{S}^4 - \mathbf{S}^7$

$(v_1, \ldots, v_k)$ are the binary expansions of u and v, respectively, is a distance r symmetry if $u_i \neq v_i$ for exactly r distinct values of i.

We shall return to a discussion of the calculation of the symmetries at the end of this section.

5.6.2 Conditions for Multivariable Symmetries for Incompletely Specified Functions

The approach here is a direct extension of that in Section 5.5. As there, we shall consider two possible coding schemes for the function. The first uses two vectors $\hat{\mathbf{Y}}$ and $\hat{\hat{\mathbf{Y}}}$ to represent the function, $\hat{\mathbf{Y}}$ having all the don't cares coded as logic zeroes while $\hat{\hat{\mathbf{Y}}}$ has them all coded as logic ones, that is,

$$\hat{y} = \begin{cases} +1 & \text{if} \quad f(X) = 0 \text{ or } f(X) \text{ not specified} \\ -1 & \text{if} \quad f(X) = 1 \end{cases}$$

$$\hat{\hat{y}} = \begin{cases} +1 & \text{if} \quad f(X) = 0 \\ -1 & \text{if} \quad f(X) = 1 \text{ or } f(X) \text{ not specified} \end{cases}$$

Recall that we had to introduce the notion of compatibility, where two functions $f(x_1, \ldots, x_n)$ and $g(x_1, \ldots, x_n)$ are compatible if there is an assignment to some or all of their don't cares such that they are equal.

Consider two subfunctions $f_u(x_1, \ldots, x_m)$ and $f_v(x_1, \ldots, x_m)$. It follows that if f_u and f_v are compatible, then there is an assignment to some or all of the don't cares that will ensure that $f(x_1, \ldots, x_n)$ possesses a multivariable symmetry of degree k ($k = n - m$) in the variables $x_{m+1}, \ldots, x_n$.

The theoretical development of Section 5.6.1 clearly carries over for $\hat{\mathbf{Y}}$ and $\hat{\hat{\mathbf{Y}}}$ so that from (5.6) we have

$$\hat{\mathbf{S}}_u = \frac{1}{2^k} \sum_{i=0}^{\beta} t_{ui} \hat{\mathbf{S}}^i, \qquad \hat{\mathbf{S}}_v = \frac{1}{2^k} \sum_{i=0}^{\beta} t_{vi} \hat{\mathbf{S}}^i$$

$$\hat{\hat{\mathbf{S}}}_u = \frac{1}{2^k} \sum_{i=0}^{\beta} t_{ui} \hat{\hat{\mathbf{S}}}^i, \qquad \hat{\hat{\mathbf{S}}}_v = \frac{1}{2^k} \sum_{i=0}^{\beta} t_{vi} \hat{\hat{\mathbf{S}}}^i \tag{5.7}$$

A necessary and sufficient condition for $\mathbf{Y}_u$ and $\mathbf{Y}_v$ to be compatible is that, for every i ($0 \leq i \leq \beta$), the ith element of either $(\hat{\mathbf{Y}}_u - \hat{\mathbf{Y}}_v)$ or $(\hat{\hat{\mathbf{Y}}}_u - \hat{\hat{\mathbf{Y}}}_v)$ is zero. A necessary condition for this result is that

$$(\hat{\mathbf{Y}}_u - \hat{\mathbf{Y}}_v)^t (\hat{\hat{\mathbf{Y}}}_u - \hat{\hat{\mathbf{Y}}}_v) = 0$$

The condition is also sufficient, since the product of corresponding entries of $(\hat{\mathbf{Y}}_u - \hat{\mathbf{Y}}_v)$ and $(\hat{\hat{\mathbf{Y}}}_u - \hat{\hat{\mathbf{Y}}}_v)$ cannot be negative.

Now from their definitions,

$$\hat{\mathbf{Y}}_u - \hat{\mathbf{Y}}_v = \frac{1}{2^k}\mathbf{T}^k(\hat{\mathbf{S}}_u - \hat{\mathbf{S}}_v), \qquad \hat{\hat{\mathbf{Y}}}_u - \hat{\hat{\mathbf{Y}}}_v = \frac{1}{2^k}\mathbf{T}^k(\hat{\hat{\mathbf{S}}}_u - \hat{\hat{\mathbf{S}}}_v)$$

so that

$$(\hat{\mathbf{Y}}_u - \hat{\mathbf{Y}}_v)^t(\hat{\hat{\mathbf{Y}}}_u - \hat{\hat{\mathbf{Y}}}_v) = \frac{1}{2^{2k}}(\hat{\mathbf{S}}_u - \hat{\mathbf{S}}_v)^t\mathbf{T}^k\mathbf{T}^k(\hat{\hat{\mathbf{S}}}_u - \hat{\hat{\mathbf{S}}}_v)$$

$$= \frac{1}{2^k}(\hat{\mathbf{S}}_u - \hat{\mathbf{S}}_v)^t(\hat{\hat{\mathbf{S}}}_u - \hat{\hat{\mathbf{S}}}_v)$$

Using (5.7) with this gives the following result.

A necessary and sufficient condition for $\mathbf{Y}_u$ and $\mathbf{Y}_v$ to be compatible is that

$$\left(\sum_{i=0}^{\beta}(t_{ui} - t_{vi})\hat{\mathbf{S}}^i\right)^t\left(\sum_{i=0}^{\beta}(t_{ui} - t_{vi})\hat{\hat{\mathbf{S}}}^i\right) = 0 \qquad (u \neq v, \quad 0 < u, v < \beta)$$

An alternative way of stating the same condition is that

$$\hat{\mathbf{S}}^t[(\mathbf{T}^k_{u*} - \mathbf{T}^k_{v*})^t(\mathbf{T}^k_{u*} - \mathbf{T}^k_{v*}) \circledast \mathbf{I}^{n-k}]\hat{\hat{\mathbf{S}}} = 0$$

However, it is probably easier to follow the former statement. Exactly as we did in Section 5.5.1, to test for the compatibility of two subfunctions of an incompletely specified function, we apply the appropriate degree-k symmetry test to the function with the don't cares coded as logic zeroes and then a second time with all the don't cares coded as logic ones. The inner product of the resulting vectors must be zero for the compatibility to exist.

As an example, consider the incompletely specified function that is illustrated in Fig. 5.25.

Initially we have

$$\hat{\mathbf{Y}} = -1\ \ -1\ \ 1\ \ -1\quad -1\ \ 1\ \ -1\ \ 1\quad 1\ \ 1\ \ 1\ \ -1\quad 1\ \ -1\ \ 1\ \ 1]^t$$

$$\hat{\hat{\mathbf{Y}}} = -1\ \ -1\ \ 1\ \ -1\quad -1\ \ 1\ \ -1\ \ 1\quad 1\ \ 1\ \ 1\ \ -1\quad -1\ \ -1\ \ 1\ \ 1]^t$$

giving the two sets of spectral coefficients

$\hat{s}_0$	$\hat{s}_1$	$\hat{s}_2$	$\hat{s}_3$	$\hat{s}_4$	$\hat{s}_{12}$	$\hat{s}_{13}$	$\hat{s}_{14}$	$\hat{s}_{23}$	$\hat{s}_{24}$	$\hat{s}_{34}$	$\hat{s}_{123}$	$\hat{s}_{124}$	$\hat{s}_{134}$	$\hat{s}_{234}$	$\hat{s}_{1234}$
2;	2	−2	−2	−6;	−2	6	−6	2	−2	−2;	−6	−2	6	−6;	2

$\hat{\hat{s}}_0$	$\hat{\hat{s}}_1$	$\hat{\hat{s}}_2$	$\hat{\hat{s}}_3$	$\hat{\hat{s}}_4$	$\hat{\hat{s}}_{12}$	$\hat{\hat{s}}_{13}$	$\hat{\hat{s}}_{14}$	$\hat{\hat{s}}_{23}$	$\hat{\hat{s}}_{24}$	$\hat{\hat{s}}_{34}$	$\hat{\hat{s}}_{123}$	$\hat{\hat{s}}_{124}$	$\hat{\hat{s}}_{134}$	$\hat{\hat{s}}_{234}$	$\hat{\hat{s}}_{1234}$
0;	0	−4	0	−4;	−4	8	−4	4	0	−4;	−4	0	4	−4;	−8

This function is very similar to that considered earlier in this section (it is the same function as that considered in Fig. 5.24 except for the presence of the single don't care). Initially it can be verified that this function cannot possess any degree 2 symmetries under any assignment to the don't cares.

x_3x_4 \ x_1x_2	00	01	11	10
00	1	0	1	1
01	0	0	1	0
11	—	0	0	1
10	1	1	0	0

Fig. 5.25. The function of Fig. 5.24, but with minterm $\bar{x}_1\bar{x}_2x_3x_4$ made don't-care.

We shall look at two of the degree 3 symmetries. In particular, we shall check whether $\mathbf{Y}_0, \mathbf{Y}_6$ are compatible and whether $\mathbf{Y}_2, \mathbf{Y}_6$ are compatible.

To check whether $\mathbf{Y}_0, \mathbf{Y}_6$ are compatible, we apply the check from row 15 of Table 5.6 to each of $\hat{\mathbf{S}}, \hat{\hat{\mathbf{S}}}$ in turn and take the inner product of the two resulting vectors. We have

$$\hat{\mathbf{S}}^2 + \hat{\mathbf{S}}^3 + \hat{\mathbf{S}}^4 + \hat{\mathbf{S}}^5 = \begin{bmatrix} -2 \\ 6 \end{bmatrix} + \begin{bmatrix} 2 \\ -6 \end{bmatrix} + \begin{bmatrix} -6 \\ -6 \end{bmatrix} + \begin{bmatrix} -2 \\ -2 \end{bmatrix} = \begin{bmatrix} -8 \\ -8 \end{bmatrix}$$

$$\hat{\hat{\mathbf{S}}}^2 + \hat{\hat{\mathbf{S}}}^3 + \hat{\hat{\mathbf{S}}}^4 + \hat{\hat{\mathbf{S}}}^5 = \begin{bmatrix} 0 \\ 8 \end{bmatrix} + \begin{bmatrix} 4 \\ -4 \end{bmatrix} + \begin{bmatrix} -4 \\ -4 \end{bmatrix} + \begin{bmatrix} 0 \\ 0 \end{bmatrix} = \begin{bmatrix} 0 \\ 0 \end{bmatrix}$$

and the inner product is obviously 0, so $\mathbf{Y}_0$ and $\mathbf{Y}_6$ are compatible. In this case one of the two tests gave the zero vector, but this will not normally happen (it happens whenever the function possesses the symmetry in the case of all the don't cares taking same value).

To check for $\mathbf{Y}_2, \mathbf{Y}_6$, we use the check from row 7 of Table 5.6, giving

$$\hat{\mathbf{S}}^4 + \hat{\mathbf{S}}^5 - \hat{\mathbf{S}}^6 - \hat{\mathbf{S}}^7 = \begin{bmatrix} -6 \\ -6 \end{bmatrix} + \begin{bmatrix} -2 \\ -2 \end{bmatrix} - \begin{bmatrix} -2 \\ 6 \end{bmatrix} - \begin{bmatrix} -6 \\ 2 \end{bmatrix} = \begin{bmatrix} 0 \\ -16 \end{bmatrix}$$

$$\hat{\hat{\mathbf{S}}}^4 + \hat{\hat{\mathbf{S}}}^5 - \hat{\hat{\mathbf{S}}}^6 - \hat{\hat{\mathbf{S}}}^7 = \begin{bmatrix} -4 \\ -4 \end{bmatrix} + \begin{bmatrix} 0 \\ 0 \end{bmatrix} - \begin{bmatrix} -4 \\ 4 \end{bmatrix} - \begin{bmatrix} -8 \\ 0 \end{bmatrix} = \begin{bmatrix} 8 \\ -8 \end{bmatrix}$$

and the inner product

$$[0 \ \ -16] \begin{bmatrix} 8 \\ -8 \end{bmatrix} = 128$$

Hence $\mathbf{Y}_2, \mathbf{Y}_6$ are not compatible.

As we saw in Section 5.5.2, there is an alternative coding scheme that can be used to represent the function, namely, $\mathbf{Y}^*$ where

$$y_i^* = \begin{cases} +1 & \text{if} \quad f(X) = 0 \\ 0 & \text{if} \quad f(X) \text{ is not specified} \\ -1 & \text{if} \quad f(X) = -1 \end{cases}$$

As before, we shall also make use of $\hat{\mathbf{Y}}^*$, which explicitly indicates the location of all the don't cares and is defined by

$$\hat{y}_i^* = 1 - (y_i^*)^2$$

If $\mathbf{S}^*$ and $\hat{\mathbf{S}}^*$ correspond to $\mathbf{Y}^*$ and $\hat{\mathbf{Y}}^*$, then

$$\mathbf{S}^* = \tfrac{1}{2}(\hat{\mathbf{S}}^* + \hat{\hat{\mathbf{S}}}), \qquad \hat{\mathbf{S}}^* = \tfrac{1}{2}(\hat{\mathbf{S}} - \hat{\hat{\mathbf{S}}})$$

Using these results, we can rewrite our condition for $\mathbf{Y}_u$ and $\mathbf{Y}_v$ to be compatible in the form

$$\left(\sum_{i=0}^{\beta} (t_{ui} - t_{vi})(\mathbf{S}^{*i} + \hat{\mathbf{S}}^{*i})^t\right)\left(\sum_{i=0}^{\beta} (t_{ui} - t_{vi})(\mathbf{S}^{*i} - \hat{\mathbf{S}}^{*i})\right) = 0$$

An easier way to state this condition is as follows:

A necessary and sufficient condition for $\mathbf{Y}_u$ and $\mathbf{Y}_v$ to be compatible is that $\mathbf{A}^t\mathbf{A} = \mathbf{B}^t\mathbf{B}$, where

$$\mathbf{A} = \sum_{i=0}^{\beta} (t_{ui} - t_{vi})\mathbf{S}^{*i}$$

and

$$\mathbf{B} = \sum_{i=0}^{\beta} (t_{ui} - t_{vi})\hat{\mathbf{S}}^{*i} \qquad (u \neq v, \quad 0 \leq u, v \leq \beta)$$

The effect of this condition is that it is a sum-of-squares condition, in that the sum of squares in **A** must be identical to the sum of squares in **B**. The procedure involved is to perform the normal symmetry test using $\mathbf{S}^*$ and repeat the procedure using $\hat{\mathbf{S}}^*$. The compatibility exists if the sums of the squares of the entries in the two resulting vectors are equal. The method is illustrated using the same example for the same two cases as above.

For the function that is illustrated in Fig. 5.25, we have

$$\mathbf{Y}^* = -1 \ -1 \ 1 \ -1 \quad -1 \ 1 \ -1 \ 1 \quad 1 \ 1 \ 1 \ -1 \quad 0 \ -1 \ 1 \ 1]^t$$

$$\hat{\mathbf{Y}}^* = \quad 0 \quad 0 \ 0 \quad 0 \quad 0 \ 0 \quad 0 \ 0 \quad 0 \ 0 \ 0 \quad 0 \quad 1 \quad 0 \ 0 \ 0]^t$$

giving the two sets of spectral coefficients,

s_0^*	s_1^*	s_2^*	s_3^*	s_4^*	s_{12}^*	s_{13}^*	s_{14}^*	s_{23}^*	s_{24}^*	s_{34}^*	s_{123}^*	s_{124}^*	s_{134}^*	s_{234}^*	s_{1234}^*
1;	1	−3	−1	−5;	−3	7	−5	3	−1	−3;	−5	−1	5	−7;	1
$\hat{s}_0^*$	$\hat{s}_1^*$	$\hat{s}_2^*$	$\hat{s}_3^*$	$\hat{s}_4^*$	$\hat{s}_{12}^*$	$\hat{s}_{13}^*$	$\hat{s}_{14}^*$	$\hat{s}_{23}^*$	$\hat{s}_{24}^*$	$\hat{s}_{34}^*$	$\hat{s}_{123}^*$	$\hat{s}_{124}^*$	$\hat{s}_{134}^*$	$\hat{s}_{234}^*$	$\hat{s}_{1234}^*$
1;	1	1	−1	−1;	1	−1	−1	−1	−1	1;	−1	−1	1	1;	1

The condition used to check for whether $\mathbf{Y}_0$ and $\mathbf{Y}_6$ are compatible comes

$\bar{x}_5$:

$x_3x_4 \backslash x_1x_2$	00	01	11	10
00	1	0	—	0
01	—	—	1	0
11	1	—	1	—
10	0	—	—	1

x_5:

$x_3x_4 \backslash x_1x_2$	00	01	11	10
00	0	0	1	—
01	—	0	—	0
11	—	1	—	1
10	0	1	1	—

Fig. 5.26. A five-variable function $f(x_1x_2x_3x_4x_5)$ with 13 don't-care minterms.

from row 15 of Table 5.6 and has the same form as we used above.

$$\mathbf{S}^{*2} + \mathbf{S}^{*3} + \mathbf{S}^{*4} + \mathbf{S}^{*5} = \begin{bmatrix} -1 \\ 7 \end{bmatrix} + \begin{bmatrix} 3 \\ -5 \end{bmatrix} + \begin{bmatrix} -5 \\ -5 \end{bmatrix} + \begin{bmatrix} -1 \\ -1 \end{bmatrix} = \begin{bmatrix} -4 \\ -4 \end{bmatrix}$$

$$\hat{\mathbf{S}}^{*2} + \hat{\mathbf{S}}^{*3} + \hat{\mathbf{S}}^{*4} + \hat{\mathbf{S}}^{*5} = \begin{bmatrix} -1 \\ -1 \end{bmatrix} + \begin{bmatrix} -1 \\ -1 \end{bmatrix} + \begin{bmatrix} -1 \\ -1 \end{bmatrix} + \begin{bmatrix} -1 \\ -1 \end{bmatrix} = \begin{bmatrix} -4 \\ -4 \end{bmatrix}$$

and in both cases the sum of squares of the entries is 32, indicating that the compatibility exists.

In the case of checking $\mathbf{Y}_2$ and $\mathbf{Y}_6$, the condition comes from row 7 of Table 5.6, giving

$$\mathbf{S}^{*4} + \mathbf{S}^{*5} - \mathbf{S}^{*6} - \mathbf{S}^{*7} = \begin{bmatrix} -5 \\ -5 \end{bmatrix} + \begin{bmatrix} -1 \\ -1 \end{bmatrix} - \begin{bmatrix} -3 \\ -5 \end{bmatrix} - \begin{bmatrix} -7 \\ 1 \end{bmatrix} = \begin{bmatrix} 4 \\ -2 \end{bmatrix}$$

$$\hat{\mathbf{S}}^{*4} + \hat{\mathbf{S}}^{*5} - \hat{\mathbf{S}}^{*6} - \hat{\mathbf{S}}^{*7} = \begin{bmatrix} -1 \\ -1 \end{bmatrix} + \begin{bmatrix} -1 \\ -1 \end{bmatrix} - \begin{bmatrix} 1 \\ 1 \end{bmatrix} - \begin{bmatrix} 1 \\ 1 \end{bmatrix} = \begin{bmatrix} -4 \\ -4 \end{bmatrix}$$

In the first case, the sum of squares is 20, while in the second, it is 32. Consequently the compatibility does not exist.

As a slightly more complex example, consider the five-variable function illustrated in Fig. 5.26. For this function we shall check for a symmetry of degree 3 using the first approach. Consider the possibility of the symmetry $f_0(x_1, x_2) = f_3(x_1, x_2)$. To see if those two subfunctions are compatible, the condition used is as specified in Table 5.6, namely, $\mathbf{S}^1 + \mathbf{S}^2 + \mathbf{S}^5 + \mathbf{S}^6$.

For this function, we have

$$\hat{\mathbf{Y}} = -1\ 1\ 1\quad 1\ \ 1\ -1\quad 1\ \ 1\quad 1\ \ 1\quad 1\ -1\quad -1\ \ 1\quad 1\ -1$$
$$1\quad 1\ \ 1\ -1\quad 1\ \ 1\quad -1\ -1\quad 1\ \ 1\quad 1\ \ 1\quad 1\ -1\ -1\quad 1]^t$$

$$\hat{\mathbf{Y}} = -1\ 1\ 1\ -1\quad 1\ -1\quad -1\ -1\quad -1\ \ 1\quad -1\ -1\quad -1\ -1\quad -1\ -1$$
$$1\ -1\quad 1\ -1\quad 1\ -1\quad -1\ -1\quad -1\ \ 1\quad 1\ -1\quad -1\ -1\ -1\ -1]^t$$

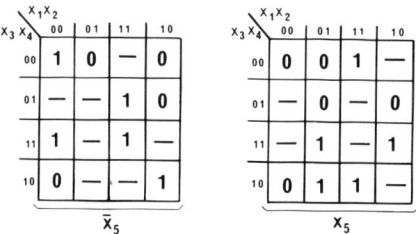

giving

$$\begin{array}{cccccccccccccccc}
\hat{s}_0 & \hat{s}_1 & \hat{s}_2 & \hat{s}_3 & \hat{s}_4 & \hat{s}_5 & \hat{s}_{12} & \hat{s}_{13} & \hat{s}_{14} & \hat{s}_{15} & \hat{s}_{23} & \hat{s}_{24} & \hat{s}_{25} & \hat{s}_{34} & \hat{s}_{35} & \hat{s}_{45} \\
12; & 4 & 4 & 8 & 0 & 0; & -4 & 0 & 0 & 0 & 0 & 0 & -8 & -4 & -4 & 4;
\end{array}$$

$$\begin{array}{cccccccccccccc}
\hat{s}_{123} & \hat{s}_{124} & \hat{s}_{125} & \hat{s}_{134} & \hat{s}_{135} & \hat{s}_{145} & \hat{s}_{234} & \hat{s}_{235} & \hat{s}_{245} & \hat{s}_{345} & \hat{s}_{1234} & \hat{s}_{1235} & \hat{s}_{1245} & \hat{s}_{1345} \\
-8 & 0 & -8 & -4 & -4 & -4 & -4 & 4 & -12 & 0; & -4 & 4 & 12 & -8;
\end{array}$$

$$\begin{array}{cc}
\hat{s}_{2345} & \hat{s}_{12345} \\
0; & -8
\end{array}$$

and

$$\begin{array}{cccccccccccccccc}
\hat{\hat{s}}_0 & \hat{\hat{s}}_1 & \hat{\hat{s}}_2 & \hat{\hat{s}}_3 & \hat{\hat{s}}_4 & \hat{\hat{s}}_5 & \hat{\hat{s}}_{12} & \hat{\hat{s}}_{13} & \hat{\hat{s}}_{14} & \hat{\hat{s}}_{15} & \hat{\hat{s}}_{23} & \hat{\hat{s}}_{24} & \hat{\hat{s}}_{25} & \hat{\hat{s}}_{34} & \hat{\hat{s}}_{35} & \hat{\hat{s}}_{45} \\
-14; & 6 & 6 & 10 & 6 & -2; & -6 & -2 & 10 & -6 & -2 & 2 & 2 & -2 & -2 & 2;
\end{array}$$

$$\begin{array}{cccccccccc}
\hat{\hat{s}}_{123} & \hat{\hat{s}}_{124} & \hat{\hat{s}}_{125} & \hat{\hat{s}}_{134} & \hat{\hat{s}}_{135} & \hat{\hat{s}}_{145} & \hat{\hat{s}}_{234} & \hat{\hat{s}}_{235} & \hat{\hat{s}}_{245} & \hat{\hat{s}}_{345} \\
-14 & 6 & -2 & 2 & -6 & -2 & -6 & 2 & -2 & 2;
\end{array}$$

$$\begin{array}{cccccc}
\hat{\hat{s}}_{1234} & \hat{\hat{s}}_{1235} & \hat{\hat{s}}_{1245} & \hat{\hat{s}}_{1345} & \hat{\hat{s}}_{2345} & \hat{\hat{s}}_{12345} \\
-2 & -2 & -6 & -2 & -2; & -6
\end{array}$$

Hence, to check whether $f_0(x_1, x_2)$ and $f_3(x_1, x_2)$ are compatible, we need

$$\hat{\mathbf{S}}^1 + \hat{\mathbf{S}}^2 + \hat{\mathbf{S}}^5 + \hat{\mathbf{S}}^6 = \begin{bmatrix} \hat{s}_3 \\ \hat{s}_{13} \\ \hat{s}_{23} \\ \hat{s}_{123} \end{bmatrix} + \begin{bmatrix} \hat{s}_4 \\ \hat{s}_{14} \\ \hat{s}_{24} \\ \hat{s}_{124} \end{bmatrix} + \begin{bmatrix} \hat{s}_{35} \\ \hat{s}_{135} \\ \hat{s}_{235} \\ \hat{s}_{1235} \end{bmatrix} + \begin{bmatrix} \hat{s}_{45} \\ \hat{s}_{145} \\ \hat{s}_{245} \\ \hat{s}_{1245} \end{bmatrix}$$

$$= \begin{bmatrix} 8 \\ 0 \\ 0 \\ -8 \end{bmatrix} + \begin{bmatrix} 0 \\ 0 \\ 0 \\ 0 \end{bmatrix} + \begin{bmatrix} -4 \\ -4 \\ 4 \\ 4 \end{bmatrix} + \begin{bmatrix} 4 \\ -4 \\ -12 \\ 12 \end{bmatrix} = \begin{bmatrix} 8 \\ -8 \\ -8 \\ 8 \end{bmatrix}$$

$$\hat{\hat{\mathbf{S}}}^1 + \hat{\hat{\mathbf{S}}}^2 + \hat{\hat{\mathbf{S}}}^5 + \hat{\hat{\mathbf{S}}}^6 = \begin{bmatrix} 10 \\ -2 \\ -2 \\ -14 \end{bmatrix} + \begin{bmatrix} 6 \\ 10 \\ 2 \\ 6 \end{bmatrix} + \begin{bmatrix} -2 \\ -6 \\ 2 \\ -2 \end{bmatrix} + \begin{bmatrix} 2 \\ -2 \\ -2 \\ -6 \end{bmatrix} = \begin{bmatrix} 16 \\ 0 \\ 0 \\ -16 \end{bmatrix}$$

and since

$$[8 \ \ -8 \ \ -8 \ \ 8] \begin{bmatrix} 16 \\ 0 \\ 0 \\ -16 \end{bmatrix} = 0$$

the compatibility exists; that is, by an appropriate choice of don't cares, we can ensure that $f_0(x_1, x_2) = f_3(x_1, x_2)$.

5.6.3 The Calculation of the Symmetries

To calculate the degree k symmetries directly, we can make use of the fast transform again, applied a number of times. We shall illustrate the procedure for the degree 3 symmetries. Looking back at Table 5.6, it will be seen that the tests can be grouped according to the particular $\mathbf{S}^i$ they use. For example, symmetries 1, 6, 9 and 12 all use $\mathbf{S}^1$, $\mathbf{S}^3$, $\mathbf{S}^5$ and $\mathbf{S}^7$, the required signs being exactly as defined by $\mathbf{T}^2$. Hence, evaluating

$$[\mathbf{S}^1\, \mathbf{S}^3\, \mathbf{S}^5\, \mathbf{S}^7]\mathbf{T}^2$$

gives the results for these four symmetries. To deduce all the symmetries requires the evaluation of seven fast transforms to give the results for the 28 possible symmetries. The details of the various tests are given in Table 5.7.

In the general case of testing for all the degree k symmetries, we shall use the T^{k-1} transform to give the results for 2^{k-1} symmetries at once. It will be necessary to evaluate $2^k - 1$ of these transforms to give complete information about all of the degree k symmetries in a particular set of k variables.

An alternative method of calculating the symmetries can be adopted if the strategy is initially to look for any degree 2 symmetries, then check for degree 3 symmetries and so on. All expressions that have to be evaluated to check for degree k symmetries are the sum or difference of the two expressions that have already been evaluated during the check for degree $(k - 1)$

Table 5.7. Tests for degree 3 symmetries in $\{x_n, x_{n-1}, x_{n-2}\}$. Each test matrix is postmultiplied by T^2, and a zero column in the resulting matrix indicates the presence of the symmetry concerned.

	Symmetry checked by			
Test matrix	Column 1	Column 2	Column 3	Column 4
$[\mathbf{S}^1\,\mathbf{S}^3\,\mathbf{S}^5\,\mathbf{S}^7]$	$\mathbf{Y}_0 = \mathbf{Y}_1$	$\mathbf{Y}_2 = \mathbf{Y}_3$	$\mathbf{Y}_4 = \mathbf{Y}_5$	$\mathbf{Y}_6 = \mathbf{Y}_7$
$[\mathbf{S}^2\,\mathbf{S}^3\,\mathbf{S}^6\,\mathbf{S}^7]$	$\mathbf{Y}_0 = \mathbf{Y}_2$	$\mathbf{Y}_1 = \mathbf{Y}_3$	$\mathbf{Y}_4 = \mathbf{Y}_6$	$\mathbf{Y}_5 = \mathbf{Y}_7$
$[\mathbf{S}^4\,\mathbf{S}^5\,\mathbf{S}^6\,\mathbf{S}^7]$	$\mathbf{Y}_0 = \mathbf{Y}_4$	$\mathbf{Y}_1 = \mathbf{Y}_5$	$\mathbf{Y}_2 = \mathbf{Y}_6$	$\mathbf{Y}_3 = \mathbf{Y}_7$
$[\mathbf{S}^1\,\mathbf{S}^2\,\mathbf{S}^5\,\mathbf{S}^6]$	$\mathbf{Y}_0 = \mathbf{Y}_3$	$\mathbf{Y}_1 = \mathbf{Y}_2$	$\mathbf{Y}_4 = \mathbf{Y}_7$	$\mathbf{Y}_5 = \mathbf{Y}_6$
$[\mathbf{S}^1\,\mathbf{S}^3\,\mathbf{S}^4\,\mathbf{S}^6]$	$\mathbf{Y}_0 = \mathbf{Y}_5$	$\mathbf{Y}_2 = \mathbf{Y}_7$	$\mathbf{Y}_1 = \mathbf{Y}_4$	$\mathbf{Y}_3 = \mathbf{Y}_6$
$[\mathbf{S}^2\,\mathbf{S}^3\,\mathbf{S}^4\,\mathbf{S}^5]$	$\mathbf{Y}_0 = \mathbf{Y}_6$	$\mathbf{Y}_1 = \mathbf{Y}_7$	$\mathbf{Y}_2 = \mathbf{Y}_4$	$\mathbf{Y}_3 = \mathbf{Y}_5$
$[\mathbf{S}^1\,\mathbf{S}^2\,\mathbf{S}^4\,\mathbf{S}^7]$	$\mathbf{Y}_0 = \mathbf{Y}_7$	$\mathbf{Y}_2 = \mathbf{Y}_5$	$\mathbf{Y}_3 = \mathbf{Y}_4$	$\mathbf{Y}_1 = \mathbf{Y}_6$

symmetries. This is true for all degree k symmetries except those of distance k. For these, the relevant expressions follow as sums or differences of expressions that have been evaluated earlier in the degree k check procedure. However, it is not clear that this approach is as computationally attractive as the method of direct calculation using the $\mathbf{T}^{k-1}$.

5.6.4 An Example

This example is an eight-variable function that is a checker for part of the Canadian postal code. The function is completely specified and Table 5.8 lists all the input assignments for which the function takes the value 1. For all other assignments, it assumes the value 0. A dash in the table indicates a variable that may take either the value 0 or 1.

For this function, there are no two-variable symmetries and no three-variable symmetries but there are four-variable symmetries, and we show how these may be used to construct a realization of the checker using four input read-only memories (ROMs).

The four-variable symmetries for x_1, x_2, x_3, x_4 and for x_5, x_6, x_7, x_8 are listed in Tables 5.9 and 5.10, respectively, these being the most useful sets of four variables. Entries of 1 in the table indicate multivariable symmetries, so, for example, in Table 5.9 the 1 in the second row and fifth column indicates that

$$f(0, 0, 0, 1, x_5, x_6, x_7, x_8) = f(0, 1, 0, 0, x_5, x_6, x_7, x_8)$$

while the 0 in the second row and fourth column indicates that

$$f(0, 0, 0, 1, x_5, x_6, x_7, x_8) \neq f(0, 0, 1, 1, x_5, x_6, x_7, x_8)$$

The most important point for circuit synthesis is that each of these tables

Table 5.8. Example eight-variable function.

x_1	x_2	x_3	x_4	x_5	x_6	x_7	x_8
0	0	0	0	—	—	—	—
—	—	—	—	1	1	1	1
1	1	1	1	—	—	—	—
—	1	0	0	—	0	0	0
0	0	0	1	—	0	0	0
1	0	0	0	—	0	0	0
1	0	0	0	0	0	0	1
1	0	0	0	1	1	0	0

only has four distinct columns so we may use a 4-input, 2-output ROM for each with appropriate coding (different columns from the table must have different assignments from the ROM). One possible assignment set is listed in Table 5.11. Of course, once we have this information, we are not limited to a ROM realization, and an AND-OR-NOT circuit is deduced very easily from the expressions:

$$x_9 = \bar{x}_1\bar{x}_2\bar{x}_3 + x_2\bar{x}_3\bar{x}_4 + x_1x_2x_3x_4$$

$$x_{10} = \bar{x}_2\bar{x}_3\bar{x}_4 + x_1x_2x_3x_4$$

$$x_{11} = \bar{x}_5\bar{x}_6\bar{x}_7 + x_5\bar{x}_7\bar{x}_8$$

$$x_{12} = \bar{x}_6\bar{x}_7\bar{x}_8 + x_5x_6x_7x_8$$

$$f(X) = x_9x_{10} + \bar{x}_{11}x_{12} + x_9x_{12} + x_{10}x_{11}$$

The ROM circuit is in Fig. 5.27.

Table 5.9. Four-variable symmetries in x_1, x_2, x_3, x_4.

			x_1	0	0	0	0	0	0	0	0	1	1	1	1	1	1	1	1
			x_2	0	0	0	0	1	1	1	1	0	0	0	0	1	1	1	1
			x_3	0	0	1	1	0	0	1	1	0	0	1	1	0	0	1	1
			x_4	0	1	0	1	0	1	0	1	0	1	0	1	0	1	0	1
x_1	x_2	x_3	x_4																
0	0	0	0	1	0	0	0	0	0	0	0	0	0	0	0	0	0	0	1
0	0	0	1	0	1	0	0	1	0	0	0	0	0	0	0	1	0	0	0
0	0	1	0	0	0	1	1	0	1	1	1	0	1	1	1	0	1	1	0
0	0	1	1	0	0	1	1	0	1	1	1	0	1	1	1	0	1	1	0
0	1	0	0	0	1	0	0	1	0	0	0	0	0	0	0	1	0	0	0
0	1	0	1	0	0	1	1	0	1	1	1	0	1	1	1	0	1	1	0
0	1	1	0	0	0	1	1	0	1	1	1	0	1	1	1	0	1	1	0
0	1	1	1	0	0	1	1	0	1	1	1	0	1	1	1	0	1	1	0
1	0	0	0	0	0	0	0	0	0	0	1	0	0	0	0	0	0	0	0
1	0	0	1	0	0	1	1	0	1	1	1	0	1	1	1	0	1	1	0
1	0	1	0	0	0	1	1	0	1	1	1	0	1	1	1	0	1	1	0
1	0	1	1	0	0	1	1	0	1	1	1	0	1	1	1	0	1	1	0
1	1	0	0	0	1	0	0	1	0	0	0	0	0	0	0	1	0	0	0
1	1	0	1	0	0	1	1	0	1	1	1	0	1	1	1	0	1	1	0
1	1	1	0	0	0	1	1	0	1	1	1	0	1	1	1	0	1	1	0
1	1	1	1	1	0	0	0	0	0	0	0	0	0	0	0	0	0	0	1

Table 5.10. Four-variable symmetries in $x_5, x_6, x_7 . x_8$.

			x_5	0	0	0	0	0	0	0	0	1	1	1	1	1	1	1	1
			x_6	0	0	0	0	1	1	1	1	0	0	0	0	1	1	1	1
			x_7	0	0	1	1	0	0	1	1	0	0	1	1	0	0	1	1
			x_8	0	1	0	1	0	1	0	1	0	1	0	1	0	1	0	1
x_5	x_6	x_7	x_8																
0	0	0	0	1	0	0	0	0	0	0	0	1	0	0	0	0	0	0	0
0	0	0	1	0	1	0	0	0	0	0	0	0	0	0	0	1	0	0	0
0	0	1	0	0	0	1	1	1	1	1	1	0	1	1	1	0	1	1	0
0	0	1	1	0	0	1	1	1	1	1	1	0	1	1	1	0	1	1	0
0	1	0	0	0	0	1	1	1	1	1	1	0	1	1	1	0	1	1	0
0	1	0	1	0	0	1	1	1	1	1	1	0	1	1	1	0	1	1	0
0	1	1	0	0	0	1	1	1	1	1	1	0	1	1	1	0	1	1	0
0	1	1	1	0	0	1	1	1	1	1	1	0	1	1	1	0	1	1	0
1	0	0	0	1	0	0	0	0	0	0	0	1	0	0	0	0	0	0	0
1	0	0	1	0	0	1	1	1	1	1	1	0	1	1	1	0	1	1	0
1	0	1	0	0	0	1	1	1	1	1	1	0	1	1	1	0	1	1	0
1	0	1	1	0	0	1	1	1	1	1	1	0	1	1	1	0	1	1	0
1	1	0	0	0	1	0	0	0	0	0	0	0	0	0	0	1	0	0	0
1	1	0	1	0	0	1	1	1	1	1	1	0	1	1	1	0	1	1	0
1	1	1	0	0	0	1	1	1	1	1	1	0	1	1	1	0	1	1	0
1	1	1	1	0	0	0	0	0	0	0	0	0	0	0	0	0	0	0	1

Table 5.11. Assignment sets for the three ROMs.

x_1	x_2	x_3	x_4	x_9	x_{10}	x_5	x_6	x_7	x_8	x_{11}	x_{12}	x_9	x_{10}	x_{11}	x_{12}	f
0	0	0	0	1	1	0	0	0	0	1	1	0	0	0	0	0
0	0	0	1	1	0	0	0	0	1	1	0	0	0	0	1	1
0	0	1	0	0	0	0	0	1	0	0	0	0	0	1	0	0
0	0	1	1	0	0	0	0	1	1	0	0	0	0	1	1	0
0	1	0	0	1	0	0	1	0	0	0	0	0	1	0	0	0
0	1	0	1	0	0	0	1	0	1	0	0	0	1	0	1	1
0	1	1	0	0	0	0	1	1	0	0	0	0	1	1	0	1
0	1	1	1	0	0	0	1	1	1	0	0	0	1	1	1	1
1	0	0	0	0	1	1	0	0	0	1	1	1	0	0	0	0
1	0	0	1	0	0	1	0	0	1	0	0	1	0	0	1	1
1	0	1	0	0	0	1	0	1	0	0	0	1	0	1	0	0
1	0	1	1	0	0	1	0	1	1	0	0	1	0	1	1	1
1	1	0	0	1	0	1	1	0	0	1	0	1	1	0	0	1
1	1	0	1	0	0	1	1	0	1	0	0	1	1	0	1	1
1	1	1	0	0	0	1	1	1	0	0	0	1	1	1	0	1
1	1	1	1	1	1	1	1	1	1	0	1	1	1	1	1	1

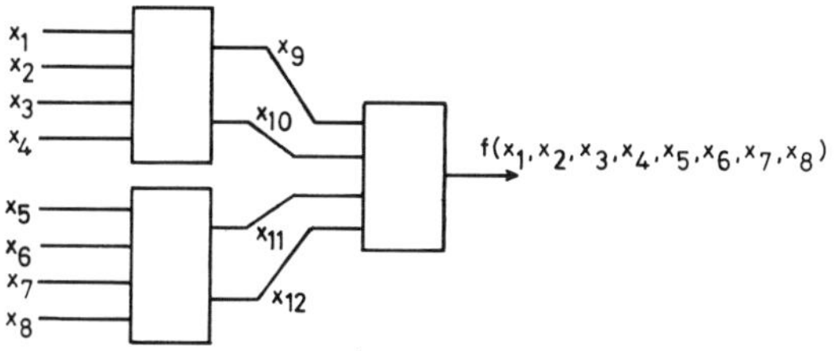

Fig. 5.27. Final example: ROM circuit realisation for Canadian Postal Code checker.

5.7 CHAPTER SUMMARY

In this chapter we have considered symmetries in Boolean functions and their detection from the spectrum. The symmetries of degree 2 have been demonstrated to be of considerable value in the synthesis of Boolean functions. Of course it is not necessary to use the spectrum for their detection, and they can be derived directly from the minterms of the function. An alternative technique based on compatibility arguments is given by Miller and Muzio.[12]

The results that we have described can be generalized to multiple-valued logic where analgous tests have been derived by Miller.[13] More complexity is involved because of the nature of the spectrum for multiple-valued logic. If the reader looks at the equations used for verifying the presence of symmetries, it will be seen that some coefficients occur more frequently than others. For example, all complete checks for the presence of single-variable degree 2 symmetries make use of the high-order coefficient. More detail as to how the symmetry information is distributed in the spectrum is given by Miller and Muzio.[14]

The synthesis methods described in this chapter complement those in the preceding chapter to give a good method. The use of the spectrum to detect properties of the underlying Boolean function that are valuable for the synthesis procedure has been demonstrated. However, the symmetries discussed in this chapter can also be detected by alternative approaches, while the methods of the last chapter based on Exclusive-OR gates are not easily performed in the Boolean domain.

The idea of higher-degree symmetries and their spectral detection is a comparatively new one and more study is required to determine the best way for them to be used in synthesis techniques. There are such a large number of possibilities that it is unlikely that a complete evaluation of all of them would ever be useful.

References

1. Dietmeyer, D. L., "Logic Design of Digital Systems." Allyn and Bacon, Boston, 1971.
2. McClusky, E. J., Detection of group invariance or total symmetry of a Boolean function, *Bell Syst. Tech. J.* **35,** 1445–1453 (1956).
3. Marcus, M. P., The detection and identification of symmetric switching functions with the use of tables of combinations, *Trans. IRE* **EC-5,** 237–239 (1956).
4. Mukhopadhyay, A., Detection of total or partial symmetry of a switching function with the use of decomposition charts, *Trans. IEEE* **EC-12,** 553–557 (1963).
5. Das, S. R., and Sheng, C. L., On detectiag total or partial symmetry of switching functions, *Trans. IEEE* **C-20,** 352–355 (1971).
6. Edwards, C. R., The design of easily tested circuits using mapping and spectral techniques, *Radio Electron. Eng.* **47,** 321–342 (1977).
7. Edwards, C. R., and Hurst, S. L., Preliminary considerations of the combinatorial and sequential digital systems under symmetry methods, *Internat. J. Electron.* **40,** 499–507 (1976).
8. Hurst, S. L., "Logical Processing of Digital Signals." Crane Russak, New York, 1978.
9. Hurst, S. L., Detection of symmetries in combinatorial functions by spectral means, *IEE J. Electronic Circs. Sys.* **1,** 173–180 (1977).
10. Edwards, C. R., and Hurst, S. L., A digital synthesis procedure under function symmetries and mapping methods, *IEEE Trans. Comput.* **C-27,** 985–997 (1978).
11. Muzio, J. C., Miller, D. M., and Hurst, S. L., Multi-variable symmetries and their detection, *IEE Proc. Part E,* **130,** 141–148 (1983).
12. Miller, D. M., and Muzio, J. C., Detection of symmetries in totally specified or partially specified combinational functions, *IEE J. Comput. Digital Techniq.* **2,** 203–209 (1979).
13. Miller, D. M., Spectral symmetry tests, *Proc. 11th Internat. Symp. Multiple-Valued Logic* 130–134 (1981).
14. Miller, D. M., and Muzio, J. C., The distribution of symmetry information in the spectrum of a Boolean function, *Electron. Lett.* **15,** 816–817 (1979).

6

Fault Detection

6.1 INTRODUCTION

Consider a combinational network designed to realize $f(X)$. A permanent functional fault is a failure that causes the network to realize a faulty function $\overset{*}{f}(X)$, which differs from $f(X)$ for one or more assignments to X. The fault detection problem is to construct and, subsequently, to execute an efficient, cost-effective test to verify that a network realizes $f(X)$ and not some $\overset{*}{f}(X)$.

A fault model is a logical abstraction describing the functional effect of physical failures. The most widely adopted fault model assumes that the physical failures cause certain network lines to be stuck-at-0 or stuck-at-1.[1–5] A single stuck-at fault is one where precisely one line is stuck, although the effect of the fault may fan out to other lines. A network with p lines has $2p$ potential single stuck-at faults.

A network with p lines has $3^p - 1$ potential multiple stuck-at faults. This large value means the detailed analysis of multiple stuck-at faults is only practical for small networks.

We shall concentrate on the single stuck-at fault model but also consider bridging faults,[6–10] which model shorts, and contact faults which correspond to missing or extra connections in the AND or OR section of a programmable logic array (PLA).[11–13]

The traditional approach to fault detection[1–5] involves a test set that is a subset of the input assignments to the network. Each assignment in the test set is termed a test vector. These vectors are selected so that at least one will produce an erroneous response for a fault in the target fault class. It has been estimated[14] that for a network with p lines, the computation involved in determining a test set to detect single stuck-at faults is on the order of p^3. While this is large, it is a one-time cost.

A more serious drawback to the test set is the sheer volume of data to be handled at test time. The test vectors must be stored and applied one at a time. The response to each must then be compared to a stored correct response. While this is practical if a computer-based test unit is used, it is not for field testing or for on-board or on-chip self-testing. As digital systems become more complex with more functionality on each chip, self-testing will become increasingly more important. The techniques presented below are suited to self-test since they require little test hardware and avoid the concept of the test set and its associated problems.

The discussion will be entirely concerned with combinational networks. However, it is now common for a sequential network to be designed in such a way that it can be tested in a purely combinational manner. In level sensitive scan design,[15–19] for example, the data latches in the feedback loops can be operated in the normal manner or in a test mode where they are controlled by two clocks. This allows the contents of each latch to be examined or loaded, effectively breaking the feedback paths. This converts the sequential network to a combinational network. The latches are connected into shift registers so that data can be scanned in or out of several latches via four pins: two data and two control clocks.

We shall begin by considering single-output networks. The extension to multiple-output networks is straightforward. The principal decision is whether to test the outputs one at a time or in parallel.

Section 6.2 will present spectral characterizations of the logical faults considered. In Sections 6.3 and 6.4 these characterizations will be used to develop a variety of fault-detection techniques. Finally, Section 6.5 will examine a variety of aspects, such as the implementation of the tests and fault equivalence.

Throughout the development we restrict our attention to irredundant networks. A network is redundant if it contains one or more lines whose values can be fixed to 0 or 1 with no effect on the functional behaviour of the network. Faults on redundant lines cannot be detected by the techniques discussed here.

6.2 ANALYSIS OF FAULTS

6.2.1 Input Stuck-at Faults

An input stuck-at fault is a failure that causes one or more network inputs to be permanently fixed at 0 or 1. If the network is designed to realize $f(X)$, such a fault will cause it to realize $\overset{*}{f}(X)$, which will be independent of the stuck inputs and will in fact simply be a subfunction of $f(X)$. Here we examine the description of such a fault in the spectral domain. The results will be presented in $\{0, 1\}$ coding since it is more closely related to the actual implementation of the tests than the alternative $\{+1, -1\}$ coding.

We shall begin by considering single-input stuck-at faults, which we denote as $x_i/0$ (x_i stuck-at-0) and $x_i/1$ (x_i stuck-at-1), $1 \leq i \leq n$. Since permuting the variables of a function simply results in an equivalent reordering of the spectral coefficients, it will be sufficient to consider $x_n/0$ and $x_n/1$.

Consider a network realizing the function $f(X)$. For $x_n/0$, the network realizes a faulty function $\overset{*}{f}(X)$ whose truthtable column vector is given by

$$\overset{*}{\mathbf{Z}} = \begin{bmatrix} \mathbf{Z}_0 \\ \mathbf{Z}_0 \end{bmatrix} \tag{6.1}$$

where $\mathbf{Z}_0$ is the minterm vector of $f_0(x_i, x_2, \ldots, x_{n-1}) = f(x_1, x_2, \ldots, x_{n-1}, 0)$. $\overset{*}{f}(X)$ is clearly independent of x_n, but we express it as a function of all n variables, since during the testing process the network must be assumed to realize a function of n variables. For $x_n/1$,

$$\overset{*}{\mathbf{Z}} = \begin{bmatrix} \mathbf{Z}_1 \\ \mathbf{Z}_1 \end{bmatrix} \tag{6.2}$$

where $\mathbf{Z}_1$ is the minterm vector of $f_1(x_1, x_2, \ldots, x_{n-1}) = f(x_1, x_2, \ldots, x_{n-1}, 1)$.

Since

$$\mathbf{R} = \begin{bmatrix} \mathbf{R}^0 \\ \mathbf{R}^1 \end{bmatrix} = \begin{bmatrix} \mathbf{T}^{n-1} & \mathbf{T}^{n-1} \\ \mathbf{T}^{n-1} & -\mathbf{T}^{n-1} \end{bmatrix} \begin{bmatrix} \mathbf{Z}_0 \\ \mathbf{Z}_1 \end{bmatrix}$$

$$\mathbf{R}_0 = [\mathbf{T}^{n-1}]\,\mathbf{Z}_0] \qquad \text{and} \qquad \mathbf{R}_1 = [\mathbf{T}^{n-1}]\,\mathbf{Z}_1]$$

we have

$$\mathbf{R}^0 = \mathbf{R}_0 + \mathbf{R}_1 \qquad \text{and} \qquad \mathbf{R}^1 = \mathbf{R}_0 - \mathbf{R}_1 \tag{6.3}$$

or, equivalently,

$$\mathbf{R}_0 = \tfrac{1}{2}(\mathbf{R}^0 + \mathbf{R}^1) \qquad \text{and} \qquad \mathbf{R}_1 = \tfrac{1}{2}(\mathbf{R}^0 - \mathbf{R}^1) \tag{6.4}$$

Therefore, for $x_n/0$,

$$\overset{*}{\mathbf{R}} = \begin{bmatrix} \overset{*}{\mathbf{R}}{}^0 \\ \overset{*}{\mathbf{R}}{}^1 \end{bmatrix} = \begin{bmatrix} \mathbf{T}^{n-1} & \mathbf{T}^{n-1} \\ \mathbf{T}^{n-1} & -\mathbf{T}^{n-1} \end{bmatrix} \begin{bmatrix} \mathbf{Z}_0 \\ \mathbf{Z}_0 \end{bmatrix}$$

and thus

$$\overset{*}{\mathbf{R}}{}^0 = 2\mathbf{R}_0 = \mathbf{R}^0 + \mathbf{R}^1 \qquad \text{and} \qquad \overset{*}{\mathbf{R}}{}^1 = \mathbf{0} \tag{6.5}$$

For $x_n/1$,

$$\overset{*}{\mathbf{R}}{}^0 = 2\mathbf{R}_1 = \mathbf{R}^0 - \mathbf{R}^1 \qquad \text{and} \qquad \overset{*}{\mathbf{R}}{}^1 = \mathbf{0} \tag{6.6}$$

For any set of subscripts $\alpha \subseteq \{1, 2, \ldots, n-1\}$, Eqs. (6.5) and (6.6) yield

$$\overset{*}{r}_\alpha = r_\alpha \pm r_{\alpha n}, \qquad \overset{*}{r}_{\alpha n} = 0 \tag{6.7}$$

where the sign is + for $x_n/0$ and − for $x_n/1$. Note that α may be empty, in which case $r_\alpha = r_0$ and $r_{\alpha n} = r_n$.

A fault is termed r_β-testable, $\beta \subseteq \{1, 2, \ldots, n\}$, if $\mathring{r}_\beta \neq r_\beta$. From Eq. (6.7), $x_n/0$ and $x_n/1$ are r_α-testable, $n \notin \alpha$, and $r_{\alpha n}$-testable if and only if $r_{\alpha n} \neq 0$. By permutation of variables, we thus have.

Criterion 6.1. $x_i/0$ and $x_i/1$ are r_α-testable, $\alpha \subseteq \{1, 2, \ldots, i-1, i+1, \ldots, n\}$, and $r_{\alpha i}$-testable if and only if $r_{\alpha i} \neq 0$.

As an example, consider the function

$$f(x) = x_1x_2x_3 + x_4(\bar{x}_2 + \bar{x}_3)$$

which has the spectrum

r_0	r_1	r_2	r_3	r_4	r_{12}	r_{13}	r_{14}	r_{23}	r_{24}	r_{34}	r_{123}	r_{124}	r_{134}	r_{234}	r_{1234}
8;	−2	0	0	−6;	2	2	0	0	−2	−2;	−2	0	0	2;	0

Applying Criterion 6.1, we find the results shown in Table 6.1.

A multiple-input stuck-at fault is a fault where one or more input lines are stuck. Single-input stuck-at faults are a special case. The fault is termed

Table 6.1. Spectral coefficients testing single-input stuck-at faults for a network realizing $f(X) = x_1x_2x_3 + x_4(\bar{x}_2 + \bar{x}_3)$.

Faulty input	Testing coefficients	Criterion
x_1	r_0, r_1	$r_1 \neq 0$
	r_2, r_{12}	$r_{12} \neq 0$
	r_3, r_{13}	$r_{13} \neq 0$
	r_{23}, r_{123}	$r_{123} \neq 0$
x_2	r_1, r_{12}	$r_{12} \neq 0$
	r_4, r_{24}	$r_{24} \neq 0$
	r_{13}, r_{123}	$r_{123} \neq 0$
	r_{34}, r_{234}	$r_{234} \neq 0$
x_3	r_1, r_{13}	$r_{13} \neq 0$
	r_4, r_{34}	$r_{34} \neq 0$
	r_{12}, r_{123}	$r_{123} \neq 0$
	r_{24}, r_{234}	$r_{234} \neq 0$
x_4	r_0, r_4	$r_4 \neq 0$
	r_2, r_{24}	$r_{24} \neq 0$
	r_3, r_{34}	$r_{34} \neq 0$
	r_{23}, r_{234}	$r_{234} \neq 0$

unidirectional if all stuck lines take the same value; otherwise it is termed mixed.

We shall consider faults where each x_{m+i}/u_i, $1 \le i \le n - m$, $m < n$, $u_i = 0$ or 1. For convenience, we shall refer to the fault as $x_{m+1}, \ldots, x_n$ stuck-at-u, denoted $(x_{m+1}, \ldots, x_n)/u$, $u = \sum_{i=1}^{n-m} u_i 2^{i-1}$. This class of multiple-input stuck-at faults is sufficient, since all others are equivalent up to permutation of variables.

Consider the partition

$$\mathbf{Z} = \begin{bmatrix} \mathbf{Z}_0 \\ \mathbf{Z}_1 \\ \vdots \\ \mathbf{Z}_\beta \end{bmatrix}, \qquad \beta = 2^{n-m} - 1$$

where each $\mathbf{Z}_i$, $0 \le i \le \beta$, has 2^m elements. $\mathbf{Z}_u$ is the truthtable column vector of

$$f_u(x_1, x_2, \ldots, x_m) = f(x_1, x_2, \ldots, x_m, u_1, u_2, \ldots, u_{n-m})$$

with $u = \sum_{i=1}^{n-m} u_i 2^{i-1}$. It defines the function that a network realizing $f(X)$ in the fault-free case will realize for the fault $(x_{m+1}, \ldots, x_n)/u$. The spectrum of $f_u(x_1, x_2, \ldots, x_n)$ is given by

$$\mathbf{R}_u = \mathbf{T}^m \mathbf{Z}_u$$

We partition $\mathbf{R}$ such that

$$\mathbf{R} = \begin{bmatrix} \mathbf{R}^0 \\ \mathbf{R}^1 \\ \vdots \\ \mathbf{R}^\beta \end{bmatrix}, \qquad \beta = 2^{n-m} - 1$$

where the $\mathbf{R}^i$, $0 \le i \le \beta$, each have 2^m coefficients. In Section 4.2, we found

$$[\mathbf{R}_0 \mathbf{R}_1 \ldots \mathbf{R}_\beta] = \frac{1}{2^{n-m}} [\mathbf{R}^0 \mathbf{R}^1 \ldots \mathbf{R}^\beta] \mathbf{T}^{n-m} \tag{6.9}$$

Expression (6.4), which dealt with single stuck-at faults, is the special case of Eq. (6.9) with $m = n - 1$.

For $(x_{m+1}, x_{m+2}, \ldots, x_n)/u$, the network realizes a faulty function $\overset{*}{f}(X)$ such that

$$\overset{*}{\mathbf{R}}_0 = \overset{*}{\mathbf{R}}_1 = \cdots \overset{*}{\mathbf{R}}_\beta = \mathbf{R}_u \tag{6.10}$$

From Eq. (6.9), we have

$$[\overset{*}{\mathbf{R}}{}^0 \overset{*}{\mathbf{R}}{}^1 \ldots \overset{*}{\mathbf{R}}{}^\beta] = [\overset{*}{\mathbf{R}}_0 \overset{*}{\mathbf{R}}_1 \ldots \overset{*}{\mathbf{R}}_\beta] \mathbf{T}^{n-m}$$

which together with Eq. (6.10) yields

$$\overset{*}{\mathbf{R}}{}^0 = 2^{n-m}\mathbf{R}_u, \qquad \overset{*}{\mathbf{R}}{}^i = \mathbf{0}, \quad 1 \le i \le \beta \tag{6.11}$$

Combining Eqs. (6.11) and (6.9) yields

$$\overset{*}{\mathbf{R}}{}^0 = [\mathbf{R}^0\mathbf{R}^1 \ldots \mathbf{R}^\beta]\mathbf{T}_{*u}^{n-m}, \qquad \overset{*}{\mathbf{R}}{}^i = \mathbf{0}, \quad 1 \le i \le \beta \tag{6.12}$$

From Eq. (6.11), it is clear that any multiple-input stuck-at fault involving p variables ($p = n - m$) causes the network to realize a faulty function with all its spectral coefficients integer multiples of 2^p. This yields the following criterion.

Criterion 6.2. A multiple-input fault involving p or more inputs is r_α-testable, $\alpha \subseteq \{1, 2, \ldots, n\}$, if r_α is not an integer multiple of 2^p.

For our previous example, $f(x) = x_1x_2x_3 + x_4(\bar{x}_2 + \bar{x}_3)$,

r_0	r_1	r_2	r_3	r_4	r_{12}	r_{13}	r_{14}	r_{23}	r_{24}	r_{34}	r_{123}	r_{124}	r_{134}	r_{234}	r_{1234}
8;	−2	0	0	−6;	2	2	0	0	−2	−2;	−2	0	0	2;	0

it is clear that $r_1, r_4, r_{12}, r_{13}, r_{24}, r_{34}, r_{123}$, and r_{234} each test all multiple faults involving two or more inputs. r_0 tests those faults involving all four inputs. The criterion says nothing about the remaining coefficients. It is important to note that Criterion 6.2 is a sufficient but not necessary condition.

It is easily shown by induction on the number of inputs that a fan-out-free network devoid of Exclusive-OR and Exclusive-NOR gates with two or more inputs must realize a function with an odd number of minterms. Since r_0 is odd for such a function, all spectral coefficients are odd.† Since all stuck-at faults, single or multiple, in a fan-out-free network are functionally equivalent to single or multiple stuck-at input faults, it follows that all stuck-at faults in a fan-out-free network are r_α-testable for all $\alpha \subseteq \{1, 2, \ldots, n\}$. This is a generalization of the result due to Tzidon *et al.*,[21] which showed the r_0-testability of such faults.

Several authors[21–23] have stated the result that all single- and multiple-input faults are r_0-testable if r_0 is odd. Criterion 6.2 generalizes this result both to the spectral coefficients other than r_0 and to the case of r_0 even.

When r_0 is even, Criterion 6.2 leaves the testability of certain multiple-input stuck-at faults an open question. For $\alpha \subseteq \{1, 2, \ldots, m\}$ Eq. (6.12) yields

$$r_\alpha = [r_{\gamma_0}, r_{\gamma_1}, \ldots, r_{\gamma_\beta}]\mathbf{T}_{*u}^{n-m}$$

where $\gamma_0 = \alpha$, $\gamma_1 = \alpha \cup \{x_{m+1}\}$, $\gamma_2 = \alpha \cup \{x_{m+2}\}$, $\gamma_3 = \alpha \cup \{x_{m+1}, x_{m+2}\}$,

† This follows since r_0 is the number of ones in the function, in which case every other coefficient is the difference between an even and an odd number.

$\ldots, \gamma_\beta = \alpha \cup \{x_{m+1}, \ldots, x_n\}$. Since a fault is r_α-testable if and only if $\mathring{r}_\alpha \neq r_\alpha$, by permutation of variables we have the following general criterion.

Criterion 6.3. $(x_{i_1}, x_{i_2}, \ldots, x_{i_q})/u$ is r_α-testable, $\alpha \cap \{i_1, i_2, \ldots, i_q\} = \phi$ if and only if

$$r_\alpha \neq [r_{\gamma_0}, r_{\gamma_1}, \ldots, r_{\gamma_\beta}]\mathbf{T}^q_{*u} \tag{6.13}$$

where $\gamma_0 = \alpha$, $\gamma_1 = \alpha \cup \{i_1\}$, $\gamma_2 = \alpha \cup \{i_2\}$, $\gamma_3 = \alpha \cup \{i_1, i_2\}, \ldots, \gamma_\beta = \alpha \cup \{i_1, i_2, \ldots, i_q\}$, $\beta = 2^q - 1$. Criterion 6.1 is the special case of Criterion 6.3, where $q = 1$, i.e., single-input faults.

As an example, consider the fault $(x_1, x_2)/2$, i.e., $x_1/1, x_2/0$, for the function

$$f(X) = x_1x_2x_3 + x_4(\bar{x}_2 + \bar{x}_3)$$

Applying Criterion 6.3, this fault is r_0-testable if and only if

$$r_0 \neq [r_0 r_1 r_2 r_{12}]\mathbf{T}^2_{*2}$$

Substituting the appropriate values yields

$$8 \neq [8 \quad -2 \quad 0 \quad 2]\begin{bmatrix} 1 \\ 1 \\ -1 \\ -1 \end{bmatrix}$$

so the fault is r_0-testable. $(x_1, x_2)/0$ is not, since

$$8 = [8 \quad -2 \quad 0 \quad 2]\begin{bmatrix} 1 \\ 1 \\ 1 \\ 1 \end{bmatrix}$$

A network with n inputs has $3^n - 1$ multiple-input stuck-at faults, including single-input stuck-at faults of which there are $2n$. For large n, there are too many faults to be considered individually. Our interest in multiple-input stuck-at faults stems from the fact that certain single stuck-at faults on internal lines are functionally equivalent to input stuck-at faults. We shall show in the next section that it is, in general, easier to determine the testability of the input fault than it is to determine the testability of the corresponding internal fault.

It is clear from Eqs. (6.7) and (6.12) that an input stuck-at fault (single or multiple) involving x_i yields $\mathring{r}_\alpha = 0$ for all α, $i \in \alpha$. We thus have the following criterion.

Criterion 6.4. Any input stuck-at fault (single or multiple) involving x_i is r_α-testable, $i \in \alpha$, if and only if $r_\alpha \neq 0$.

Input stuck-at faults are "functional" faults in the sense that their effect depends only on the function in question or, in our context, its spectrum. They are independent of the details of the network used to realize the function. This is of course not true for faults on internal network lines. The latter are thus more complex to analyze.

6.2.2 Internal Stuck-at Faults

A single internal stuck-at fault is a stuck-at fault on any network line except a primary input. The network output is considered an internal line. For a particular line g in a network, we adopt the functional model

$$f(X) = h(X, g(X)) \tag{6.14}$$

where $g(X)$ is the function realized by line g and $h(X, g)$ is the function realized by the residual network found by treating g as an input.

It is important to note that $g(X)$ and $h(X, g)$ are defined by the network in question and not by $f(X)$ alone. Either $g(X)$ or $h(X, g)$ may be independent of one or more x_i, $1 \leq i \leq n$.

As an example, consider the network of Fig. 6.1. For the line labeled g.

$$g(X) = \overline{x_2 x_3} \qquad \text{and} \qquad h(X, g) = x_1\bar{g} + x_4(\bar{x}_2 + \bar{x}_3)$$

Note that, as this example shows, $g(X)$ and $h(X, g)$ may have common inputs and their realization may share gates.

Eqι ιtion (6.14) may be written as

$$f(X) = h_0(X)\overline{g(X)} + h_1(X)g(X) \tag{6.15}$$

where $h_0(X) = h(X, 0)$ and $h_1 = h(X, 1)$. Let $\hat{\mathbf{R}}$ be the spectrum of $h(X, g)$ and let $\mathbf{R}^g$ be the spectrum of $g(X)$. In a similar manner to $f(X)$, we have $\hat{\mathbf{R}}_0$ and $\hat{\mathbf{R}}_1$ the spectra of $h_0(X)$ and $h_1(X)$, respectively. From (6.9),

$$\hat{\mathbf{R}}_0 = \tfrac{1}{2}(\hat{\mathbf{R}}^0 + \hat{\mathbf{R}}^1) \qquad \text{and} \qquad \hat{\mathbf{R}}_1 = \tfrac{1}{2}(\hat{\mathbf{R}}^0 - \hat{\mathbf{R}}^1) \tag{6.16}$$

where

$$\hat{\mathbf{R}} = \begin{bmatrix} \hat{\mathbf{R}}^0 \\ \hat{\mathbf{R}}^1 \end{bmatrix}$$

Transforming Eq. (6.15) to the spectral domain using the computational procedures discussed in Section 2.5 gives

$$\mathbf{R} = \frac{1}{2^n}\hat{\mathbf{R}}_0 * (\mathbf{J} - \mathbf{R}^g) + \frac{1}{2^n}\hat{\mathbf{R}}_1 * \mathbf{R}^g$$

$$= \hat{\mathbf{R}}_0 - \frac{1}{2^n}\hat{\mathbf{R}}_0 * \mathbf{R}^g + \frac{1}{2^n}\hat{\mathbf{R}}_1 * \mathbf{R}^g$$

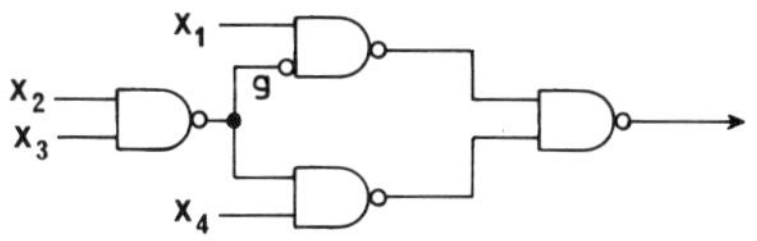

Fig. 6.1. Network realizing $f(X) = x_1x_2x_3 + x_4(\bar{x}_2 + \bar{x}_3)$. g identifies the faulty line in question.

Substituting in Eq. (6.16) and rearranging gives

$$\mathbf{R} = \frac{1}{2}\hat{\mathbf{R}}^0 + \frac{1}{2}\hat{\mathbf{R}}^1 - \frac{1}{2^n}\hat{\mathbf{R}}^1 * \mathbf{R}^g \tag{6.17}$$

Suppose g is stuck, in which case Eq. (6.17) becomes

$$\overset{*}{\mathbf{R}} = \frac{1}{2}\hat{\mathbf{R}}^0 + \frac{1}{2}\hat{\mathbf{R}}^1 - \frac{1}{2^n}\hat{\mathbf{R}}^1 * \overset{*}{\mathbf{R}}{}^g \tag{6.18}$$

where $\overset{*}{\mathbf{R}}{}^g$ is the spectrum of the faulty $g(X)$. For $g/0$, $\overset{*}{\mathbf{R}}{}^g = \mathbf{0}$ and

$$\overset{*}{\mathbf{R}} = \tfrac{1}{2}(\hat{\mathbf{R}}^0 + \hat{\mathbf{R}}^1)$$

For $g/1$, $\overset{*}{\mathbf{R}}{}^g = \mathbf{J}$ and

$$\overset{*}{\mathbf{R}} = \tfrac{1}{2}(\hat{\mathbf{R}}^0 - \hat{\mathbf{R}}^1)$$

For $\alpha \subseteq \{1, 2, \ldots, n\}$, these yield

$$\overset{*}{r}_\alpha = \tfrac{1}{2}(\hat{r}_\alpha + \hat{r}_{\alpha n+1}), \quad g/0; \qquad \text{and} \qquad \overset{*}{r}_\alpha = \tfrac{1}{2}(\hat{r}_\alpha - \hat{r}_{\alpha n+1}), \quad g/1$$

including the case of α as the empty set.

Since a fault is r_α-testable if and only if $r_\alpha \neq \overset{*}{r}_\alpha$, we have

Criterion 6.5. $g/0$ is r_α-testable if and only if

$$r_\alpha \neq \tfrac{1}{2}(\hat{r}_\alpha + \hat{r}_{\alpha n+1}) \tag{6.19a}$$

$g/1$ is r_α-testable if and only if

$$r_\alpha \neq \tfrac{1}{2}(\hat{r}_\alpha - \hat{r}_{\alpha n+1}) \tag{6.19b}$$

Equation (6.19) depends on both $\mathbf{R}$, the spectrum of the function realized by the network, and $\hat{\mathbf{R}}$, the spectrum of the function realized by the residual network found by treating g as a primary input. The factor of $\frac{1}{2}$ arises from the fact $h(X, g)$ is a function of $n + 1$ variables, whereas $h(X, 0)$ and $h(X, 1)$ are functions of only n.

For an example, consider the network in Fig. 6.1. Here

$$f(X) = x_1x_2x_3 + x_4(\bar{x}_2 + \bar{x}_3)$$

with $r_0 = 8$. For the line marked g,

$$h(X, g) = x_1\bar{g} + x_4(\bar{x}_2 + \bar{x}_3)$$

which yields $\hat{r}_0 = 16$, $\hat{r}_5 = 4$ (recall g corresponds to x_{n+1}). Since $r_0 \neq \frac{1}{2}(\hat{r}_0 \pm \hat{r}_5)$, $g/0$ and $g/1$ are both r_0-testable.

As a second example, consider the network in Fig. 6.2, which is the same as in Fig. 6.1 except for the line chosen as g. In this case,

$$h(X, g) = x_1\bar{g} + x_4 g$$

$\hat{r}_0 = 16$ and $\hat{r}_5 = 0$. Since $r_0 = \frac{1}{2}(\hat{r}_0 \pm \hat{r}_5)$, $g/0$ and $g/1$ are not r_0-testable.

Note that in this second example, $h(X, g)$ is independent of x_2 and x_3; however, $h(X, g)$ is still treated as a function of five variables, as is $\hat{\mathbf{R}}$. This is necessary since $h(X, g)$ is treated as a function of all $n + 1$ variables in the derivation of Criterion 6.5.

Two faults are functionally equivalent if they cause the network to realize the same faulty function. Only one of a class of functionally equivalent faults need be explicitly considered. When possible, it is advantageous to choose an input fault, single or multiple, in preference to an internal stuck-at fault since the testability of an input fault depends only on **R**. No $\hat{\mathbf{R}}$ need be computed, so considerable computation is saved. For an equivalence class containing no input fault, it is best to consider the fault closest to the network output since this will, in general, involve a simpler $\hat{\mathbf{R}}$. The handling of equivalent faults is discussed in detail in Section 6.5.

The notion of fault dominance used to reduce the computation required in traditional test set generation is not applicable to spectral coefficient testing. Hence, neither is the notion of minimal checkpoints.[1]

In Section 6.2.1 we showed that any multiple-input stuck-at fault involving p inputs causes the network to realize a faulty function $\overset{*}{f}(X)$, where every $\overset{*}{r}_\alpha$ is an integer multiple of 2^p. Clearly, any internal stuck-at fault that causes the network to realize a faulty function $\overset{*}{f}(X)$ independent of p variables will also have all $\overset{*}{r}_\alpha$ integer multiples of 2^p. This yields the following criterion.

Criterion 6.6. A single- or multiple-internal stuck-at fault that causes a network to realize an $\overset{*}{f}(X)$ independent of p variables is r_α-testable, $\alpha \subseteq \{1, 2, \ldots, n\}$, if r_α is not an integer multiple of 2^p.

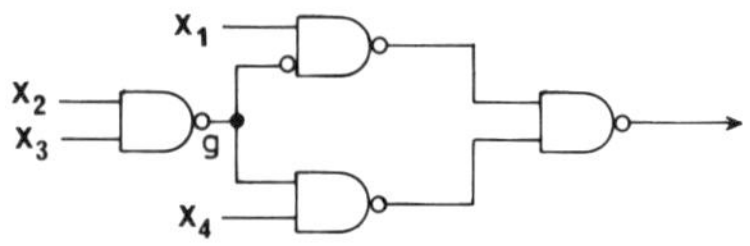

Fig. 6.2. Identical to the network in Fig. 6.1 except for the line designated g.

Note that the fault need not be functionally equivalent to an input stuck-at fault. For example, consider the network in Fig. 6.3, which realizes

$$f(X) = x_1x_4 \oplus x_2x_4 + x_2x_3$$

for which $r_0 = 7$. $g/1$ causes the network to realize

$$\overset{*}{f}(X) = \bar{x}_2 + \bar{x}_4 + x_2x_3$$

which is independent of x_1. Since r_0 is odd, $g/1$ is r_α-testable for all $\alpha \subseteq \{1, 2, 3, 4\}$. $g/1$ is not equivalent to an input fault.

Consider any internal line g such that

$$f(X) = h(X_1, g(X_2))$$

where $X = X_1 \cup X_2$, $X_1 \cap X_2 = \phi$. Recall Eqs. (6.17) and (6.18):

$$\mathbf{R} = \frac{1}{2}\hat{\mathbf{R}}^0 + \frac{1}{2}\hat{\mathbf{R}}^1 - \frac{1}{2^n}\hat{\mathbf{R}}^1 * \mathbf{R}^g \tag{6.17}$$

$$\overset{*}{\mathbf{R}} = \frac{1}{2}\hat{\mathbf{R}}^0 + \frac{1}{2}\hat{\mathbf{R}}^1 - \frac{1}{2^n}\hat{\mathbf{R}}^1 * \overset{*}{\mathbf{R}}^g \tag{6.18}$$

Since $h(X_1, g)$ and $g(X_2)$ have no common variables, either $\hat{r}^1_\alpha = 0$ or $r^g_\alpha = 0$ (and thus $\overset{*}{r}{}^g_\alpha = 0$) for all $\alpha \neq 0$. It follows from the definition of the convolution operation (see Eq. (2.28)) that

$$r_{\alpha\beta} = \frac{1}{2}\hat{r}^0_{\alpha\beta} + \frac{1}{2}\hat{r}^1_{\alpha\beta} - \frac{1}{2^n}\hat{r}^1_\alpha r^g_\beta$$

$$\overset{*}{r}_{\alpha\beta} = \frac{1}{2}\hat{r}^0_{\alpha\beta} + \frac{1}{2}\hat{r}^1_{\alpha\beta} - \frac{1}{2^n}\hat{r}^1_\alpha \overset{*}{r}{}^g_\beta$$

where α identifies a subset of X_1 and β identifies a subset of X_2. Either subset may be empty ($\alpha = 0$ or $\beta = 0$).

The following criterion follows immediately.

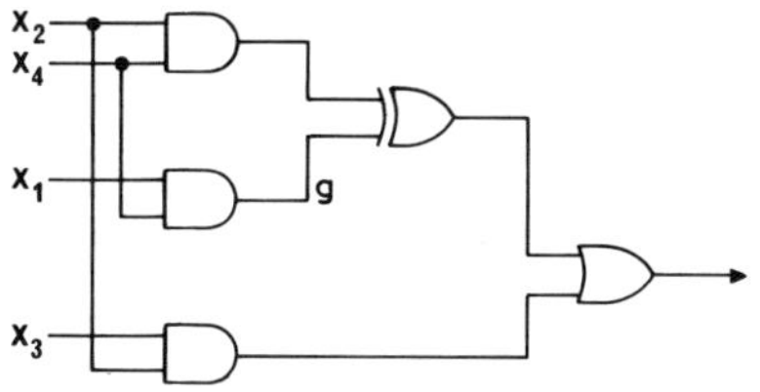

Fig. 6.3. A network to illustrate the application of Criterion 6.6. The fault $g/1$ is not equivalent to an input stuck-at fault, but is testable since it results in a faulty network independent of x_1.

Criterion 6.7. Given a network realizing $f(X)$ with an internal line g such that

$$f(X) = h(X_1, g(X_2))$$

$X_1 \cup X_2 = X$, $X_1 \cap X_2 = \phi$, any fault contained within the subnetwork realizing $g(X_2)$, which is r_β^g-testable with respect to $g(X_2)$, is $r_{\alpha\beta}$-testable with respect to $f(X)$ if and only if $\hat{r}_{\alpha n+1} \neq 0$.

We defer examples of this criterion till later. Its benefit is it allows us to partition certain networks into subnetworks, the testability of which is more readily determined, and which can then be combined to give the testability of the original network.

It is mathematically straightforward to extend these results to multiple-internal stuck-at faults, but as mentioned, there are too many faults for explicit testing to be practical.

6.2.3 Unate Lines

Unate functions were discussed at some length in Section 3.4.1. Here we examine the effect unateness has on the testability of stuck-at faults in combinational networks. $f(X)$ is said to be positive in x_i if

$$f(x_1, x_2, \ldots, x_{i-1}, 0, x_{i+1}, \ldots, x_n) \leq f(x_1, x_2, \ldots, x_{i-1}, 1, x_{i+1}, \ldots, x_n)$$

and negative in x_i if

$$f(x_1, x_2, \ldots, x_{i-1}, 0, x_{i+1}, \ldots, x_n) \geq f(x_1, x_2, \ldots, x_{i-1}, 1, x_{i+1}, \ldots, x_n)$$

where $\leq$ denotes the partial ordering of the Boolean lattice, i.e., $0 \leq 1$. $f(X)$ is unate in x_i if it is positive or negative in x_i. A unate function is unate in all its variables.

Clearly if $f(X)$ is positive in x_i, $r_i < 0$, and if $f(X)$ is negative in x_i, $r_i > 0$. When $r_i = 0$, $f(X)$ is neither positive or negative in x_i unless x_i is a redundant variable, a case that we ignore. It follows immediately that a network realizing an $f(X)$ that is unate in x_i is r_0-testable for $x_i/0$ and $x_i/1$.

Unfortunately, this result does not extend to multiple-input stuck-at faults. Consider the function

$$f(X) = x_1x_2 + \bar{x}_2x_3$$

with the spectrum

$$\begin{array}{cccccccc} r_0 & r_1 & r_2 & r_3 & r_{12} & r_{13} & r_{23} & r_{123} \\ 4; & -2 & 0 & -2; & 2 & 0 & -2; & 0 \end{array}$$

For the fault $(x_1/0, x_3/1)$, the faulty function is

$$\overset{*}{f}(X) = \bar{x}_2$$

which has the spectrum

$$\begin{array}{cccccccc} \overset{*}{r}_0 & \overset{*}{r}_1 & \overset{*}{r}_2 & \overset{*}{r}_3 & \overset{*}{r}_{12} & \overset{*}{r}_{13} & \overset{*}{r}_{23} & \overset{*}{r}_{123} \\ 4; & 0 & 4 & 0; & 0 & 0 & 0; & 0 \end{array}$$

Since $r_0 = \overset{*}{r}_0$, the fault is not r_0-testable. Neither is $(x_1/1, x_3/0)$.

Consider an internal line g in a network realizing $f(X)$ and let $g(X)$ and $h(X, g)$ be defined as above. g will be termed a unate internal line if $h(X, g)$ is unate in g.

Suppose $h(X, g)$ is positive in g, whence

$$h(X, 0) \leq h(X, 1)$$

For the fault-free case,

$$f(X) = h(X, g(X))$$

For $g/0$,

$$\overset{*}{f}(X) = h(X, 0)$$

It follows immediately that

$$\overset{*}{f}(X) \leq f(X) \tag{6.20}$$

Since Eq. (6.20) cannot be equal for all assignments to X (we preclude the case of g a redundant line), $\overset{*}{r}_0 \neq r_0$, and $g/0$ is r_0-testable. Similar arguments hold for $g/1$ and for $h(X, g)$ negative in g.

The following simple results are useful in identifying the unate line in a network. The results are for convenience stated in terms of x_{n-1}, x_n and x_{n+1}. They clearly hold for all selections of variables.

1. If $f(X)$ is positive (negative) in x_n, $f(x_1, x_2, \ldots, \bar{x}_n)$ is negative (positive) in x_n.

2. If $f(X)$ is positive (negative) in x_n,

$$\mathring{f}(x_1, x_2, \ldots, x_n, x_{n+1}) = f(x_1, x_2, \ldots, x_{n-1}, x_n \cdot x_{n+1})$$

is positive (negative) in x_n and x_{n+1}.

3. If $f(X)$ is positive (negative) in x_n,

$$\mathring{f}(x_1, x_2, \ldots, x_n, x_{n+1}) = f(x_1, x_2, \ldots, x_{n-1}, x_n + n_{n+1})$$

is positive (negative) in x_n.

4. The function $f(x_1, x_2, \ldots, x_{n-1}, x_n \oplus x_{n+1})$ is nonunate in x_n and x_{n+1}.
5. The function

$$\mathring{f}(x_1, x_2, \ldots, x_{n-1}) = f(x_1, x_2, \ldots, x_{n-1}, x_{n-1})$$

is positive (negative) in x_{n-1} if and only if $f(X)$ is positive (negative) in both x_{n-1} and x_n.

The procedure below is based on these results. It identifies most unate lines by a simple scan of the network beginning at the output and working back toward the inputs. No backtracking is required. The following symbols are used:

P denotes a positive line;
N denotes a negative line; and
C denotes a candidate line.

A candidate line is thus not *à priori* r_0-testable and must be analyzed using the methods discussed earlier.

Procedure 6.1.

1. Label the network output *P*.
2. For each gate whose output is labeled but whose inputs are not, do one of the following as appropriate:

AND or OR: label all inputs identical to the output;
NOT, NAND, or NOR: label each input as follows:

Output	Input
P	*N*
N	*P*
C	*C*

Exclusive-OR and Exclusive-NOR: label each input *C*.

3. For each fan-out point whose branches are all labeled but whose input stem is not, construct the stem label from the branch labels using the following

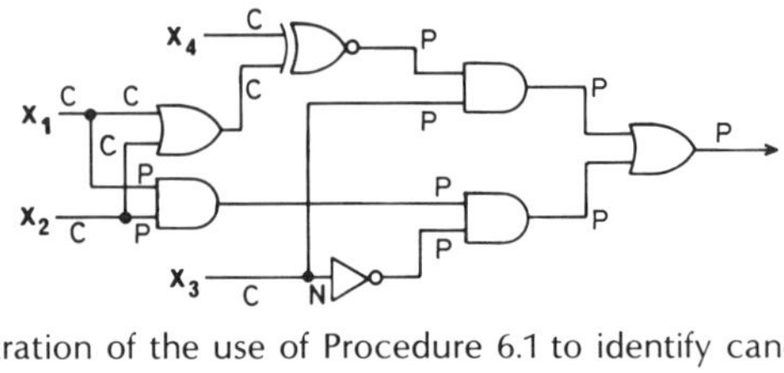

Fig. 6.4. An illustration of the use of Procedure 6.1 to identify candidate r_0-untestable network lines.

commutative and associative operator:

	P	N	C
P	P	C	C
N	C	N	C
C	C	C	C

4. Repeat steps 2 and 3 as required until all network lines are labeled.

Note that this procedure identifies both the internal and input unate lines.†

Figure 6.4 shows a network as labeled by Procedure 6.1. It is clear that the candidate r_0-untestable lines are

(i) inputs to Exclusive-OR and Exclusive-NOR gates,
(ii) lines that fan out to reconvergent paths with unequal inversion parities, and
(iii) lines that feed lines of types (i) and (ii) directly or indirectly.

Procedure 6.1 is both time and space linear in the number of network lines, and thus is as efficient as can be expected.

6.2.4 Bridging Faults

Galiay *et al.*[24] have shown that under rather strict layout rules all physical failures in an integrated circuit implementation are covered by the stuck-at fault model. In general, this is not the case. A short between two or more network lines usually does not correspond to a stuck-at fault.

A bridging fault assumes that whenever two or more network lines are shorted, a spurious AND or OR is introduced and each line involved takes the resulting value. Here we consider AND-bridging. The results are readily adapted to the OR-bridging case.

Consider a combinational network realizing $f(X)$. An input bridging fault causes each of a certain subset of X to be replaced by the AND of the variables in the subset. This causes the network to realize a faulty function $\overset{*}{f}(X)$. Without loss of generality, we can consider input bridging faults involving $x_{m+1}, x_{m+2}, \ldots, x_n$. All others are equivalent under permutation of variables. This situation is depicted in Fig. 6.5.

As before, let

$$f_u(x_1, x_2, \ldots, x_m) = f(x_1, x_2, \ldots, x_m, u_1, \ldots, u_{n-m})$$

where $u = \sum_{i=1}^{n-m} u_i 2^{i-1}$, $0 \leq u \leq \beta = 2^{n-m} - 1$. $\mathbf{R}_u$ will denote the spectrum of $f_u(x_1, x_2, \ldots, x_m)$.

† We exclude networks containing an exclusive-OR gate which could be replaced directly by a vertex gate and inverter, as, i.e., $x_i \oplus x_j$ by $x_i\bar{x}_j$. In such a network, half the exclusive-OR gate is redundant and Procedure 6.1 may label certain unate lines with C.

The nature of the bridging fault is that all the subfunctions $f_1, f_2, \ldots, f_{\beta-1}$ are replaced by f_0. For the fault-free network, we have

$$[\mathbf{R}_0\,\mathbf{R}_1 \cdots \mathbf{R}_\beta] = \frac{1}{2^{n-m}}[\mathbf{R}^0\,\mathbf{R}^1 \cdots \mathbf{R}^\beta]\mathbf{T}^{n-m} \tag{6.9}$$

where

$$\mathbf{R} = \begin{bmatrix} \mathbf{R}^0 \\ \mathbf{R}^1 \\ \vdots \\ \mathbf{R}^\beta \end{bmatrix}$$

with 2^m coefficients in each $\mathbf{R}^u$, $0 \leq u \leq \beta$. For the faulty network

$$[\mathbf{R}_0\,\mathbf{R}_0 \cdots \mathbf{R}_0\,\mathbf{R}_\beta] = \frac{1}{2^{n-m}}[\overset{*}{\mathbf{R}}{}^0\,\overset{*}{\mathbf{R}}{}^1 \cdots \overset{*}{\mathbf{R}}{}^\beta]\mathbf{T}^{n-m} \tag{6.21}$$

$\overset{*}{\mathbf{R}}$ is the spectrum of $\overset{*}{f}(X)$.

It follows that

$$[\overset{*}{\mathbf{R}}{}^0\,\overset{*}{\mathbf{R}}{}^1 \cdots \overset{*}{\mathbf{R}}{}^\beta] = \frac{1}{2^{n-m}}[\mathbf{R}^0\,\mathbf{R}^1 \cdots \mathbf{R}^\beta][\mathbf{T}^{n-m}_{*0}\,\mathbf{T}^{n-m}_{*0} \cdots \mathbf{T}^{n-m}_{*0}\,\mathbf{T}^{n-m}_{*\beta}]\mathbf{T}^{n-m} \tag{6.22}$$

For the fault to be r_α-testable, it is necessary to show $\overset{*}{r}_\alpha \neq r_\alpha$. For a particular coefficient, this can be checked using Eq. (6.22).

The case of a pair of inputs bridged leads to the following criteria, which come directly from Eq. (6.22).

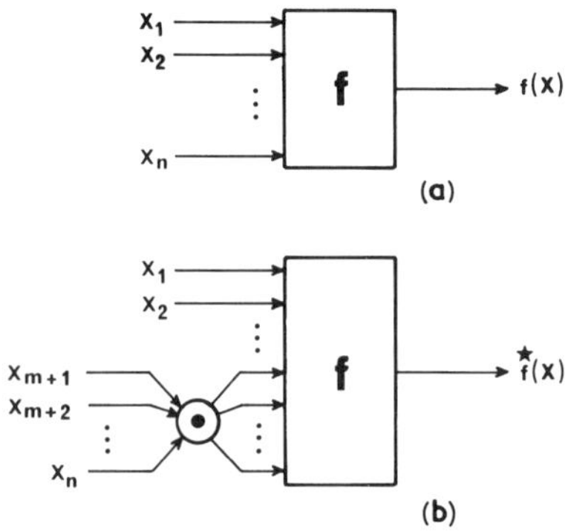

Fig. 6.5. (a) A combinational network free of input bridging. (b) The logical structure of AND-input bridging between $x_{m+1}, \ldots, x_n$.

Criterion 6.8. An AND-bridging fault between x_i and x_j is both r_α- and $r_{\alpha ij}$-testable, $i, j \notin \alpha$, if and only if

$$r_{i\alpha} + r_{j\alpha} + 2r_{ij\alpha} \neq 0 \tag{6.23}$$

Criterion 6.9. An AND-bridging fault between x_i and x_j is $r_{i\alpha}$-testable, $i, j \notin \alpha$, if and only if

$$r_{i\alpha} - r_{j\alpha} \neq 0 \tag{6.24}$$

Both criteria allow $\alpha = 0$, in which case $r_{i\alpha}$ is r_i, etc.

Consider the function with spectrum

r_0	r_1	r_2	r_3	r_4	r_{12}	r_{13}	r_{14}	r_{23}	r_{24}	r_{34}	r_{123}	r_{124}	r_{134}	r_{234}	r_{1234}
7;	1	3	−1	3;	1	−3	1	−1	−1	−1;	1	1	1	−1;	5

A network realizing this function will be r_0-testable for AND-bridging between variable pairs except for (x_2, x_3) and (x_3, x_4).

For (x_2, x_3), $r_{24} + r_{34} + 2r_{234} \neq 0$, so the bridge is r_4-testable. For (x_3, x_4), $r_{23} + r_{24} + 2r_{234} \neq 0$, so the bridge is r_2-testable. Both these bridges are also r_{13}-testable since $r_{13} \neq r_{12}$ and $r_{13} \neq r_{14}$.

An AND feedback bridging fault covers the case of the network output shorted with certain network inputs. Once again it is sufficient to consider the inputs $x_{m+1}, \ldots, x_n$ as shown in Fig. 6.6.

Such a fault converts the combinational network to a sequential one and the possibility of oscillation arises. Karpovsky and Su[6] have shown that the output will oscillate if

$$x_{m+1}x_{m+2}\cdots x_n f(x_1, \ldots, x_m, 0, 0, \ldots, 0)\bar{f}(x_1, x_2, \ldots, x_m, 1, 1, \ldots, 1) = 1$$

The following criterion is a direct consequence of this.

Criterion 6.10. A circuit cannot oscillate under feedback bridging if and only if the function it realizes is positive in all its variables.

If a circuit oscillates, it has essentially become sequential and is not subject to functional analysis techniques designed for combinational networks.

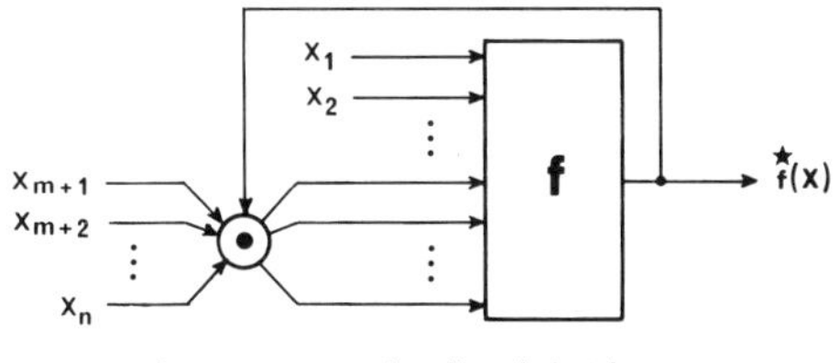

Fig. 6.6. AND feedback bridging.

The following criterion demonstrates the limited scope of spectral coefficient analysis applied to feedback bridging faults.

Criterion 6.11. A network realizing a function that is positive in all its variables is r_0-testable for all feedback bridging unless it realizes the AND of all its inputs.

Having raised the problem of feedback bridgings and having shown that the spectral coefficients are of little help, we include the following approach due to Xu Shiyi[7] based on the work of Karpovsky and Su.[6]

Procedure 6.2.

1. Apply $(0, 0, \ldots, 0)$ to the inputs. If $f(0, 0, \ldots, 0) = 1$, the fault is detected since for the AND bridging being considered $\overset{*}{f}(0, 0, \ldots, 0) = 0$, and the output cannot oscillate.
2. If $f(0, 0, \ldots, 0) = 0$, then apply a test vector $(v_1, v_2, \ldots, v_n)$ such that $f(v_1, v_2, \ldots, v_n) = 1$ but $f(\hat{v}_1, \hat{v}_2, \ldots, \hat{v}_n) = 0$ for all $(\hat{v}_1, \hat{v}_2, \ldots, \hat{v}_n)$ that contain fewer ones than $(v_1, v_2, \ldots, v_n)$.

These are two cases to consider:

(a) If any of $v_{m+1}, \ldots, v_n = 0$, $\overset{*}{f}(v_1, v_2, \ldots, v_n) = 0$, and the fault is detected since the output cannot oscillate.

(b) Suppose $v_{m+1} = \cdots = v_n = 1$, after step 1 we have $f(0, 0, \ldots, 0) = 0$ and the AND bridge will hold all $x_{m+1}, \ldots, x_n$ at 0. Hence some $f(\hat{v}_1, \hat{v}_2, \ldots, \hat{v}_n)$, where $(\hat{v}_1, \hat{v}_2, \ldots, \hat{v}_n)$ has fewer ones than $(v_1, v_2, \ldots, v_n)$, will be evaluated holding the output at 0 and the fault is detected.

This method is remarkably simple, detecting all AND feedback bridgings with the application of at most two test vectors.

6.2.5 Programmable Logic Arrays

The final fault class we shall consider is the set of contact faults[25–26] peculiar to programmable logic arrays (PLAs). The basic structure of a PLA is shown in Fig. 6.7. An n input array has the $2n$ literals $x_1, \bar{x}_1, \ldots, x_n, \bar{x}_n$ running horizontally through its AND array. Each column of the array forms a prime implicant by making "contact" with the appropriate literals. These columns run through the OR array. The output functions are formed along horizontal paths that contact the appropriate prime implicants. The nature of these contacts is technology-dependent and not important to the current discussion.

A contact fault models a physical failure that causes a required contact to be missing or an erroneous contact to be made. Smith[27] has classified contact faults as follows:

growth fault: a missing contact in the AND array causes a prime implicant to loose a literal and hence include twice as many minterms;
shrinkage fault: an erroneous contact in the AND array causes a prime implicant to gain a literal and thus include either half its original minterms or none;
disappearance fault: a missing contact in the OR array causes a prime implicant to be dropped from the corresponding output function;
appearance fault: an erroneous contact in the OR array causes a prime implicant to be added to the corresponding output function.

A contact fault is termed a functional contact fault if it alters the function realized by at least one output of the PLA. This qualification is necessary since certain contact faults have no functional effect. An appearance fault, for example may introduce a redundant prime implicant.

A functional growth or appearance fault always results in $\overset{*}{r}_0 > r_0$, while a functional shrinkage or disappearance fault always results in $\overset{*}{r}_0 < r_0$. Clearly, all single-functional contact faults are r_0-testable. In addition, all

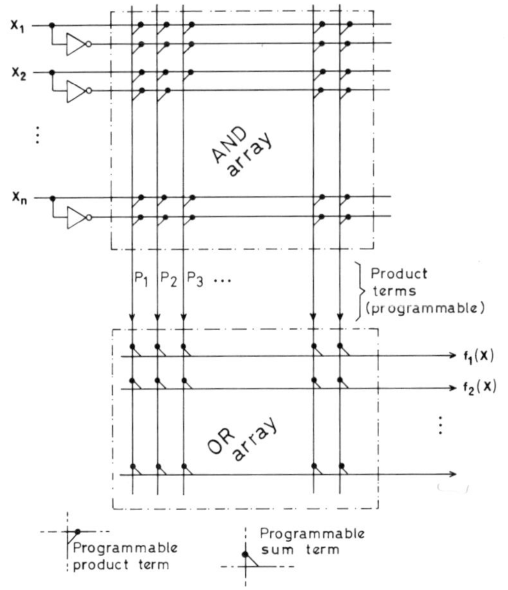

Fig. 6.7. The basic structure of a PLA.

multiple faults composed of growth and appearance faults, and all multiple faults composed of shrinkage and disappearance faults, are r_0-testable. Other multiple faults would have to be explicitly examined to ensure the number of minterms added and removed does not balance leaving $\mathring{r}_0 = r_0$. For all but the smallest arrays, such an examination would be impractical.

The degree to which these multiple-functional contact fault results are applicable is technology-dependent and beyond the scope of this discussion. A detailed examination of PLA fault mechanisms can be found in Smith.[27]

6.3 SYNDROME TESTING

Having established a spectral framework for the analysis of faults, we turn our attention to the development of fault detection tests for combinational networks. The discussion will focus on the single stuck-at fault model, but the techniques developed also apply to contact faults in PLAs. The extension to bridging faults is not considered.

In this section, we show the connection of the spectral results to the method of syndrome testing as developed by Savir.[28–31] Syndrome testing verifies the performance of a network by counting the number of ones in the truth-table column vector of the function that it realizes. This value is simply r_0, so syndrome testing is closely related to the notion of r_0-testability. In Section 6.4, we consider the extension of these ideas to the other spectral coefficients.

6.3.1 Syndrome-Testable Networks

Savir[28–31] has defined the syndrome of a function as

$$\sigma(f) = \frac{1}{2^n} W(f)$$

where $W(f) = \sum_{v=0}^{2^n-1} f(v)$ is the weight of $f(X)$. Since $r_0 = 2^n\sigma(f)$, the notion of a syndrome-testable fault and an r_0-testable fault are synonymous.

To characterize the syndrome-testability of networks, we shall first consider a number of topological classes:

fan-out-free: a network where every network input and gate output is an input to at most one gate;

internally fan-out-free: a network where the fan-out restriction applies only to internal lines;

unate: a network where every line is unate;

internally unate: a network where every internal line is unate;

two-level: a network that is a direct implementation of a sum of products or product of sums expression; and

internally nonunate: a network with one or more nonunate internal lines.

These classes are not disjoint. In particular, fan-out-free networks are internally fan-out-free and unate networks and two-level networks are internally unate. When devoid of Exclusive-OR and Exclusive-NOR gates, a fan-out-free network is unate and an internally fan-out-free network is internally unate. A unate network or two-level network can not have an Exclusive-OR or Exclusive-NOR gate. An internally unate network can only have an Exclusive-OR or Exclusive-NOR gate if all of its inputs are fan-out-free network inputs.

By an input line we mean the primary network input itself. Fan-out branches emanating from such a line are considered internal lines.

A fan-out-free function devoid of Exclusive-OR and Exclusive-NOR gates realizes a function $f(X)$ such that

$$f(X) = f_1(X_1)f_2(X_2), \qquad X = X_1 \cup X_2, \quad X_1 \cap X_2 = \phi$$

or

$$f(X) = f_1(X_1) + f_2(X_2)$$

Note that if the final gate in the realization of $f(X)$ is a NAND or NOR, the inverter can be removed by applying one of De Morgan's laws. Clearly, either

$$W(f) = W(f_1)W(f_2) \tag{6.25}$$

or

$$W(f) = 2^{n_2}W(f_1) + 2^{n_1}W(f_2) - W(f_1)W(f_2) \tag{6.26}$$

where $n_1 = |X_1|$ and $n_2 = |X_2|$. In either case, $W(f)$ is odd if and only if $W(f_1)$ and $W(f_2)$ are both odd.

The subnetworks realizing $f_1(X_1)$ and $f_2(X_2)$ are fan-out-free. Since $W(x_i) = W(\bar{x}_i) = 1$, $1 \le i \le n$, it follows by induction that $W(f)$ is odd.

In a fan-out-free network, there is a single path from each input to the output, and every single or multiple stuck-at fault either is, or is equivalent to, a single- or multiple-input stuck-at fault. It follows from Criterion 6.2 that all single and multiple stuck-at faults in a fan-out-free network devoid of Exclusive-OR and Exclusive-NOR gates are syndrome-testable.

Note that stuck-at faults in fan-out-free networks devoid of Exclusive-OR and Exclusive-NOR gates are, in fact, r_α-testable for all $\alpha \subseteq \{1, 2, \ldots, n\}$. This fact is of little practical use since, as we shall show later when given the choice, r_0 is the preferred spectral coefficient.

Exclusive-OR and Exclusive-NOR gates present difficulty since for

$$f(X) = f_1(X_1) \oplus f_2(X_2), \qquad X_1 \cup X_2 = X, \quad X_1 \cap X_2 = \phi$$

we have

$$W(f) = 2^{n_2}W(f_1) + 2^{n_1}W(f_2) - 2W(f_1)W(f_2) \tag{6.27}$$

Thus a fan-out-free network, which includes one or more Exclusive-OR or Exclusive-NOR gates, will realize a function with even weight, since Eq. (6.27) always yields an even value as do Eqs. (6.25) and (6.26) when either $W(f_1)$, $W(f_2)$ or both are even. All stuck-at faults are still equivalent to input stuck-at faults and Criterion 6.2 still applies.

For example, the network in Fig. 6.8 realizes

$$f(X) = x_1x_2 \oplus x_3x_4$$

which has $r_0 = 6$ with all other spectral coefficients equal to ± 2. By Criterion 6.2, all multiple-input stuck-at faults involving two or more inputs are syndrome-testable. Since the spectral coefficients are all multiples of 2, Criterion 6.2 says nothing about single-input stuck-at faults for this example.

However, $r_i \neq 0$, $1 \leq i \leq 4$; hence by Criterion 6.1 all single-input stuck-at faults are syndrome-testable. All stuck-at faults in the network of Fig. 6.8 are syndrome-testable even though it realizes a function with even weight. A single Exclusive-OR gate is an example of a fan-out-free network for which this is not true.

The network in Fig. 6.9 is a simple internally fan-out-free network that is not also internally unate. The latter class will be discussed shortly. This network realizes

$$f(X) = (x_1x_2 + x_2x_3) \oplus x_1x_3$$

which has $r_0 = 3$. Since r_0 is odd, this network is syndrome-testable for all input stuck-at faults.

However, $g/0$ is not equivalent to an input stuck-at fault. This fault causes the network to realize

$$\overset{*}{f}(X) = x_1x_2 + x_2x_3$$

which has $\overset{*}{r}_0 = 3$. Hence, $g/0$ is not syndrome-testable even though all input stuck-at faults are.

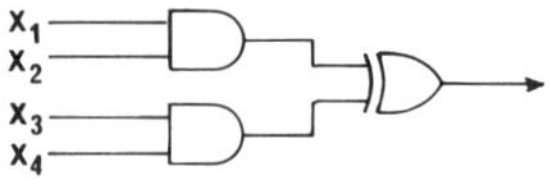

Fig. 6.8. A fan-out-free network that is syndrome-testable despite the Exclusive-OR gate.

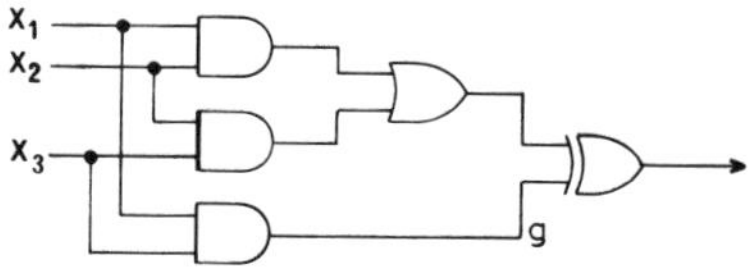

Fig. 6.9. An internally fan-out-free network that is not internally unate. The fault g/0 is not syndrome-testable.

A unate network must clearly realize a unate function. Conversely, an irredundant realization of a unate function must be a unate network. It is clear from the development of Procedure 6.1 that a unate network cannot contain an Exclusive-OR or an Exclusive-NOR gate, and that every pair of reconvergent fan-out paths must have equal inversion parities. The latter condition is equivalent to requiring that the number of inversions around every loop in the network be even, where the term loop is interpreted in the graph theoretic sense.

Since all single stuck-at faults on unate lines are r_0-testable, a unate network is syndrome-testable with respect to single stuck-at faults.

The network of Fig. 6.10 realizes

$$f(X) = x_1x_2 + x_1x_3 + x_2x_3$$

which has $r_0 = 4$. As shown, every line of this network is positive (found by applying Procedure 6.1). The multiple stuck-at fault $\{x_1/0, x_2/1\}$ yields

$$\overset{*}{f}(X) = x_3$$

which has $\overset{*}{r}_0 = 4$. This fault is syndrome-untestable.

This example shows that even the relatively simple class of unate networks has syndrome-untestable multiple stuck-at faults.

An internally unate network is clearly syndrome-testable with respect to single stuck-at faults on its internal lines. Applying Criterion 6.1, we find a single-input stuck-at fault on x_i is syndrome-testable if and only if $r_i \neq 0$. Hence an internally unate network is syndrome-testable with respect to single stuck-at faults if and only if it realizes a function with $r_i \neq 0$, $1 \leq i \leq n$.

Two-level networks are internally unate. An internally fan-out-free network is also internally unate if it is either devoid of Exclusive-OR and

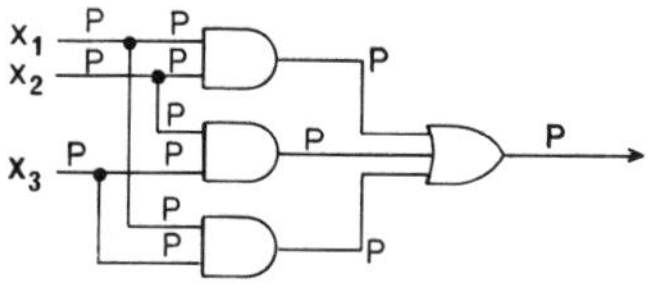

Fig. 6.10. A unate network for which the multiple-input stuck-at fault $\{x_1/0, x_2/1\}$ is not syndrome-testable.

Exclusive-NOR gates, or, if when these gates occur, they have only primary network inputs as their inputs.

Contact faults in PLAs were discussed in Section 6.2.5. A PLA is an implementation of a two-level network and, with respect to single stuck-at faults, behaves as such. Syndrome-testing thus offers good protection against faults in PLAs.

An internally nonunate network has one or more nonunate internal lines. These lines are readily identified using Procedure 6.1. Note that an internally nonunate network may also have nonunate inputs. These are handled as described for internally unate networks.

It is clear from the development of Procedure 6.1 that an internally nonunate network must have an Exclusive-OR or Exclusive-NOR gate with at least one input an internal line, or must contain a loop with an odd number of inversions.

Once the internal nonunate lines have been determined by Procedure 6.1 (the internal lines labeled C), the syndrome-testability of faults on each of these lines must be verified. Criterion 6.5 could be applied to each line in turn, but this is undesirable since a unique $\hat{\mathbf{R}}$ must be computed for each line. Alternative procedures based on the results of Section 6.2 will be illustrated in Section 6.4.3.

6.3.2 Syndrome Testing by Hardware Modification

Markowsky[32] has shown that by adding control lines and gates, a network can always be made syndrome-testable with respect to single stuck-at faults. Unfortunately, his proof does not define a practical method. Savir,[28] on the other hand, has presented a heuristic approach that does quite well in practice. Savir's method yields a near-minimal addition. It has not been formally justified. In addition, it can require considerable computation.

For two-level AND-OR networks, Savir's method proceeds as follows (an analogous method exists for OR-AND networks):

1. A control variable is added to one prime implicant of the function. This is equivalent to adding an input to one AND gate in the two-level network. The prime implicant is selected by trying each in turn and choosing the addition that makes the greatest number of previously syndrome-untestable inputs syndrome-testable. In the case of a tie, an arbitrary choice is made.
2. On each subsequent iteration, the method attempts to add an existing control variable to another prime implicant so as to further reduce the number of untestable inputs. As before, all possibilities are searched, and the addition causing the greatest improvement is chosen.

3. If step 2 fails to find any improvement, an additional control variable is introduced and the procedure proceeds as in step 1.

4. The procedure iterates until all inputs are syndrome-testable with respect to single stuck-at faults.

The method will always terminate, but the number of control lines added will not always be minimal, nor will the increase in AND gate fan-in. Essentially, the method is performing a local optimization on each iteration. Local optimization does not lead, in general, to global optimization.

As an example of this method, consider

$$f(X) = x_1\bar{x}_2 + \bar{x}_1 x_3 + x_2\bar{x}_3 + x_4 x_5 + \bar{x}_4\bar{x}_5$$

It is easily verified that a two-level realization of this function is syndrome-untestable for all single-input stuck-at faults.

The first pass of Savir's procedure yields the expression

$$c_1 x_1\bar{x}_2 + \bar{x}_1 x_3 + x_2\bar{x}_3 + x_4 x_5 + \bar{x}_4\bar{x}_5$$

for which x_3, x_4, x_5 remain untestable inputs. The second pass uses c_1 a second time, giving

$$c_1 x_1\bar{x}_2 + \bar{x}_1 x_3 + x_2\bar{x}_3 + c_1 x_4 x_5 + \bar{x}_4\bar{x}_5$$

which has a single untestable input x_3. The final pass yields

$$c_1 x_1\bar{x}_2 + c_2\bar{x}_1 x_3 + x_2\bar{x}_3 + c_1 x_4 x_5 + \bar{x}_4\bar{x}_5$$

A two-level implementation of this expression is syndrome-testable for all single stuck-at faults. Note that the control variables are only added in positive form, so faults on the corresponding input lines are by definition syndrome-testable.

From Criterion 6.2, it is clear that a network realizing a function with an odd number of ones is syndrome-testable for all input stuck-at faults. This suggests that a network that does not have this property (and therefore realizes a function with an even number of ones) be modified to realize a function with an odd number of ones. The modification must be made so that the new network can be made to function as the original. This was of course the case for Savir's method where all control lines must be set to 1 to obtain the original function.

Susskind[22] and Carter[23] have suggested a single minterm addition as follows:

$$\hat{f}(X, c) = \bar{c}f(X) + cm_j \tag{6.28}$$

where $f(X)$ is the original function, c is a control variable and m_j is an

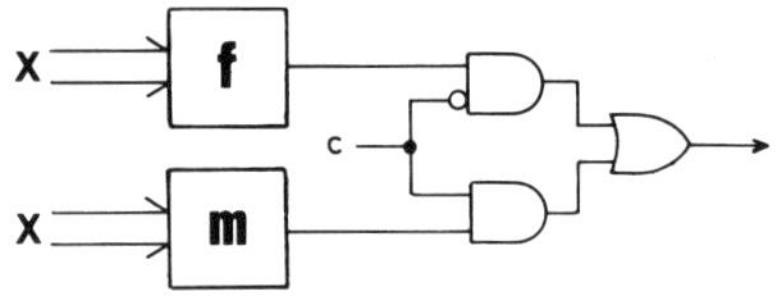

Fig. 6.11. The addition of a single minterm m to make $f(X)$ syndrome-testable for single stuck-at faults.

arbitrary minterm. Clearly, if $W(f)$ is even, $W(\hat{f})$ is odd. An implementation of $\hat{f}(X)$ is made to realize $f(X)$ by setting $c = 0$.

Equation (6.28) has the advantage that if a fixed realization of $f(X)$ is available, it can be made testable for stuck-at input faults by the addition of relatively simple hardware as illustrated in Fig. 6.11. The additions can be made so that every line added, except the input c, is unate and hence syndrome-testable. c is syndrome-testable since $W(\hat{f})$ is odd. In addition, any unate internal lines in the realization of $f(X)$ are also unate in the realization of $\hat{f}(X)$. This modification thus ensures the single stuck-at fault syndrome-testability of an internally unate network and is not restricted to two-level networks as is Savir's method.

If the realization of $f(X)$ can be altered, the direct implementation of

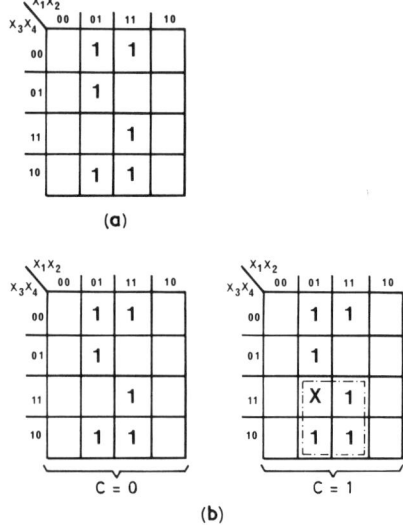

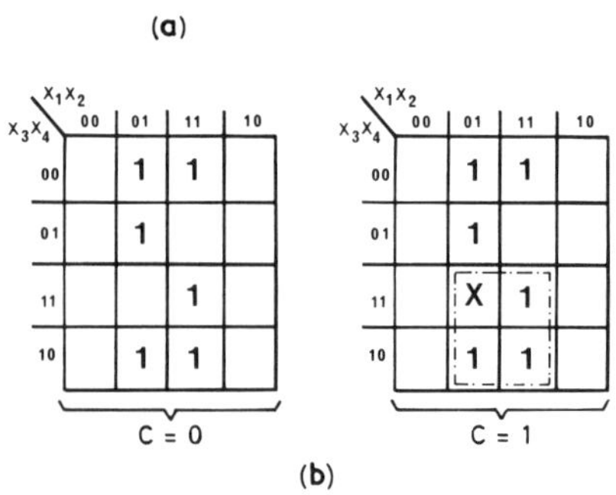

Fig. 6.12. (a) $f(X) = x_2\bar{x}_4 + \bar{x}_1x_2\bar{x}_3 + x_1x_2x_3$. (b) A testable design found by duplicating $f(X)$ over c and adding an odd number of minterms to the $c = 1$ copy.

Eq. (6.28) is not the best way to proceed. Rather, an internally unate realization of $\hat{f}(X)$ could be constructed. A simpler realization is usually found by considering

$$\hat{\hat{f}}(X, c) = f(X) + cm_j \tag{6.29}$$

where $f(X)\, m_j = 0$, i.e., m_j is a minterm that does not belong to $f(X)$. This is particularly true for a two-level realization. Every prime implicant except one of $\hat{\hat{f}}(X, c)$ involves $\bar{c}$ and the exception may involve c. All but one of the prime implicants of $\hat{f}(X, c)$ are independent of c. A two-level realization of $\hat{\hat{f}}(X)$ is simply a two-level realization of $f(X)$ with an additional AND gate realizing a prime implicant that covers cm_j.

The choice of m_j in Eq. (6.29) should not be made arbitrarily. Rather, the minterm leading to the simplest additional prime implicant should be chosen.

Consider the function

$$f(X) = x_2\bar{x}_4 + \bar{x}_1x_2\bar{x}_3 + x_1x_2x_3$$

whose Karnaugh map is shown in Fig. 6.12a. Figure 6.12b shows this function "duplicated" over c with one minterm added to yield the two-level realization

$$\hat{\hat{f}}(X) = x_2\bar{x}_4 + \bar{x}_1x_2\bar{x}_3 + x_1x_2x_3 + cx_2x_3$$

which is syndrome-testable for all single stuck-at faults and all multiple-input stuck-at faults.

Consider the function in Fig. 6.13a. Here, as depicted in Fig. 6.13b, it is advantageous to add three minterms. The approach is valid as long as an odd number of minterms are added.

Conversely, a function can be modified by removing minterms from the duplicate map. This is appropriate when the original function has a prime implicant that is the sole cover of an odd number of minterms. For example, the function in Fig. 6.12a can be modified as shown in Fig. 6.14. This yields the function

$$\hat{\hat{f}}(X, c) = x_2\bar{x}_4 + c\bar{x}_1x_2\bar{x}_3 + x_1x_2x_3$$

Note that in this approach the original function is associated with $c = 1$, so that c will be positive in the resulting expression.

We have illustrated these methods using two-level realizations because of their easy visual interpretation on the maps. Clearly, any internally unate realization of Eq. (6.29) will suffice, but it is not so clear which minterms should be added or removed. Certainly choosing the modification that leads to the simplest two-level network is a good heuristic.

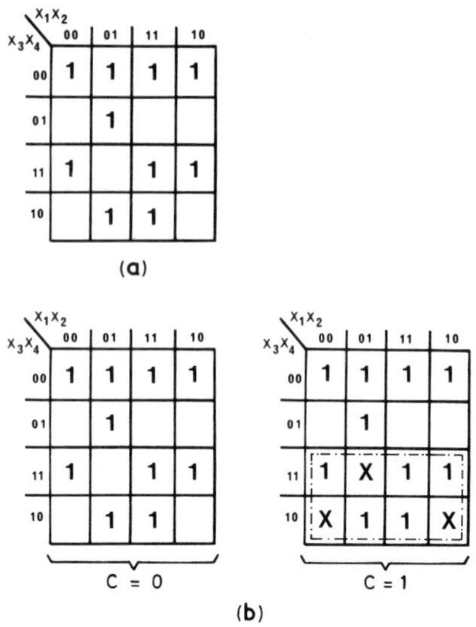

Fig. 6.13. (a) $f(X) = \bar{x}_3\bar{x}_4 + \bar{x}_1x_2\bar{x}_3 + x_2x_3\bar{x}_4 + \bar{x}_2x_3x_4 + x_1x_3x_4$. (b) A testable design for $f(X)$ found by adding three minterms giving the new prime implicant x_3.

Savir's method applied to

$$f(X) = x_1\bar{x}_2 + \bar{x}_1x_3 + x_2\bar{x}_3 + x_4x_5 + \bar{x}_4\bar{x}_5$$

gave

$$c_1x_1\bar{x}_2 + c_2\bar{x}_1x_3 + x_2\bar{x}_3 + c_1x_4x_5 + \bar{x}_4\bar{x}_5$$

The map of this function is given in Fig. 6.15a. No prime implicant of this function is the sole cover of an odd number of ones, so we choose to add minterms. The best choice is to add one minterm. The selection indicated

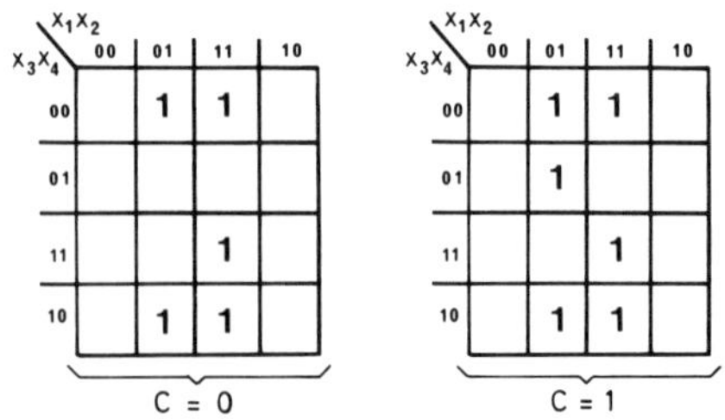

Fig. 6.14. Making the function in Fig. 6.12a syndrome-testable by the removal of minterms.

in Fig. 6.12b yields

$$x_1\bar{x}_2 + \bar{x}_1x_3 + x_2\bar{x}_3 + x_4x_5 + \bar{x}_4\bar{x}_5 + cx_1x_4$$

This solution increases the total fan-in count by 4 as opposed to an increase of 3 in Savir's method. In addition, an AND gate is added. However, only one control input is required. Since pin count is an extremely important design constraint, the method presented here, which always introduces a single control input, will in most cases be preferred. The minterm method also requires less computation than Savir's exhaustive approach.

Savir's technique can be extended to handle arbitrary combinational networks. The prime implicants in the two-level method are replaced by the set of AND, OR, NAND and NOR gates in the network. Rather than just the syndrome-untestable inputs, one must consider all lines with syndrome-untestable faults. Syndrome-untestable faults on the inputs to Exclusive-OR

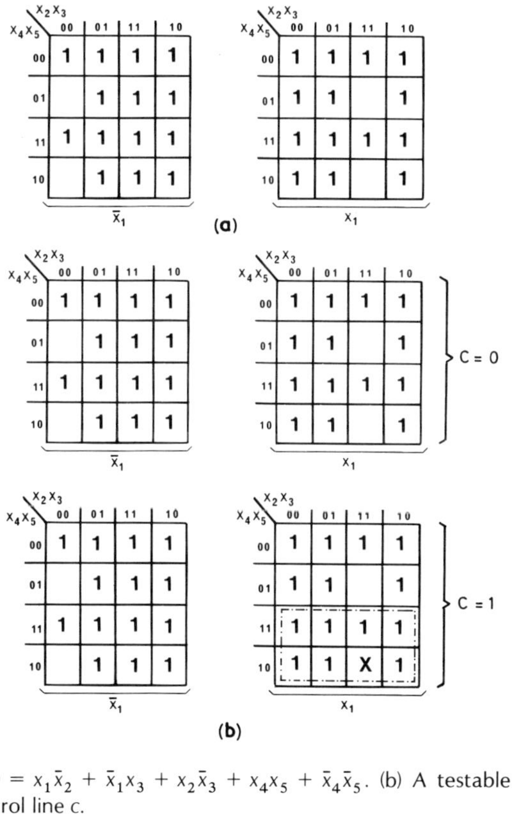

Fig. 6.15. (a) $f(X) = x_1\bar{x}_2 + \bar{x}_1x_3 + x_2\bar{x}_3 + x_4x_5 + \bar{x}_4\bar{x}_5$. (b) A testable implementation requiring one control line c.

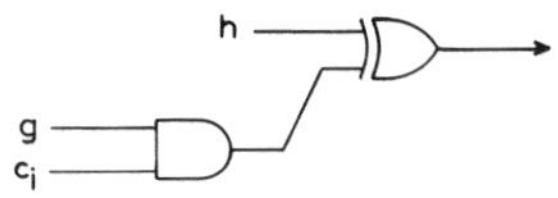

Fig. 6.16. The syndrome-testing of an Exclusive-OR gate.

(or Exclusive-NOR) gates are handled as shown in Fig. 6.16. The AND gate is an additional gate.

The computation required by this approach can be extremely high, since it is once again an exhaustive search.

6.3.3 Constrained Syndrome Testing

A constrained syndrome test[29] is one that is performed on a subset of the inputs while the other inputs are held constant. For a network realizing $f(X)$, a constrained syndrome-test verifies the behaviour of a subnetwork realizing a subfunction of $f(X)$. It is always possible to find a set of constrained syndrome tests such that correct behaviour of the network for these tests ensures it is free of single stuck-at faults. The problem is to identify a minimal, or nearly minimal, set of tests.

Constrained syndrome-testing is an alternative to hardware modification. It has the advantage that the network need not be altered to make it testable. Also no additional control inputs, which require extra pins, are required. It is thus particularly well suited to testing existing networks or networks embedded within larger systems.

Consider a network realizing $f(X)$ with a syndrome-untestable input x_i. Expanding about x_i yields

$$f(X) = Ax_i + B\bar{x}_i + C \tag{6.30}$$

where A, B and C are Boolean expressions independent of x_i. Since faults on x_i are syndrome-untestable, $f(X)$ is nonunate in x_i, so $A \neq 0$ and $B \neq 0$.

Let $\hat{X}$ denote a constraint of X found by setting certain x_i to constants. Let $\hat{A}$, $\hat{B}$ and $\hat{C}$ be the expressions that $\hat{X}$ produces from A, B and C, respectively.

There must exist an $\hat{X}$ such that $\hat{B} = 0$, since if there does not, $B = 1$ and Eq. (6.30) reduces to

$$f(X) = A + \bar{x}_i + C$$

which is unate in x_i.

Suppose for every $\hat{X}$ such that $\hat{B} = 0$, $\hat{A} = 0$. In this case, $AB = A$ and

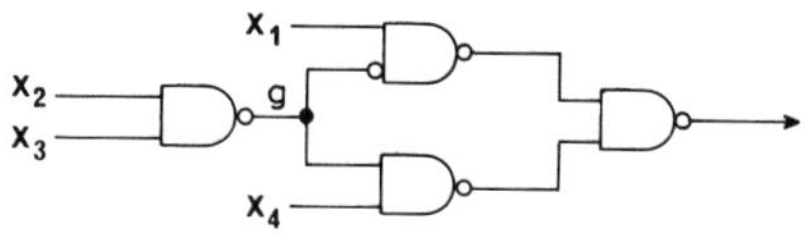

Fig. 6.17. Identical to the network in Fig. 6.1 except for the line designated g.

Eq. (6.30) becomes

$$\begin{aligned} f(X) &= ABx_i + B\bar{x}_i + C \\ &= AB + B\bar{x}_i + C \end{aligned}$$

which is once again unate in x_i.

Therefore, $\hat{X}$ exists such that

$$f(\hat{X}) = \hat{A}x_i + \hat{C}$$

Similarly, a constraint exists to make x_i a negative line.

The network in Fig. 6.17 realizes the function

$$f(X) = x_1x_2x_3 + x_4(\bar{x}_2 + \bar{x}_3)$$

which is syndrome-untestable for faults on x_2 and x_3. $f(X)$ may be written as

$$f(X) = (x_1x_3)x_2 + (x_4)\bar{x}_2 + (\bar{x}_3x_4)$$

Either of the constraints $x_1 = 0$ or $x_4 = 0$ yield a subfunction that is unate in x_2. Note that $x_3 = 0$ results in a subfunction that is independent of x_2. The case of x_3 is identical.

This development is easily extended to show that an input constraint exists to make an internal line unate and thus syndrome-testable. However, if an input line is unate, so are all the internal lines that comprise the paths from that input to the network output. Rather than treat the syndrome-untestable internal lines separately, it is better to handle them in conjunction with the inputs.

For example, $g/0$ and $g/1$ are syndrome-untestable for the network in Fig. 6.17. Since $x_1 = 0$ (or $x_4 = 0$) make x_2 and x_3 unate inputs, they also make g a unate and therefore syndrome-testable line.

A line need not be unate to be syndrome-testable. Constrained syndrome tests that insist on unateness are thus, in general, more stringent than required.

Consider a syndrome-untestable input x_i, $1 \leq i \leq m$. From Section 4.2 we have

$$\mathbf{R}_u = \frac{1}{2^{n-m}}[\mathbf{R}^0\,\mathbf{R}^1 \ldots \mathbf{R}^\beta]\mathbf{T}_{*u}^{n-m} \tag{6.31}$$

where $\beta = 2^{n-m} - 1, 0 \leq u \leq \beta$. For i this yields

$$r_{ui} = \frac{1}{2^{n-m}}[r_i^0 \, r_i^1 \cdots r_i^\beta]\mathbf{T}_{*u}^{n-m}$$

Hence $x_i/0$ and $x_i/1$ are syndrome-testable with respect to $f_u(x_1, \ldots, x_m)$ if and only if

$$[r_i^0 \, r_i^1 \cdots r_i^\beta]\mathbf{T}_{*u}^{n-m} \neq 0 \tag{6.32}$$

By permutation of variables, Eq. (6.32) yields the following.

Criterion 6.12. $x_i/0$ and $x_i/1$ are syndrome-testable with respect to the input constraint $x_{j_k} = u_k$, $1 \leq k \leq n - m$, if and only if

$$[r_i \, r_{\gamma_1 i} \, r_{\gamma_2 i} \cdots r_{\gamma_\beta i}]\mathbf{T}_{*u}^{n-m} \neq 0 \tag{6.33}$$

where $u = \sum_{k=1}^{m} u_k 2^{k-1}$ and γ_p, $1 \leq p \leq \beta$, are the nonempty subsets of $\{j_1, j_2, \ldots, j_{n-m}\}$, ordered so that $\gamma_1 = \{j_1\}$, $\gamma_2 = \{j_2\}$, $\gamma_3 = \{j_1, j_2\}$, etc.

Setting $m = 0$ in Eq. (6.33) yields the condition for the faults $x_i/0$ and $x_i/1$ to be syndrome-testable with respect to $f(X)$ itself. Criterion 6.12 is thus a partial generalization of Criterion 6.1.

Consider g an internal line in a network realizing $f(X)$. By Criterion 6.5 we know that $g/0$ is syndrome-testable with respect to $f_u(x_1, x_2, \ldots, x_m)$ if and only if

$$r_{u0} \neq \tfrac{1}{2}(\hat{r}_{u0} + \hat{r}_{un+1}) \tag{6.34}$$

where $\hat{R}_u$ is the spectrum of

$$h_u(x_1, x_2, \ldots, x_m, x_{n+1}) = h(x_1, \ldots, x_m, u_1, u_2, \ldots, u_{n-m}, x_{n+1})$$

Similarly, $g/1$ is syndrome-testable with respect to $f_u(x_1, x_2, \ldots, x_m)$ if and only if

$$r_{u0} \neq \tfrac{1}{2}(\hat{r}_{u0} - \hat{r}_{un+1}) \tag{6.35}$$

From Eq. (6.31) and its extension to $\hat{\mathbf{R}}$, we have

$$r_{u0} = \frac{1}{2^{n-m}}[r_0^0 \, r_0^1 \cdots r_0^\beta]\mathbf{T}_{*u}^{n-m}$$

$$\hat{r}_{u0} = \frac{1}{2^{n-m}}[\hat{r}_0^0 \, \hat{r}_0^1 \cdots \hat{r}_0^\beta]\mathbf{T}_{*u}^{n-m}$$

and

$$\hat{r}_{un+1} = \frac{1}{2^{n-m}}[\hat{r}_{n+1}^0 \, \hat{r}_{n+1}^1 \cdots \hat{r}_{n+1}^\beta]\mathbf{T}_{*u}^{n-m}$$

Adopting the same notation as in Criterion 6.12, Eqs. (6.34) and (6.35) yield the following.

Criterion 6.13. $g/0$ is syndrome-testable with respect to the input constraint $x_{j_k} = u_k$, $1 \leq k \leq n - m$, if and only if

$$[\hat{r}_0 + \hat{r}_{n+1} - 2r_0, \hat{r}_{\gamma_1} + \hat{r}_{\gamma_1 n+1} - 2r_{\gamma_1}, \ldots, \hat{r}_{\gamma_\beta} + \hat{r}_{\gamma_\beta n+1} - 2r_{\gamma_\beta}]\mathbf{T}^{n-m}_{*u} \neq 0 \tag{6.36a}$$

$g/1$ is syndrome-testable for this constraint if and only if

$$[\hat{r}_0 - \hat{r}_{n+1} - 2r_0, \hat{r}_{\gamma_1} - \hat{r}_{\gamma_1 n+1} - 2r_{\gamma_1}, \ldots, \hat{r}_{\gamma_\beta} - \hat{r}_{\gamma_\beta n+1} - 2r_{\gamma_\beta}]\mathbf{T}^{n-m}_{*u} \neq 0 \tag{6.36b}$$

Criterion 6.13 is a partial generalization of Criterion 6.5.

Consider the network in Fig. 6.17 that has the spectrum

r_0	r_1	r_2	r_3	r_4	r_{12}	r_{13}	r_{14}	r_{23}	r_{24}	r_{34}	r_{123}	r_{124}	r_{134}	r_{234}	r_{1234}
8;	−2	0	0	−6;	2	2	0	0	−2	−2;	−2	0	0	2;	0

$x_2/0$ and $x_2/1$ are syndrome-testable under the constraint $x_4 = 0$ since

$$[r_2 \quad r_{12}]\mathbf{T}^1_{*0} = [0 \quad 2]\begin{bmatrix}1\\1\end{bmatrix} = 2$$

The constraint $x_1 = 1$ also makes these faults syndrome-testable, as do $x_4 = 0$ and $x_4 = 1$.

The constraint $x_1 = x_4 = 0$ does not make $x_2/0$ and $x_2/1$ syndrome-testable since

$$[r_2\, r_{12}\, r_{24}\, r_{124}]\mathbf{T}^2_{*0} = [0 \quad 2 \quad -2 \quad 0]\begin{bmatrix}1\\1\\1\\1\end{bmatrix} = 0$$

This is easily verified by inspection of the network.

$g/0$ and $g/1$ are also syndrome-untestable for the network in Fig. 6.17. For this line, we have

$\hat{r}_0$	$\hat{r}_1$	$\hat{r}_4$	$\hat{r}_5$	$\hat{r}_{15}$	$\hat{r}_{45}$
16	−8	−8	0	8	−8

(all $\hat{r}_\alpha$ involving x_2 or x_3 are, of course, zero). The constraint $x_1 = 1$ makes g syndrome-testable since

$$[\hat{r}_0 \pm \hat{r}_5 - 2r_0, \hat{r}_1 \pm \hat{r}_{15} - 2r_1]\mathbf{T}^1_{*0} = [0, -8 \quad \pm 8 \quad +4]\begin{bmatrix}1\\1\end{bmatrix} \neq 0$$

Suppose x_i is a syndrome-untestable input. Let $r_{\alpha i}$ be a non-zero spectral coefficient such that $|\alpha|$ is minimal. Such a coefficient must exist since, if it does not, faults on x_i are not testable for any r_α; hence x_i is redundant. A constraint that fixes the variables of α must satisfy

$$[r_i r_{\alpha_1 i} r_{\alpha_2 i} \cdots r_{\alpha_\beta i}]\mathbf{T}_{*u}^{|\alpha|} \neq 0 \tag{6.37}$$

where α_p, $1 \leq p \leq \beta$, are the nonempty subsets of α, $\alpha_\beta = \alpha$, if it is to make x_i syndrome-testable. Since $|\alpha|$ is minimal, $r_i = 0$ and $r_{\alpha_p i} = 0$, $1 \leq p < \beta$, and Eq. (6.37) holds for all u.

Criterion 6.14. If $x_i/0$ and $x_i/1$ are r_α-testable with $|\alpha|$ minimal, $i \notin \alpha$. they are also syndrome-testable for every constraint that fixes the variables indicated by α.

The following criterion follows in an analogous manner.

Criterion 6.15. If $g/0$ ($g/1$) is r_α-testable with $|\alpha|$ minimal, $g/0$ ($g/1$) is also syndrome-testable for every constraint that fixes the variables indicated by α.

These criteria provide simple methods for identifying which inputs to constrain to make a particular fault syndrome-testable. On an individual basis, the choice of constraints is arbitrary. In combination, the constraints must be selected so that every fault is tested.

Consider the network in Fig. 6.18 that realizes

$$f(X) = x_1\bar{x}_3\bar{x}_4 + x_2x_3\bar{x}_4 + \bar{x}_1\bar{x}_3x_4 + \bar{x}_2x_3x_4$$

with the spectrum

r_0	r_1	r_2	r_3	r_4	r_{12}	r_{13}	r_{14}	r_{23}	r_{24}	r_{34}	r_{123}	r_{124}	r_{134}	r_{234}	r_{1234}
8;	0	0	0	0;	0	0	−4	0	−4	0;	0	0	−4	4;	0

All single-input stuck-at faults are syndrome-untestable ($r_i = 0$, $1 \leq i \leq 4$).

Since r_{14} and r_{24} are the lowest-order nonzero coefficients involving x_1 and x_2, constraining the value of x_4 will test single faults on x_1 and x_2.

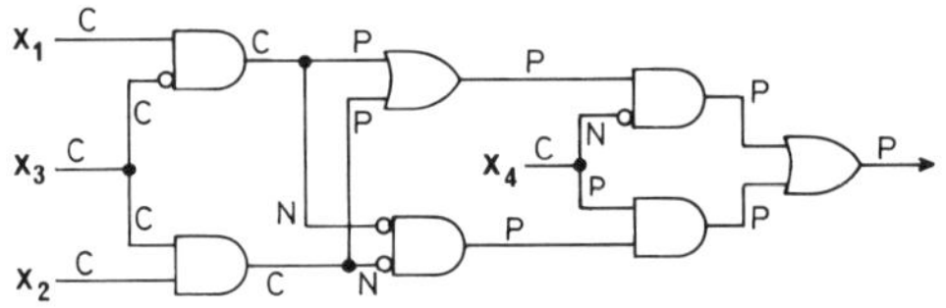

Fig. 6.18. A network to be used in illustrating constrained syndrome testing.

Faults on x_3 can be tested by constraining x_1 and x_4 ($r_{134} \neq 0$) or x_2 and x_4 ($r_{234} \neq 0$). Faults on x_4 are testable by constraining x_1 or x_2.

Suppose we choose the constraint $x_4 = 0$ as illustrated in Fig. 6.19a. x_1 and x_2 become positive and therefore syndrome-testable. Four internal lines also become testable by becoming unate. Figure 6.19b shows the effect of the constraint $x_1 = x_4 = 0$. x_3 is testable as required.

Finally, x_1 or x_2 must be constrained to test x_4. Care must be taken to select a test so that all internal lines not tested by the previous constrained tests are covered. Figure 6.19c shows an appropriate choice; $x_2 = 1$ is also acceptable, but $x_1 = 0$ or $x_2 = 0$ is not.

The assignment in Fig. 6.19c is interesting in that it makes faults on x_4 syndrome-testable without making x_4 unate. Making x_4 unate requires two inputs be fixed, for example, $x_1 = x_2 = 0$ or $x_1 = x_3 = 0$.

The problem of finding a set of constrained syndrome tests that cover all single faults in a network is closely related to the spectral signatures discussed in the next section. We defer a detailed discussion of this problem to Section 6.5.

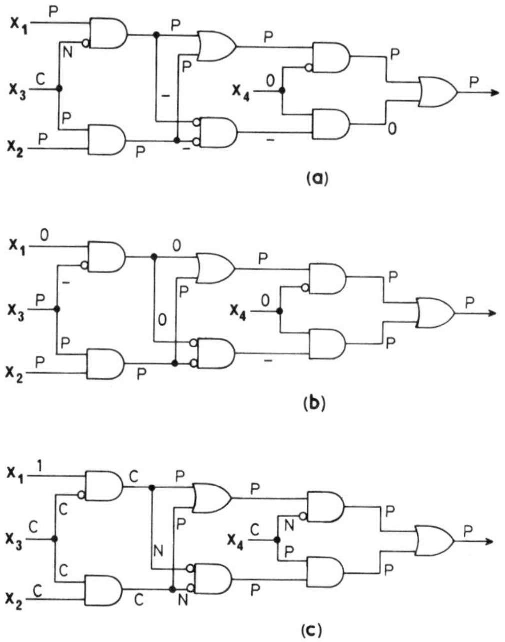

Fig. 6.19. (a) Syndrome testing the network of Fig. 6.18 for the constraint $x_4 = 0$. (b) The effect of the constraint $x_1 = x_4 = 0$. (c) The effect of the constraint $x_1 = 1$. Note that faults on x_4 are made syndrome-testable even though x_4 does not become unate.

6.4 SPECTRAL FAULT DETECTION SIGNATURES

6.4.1 Unique Signatures

As an alternative to hardware modification and constrained syndrome testing, we shall now examine fault detection procedures that make use of all the spectral coefficients and not just r_0. The idea is to identify a subset of the coefficients called a signature, such that verification of these coefficients ensures correct behaviour of the network. We first consider unique signatures. Such a signature distinguishes the function realized by a network from all others and can thus detect all permanent functional failures.

While the full 2^n spectral coefficients uniquely define any given function, it is well known (see Chapter 3) that the zero-order and n first-order coefficient values uniquely define a linearly separable (threshold) function. These coefficients thus form a unique spectral signature for any network realizing a linearly separable function, since any permanent functional failure will modify at least one of these coefficients. Unfortunately, the fraction of functions that are linearly separable tends rapidly to zero as n increases beyond 3.

Spectral techniques for function classification were also discussed in Chapter 3. It was shown that, given any function $f(X)$ and its spectrum, the five following operations can be used to reorder and sign-change the spectral coefficients such that a positive, canonic ordering results:

(i) negation of a variable,
(ii) negation of the function,
(iii) permutation of variables,
(iv) spectral translation, and
(v) disjoint spectral translation,

These operations divide the set of Boolean functions into equivalence classes with each class being uniquely defined by its canonic function.

If the canonic function in an equivalence class is linearly separable, it has a unique spectral signature containing $n + 1$ coefficients. So, in fact, do all the other functions within that class. The coefficients of the signature are precisely those that become the zero- and first-order coefficients when the function is put into canonic form. Thus for any function in a class where the canonic function is linearly separable, its unique spectral signature is completely defined by knowledge of the operations required to put it into canonic form and the $n + 1$ coefficients defining the canonic function.

For $n = 2, 3$, all the canonic functions are threshold. For $n = 4$, seven of the eight classes have linearly separable canonic functions. These seven classes contain 64,640 of the 65,536 four-variable functions (98.8%). For $n = 5$, 21 of 48 classes have linearly separable canonic functions. The number of functions they include is not known.

The eighth four-variable class has the canonic function whose spectrum is†

s_0	s_1	s_2	s_3	s_4	s_{12}	s_{13}	s_{14}	s_{23}	s_{24}	s_{34}	s_{123}	s_{124}	s_{134}	s_{234}	s_{1234}
4;	4	4	4	4;	4	4	−4	−4	4	−4;	−4	−4	−4	4;	−4

This function is unate but not linearly separable. The class has 896 functions. Since the spectral coefficients all have magnitude 4, the information content of the spectrum is essentially binary. Each function in this class must therefore have at least 10 coefficients in a unique spectral signature.

A signature consisting of 10 out of 16 coefficients is not encouraging. While no detailed study has been performed, the decreasing fraction of functions that are linearly separable would seem to indicate the size of a unique signature will grow with n.

One must remember that not all functions are used in practice. Designers favour simple functions where possible. The above result may thus be more frequently of interest than first appears. Further, for functions with don't-care conditions, we may suggest that, when possible, the don't cares be completed so that the function falls within a class with a linearly separable canonic function. A function thus completed has a unique $n + 1$ spectral coefficient signature.

6.4.2 Internally Unate Networks

When a unique signature involves too many coefficients to be practical, it is necessary to pursue a spectral signature that is only unique under certain fault assumptions. For example, $\{r_0\}$ in a multiple stuck-at fault signature for a fan-out-free network devoid of Exclusive-OR and Exclusive-NOR. $\{r_0\}$ is also a single stuck-at fault signature for a unate network, and for an internally unate network that realizes a function where $r_i \neq 0$, $1 \leq i \leq n$. We now consider the choice of single stuck-at fault signatures for internally unate networks that realize functions with one or more zero first-order coefficients. Note that such a function must of necessity have r_0 even.

The most straightforward approach involves the direct application of Criterion 6.1. r_0 covers all single stuck-at faults on internal lines and on all

† This example is presented in $\langle +1, -1 \rangle$ coding to more clearly demonstrate its binary nature. The choice of $\langle +1, -1 \rangle$ or $\langle 0, 1 \rangle$ coding is of no relevance since the transformation between the two is unique and invertible.

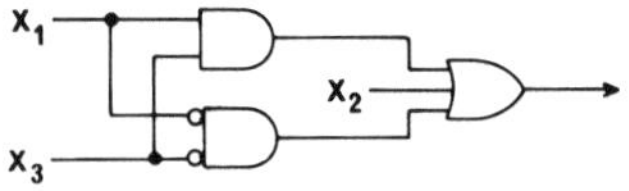

Fig. 6.20. $f(X) = x_1x_3 + \bar{x}_1\bar{x}_3 + x_2$.

inputs x_i such that $r_i \neq 0$, $1 \leq i \leq n$. For those x_j, $1 \leq j \leq n$, such that $r_j = 0$, we construct a covering table with one column per x_j and one row for each spectral coefficient other than r_0. A mark is placed in each entry where the corresponding x_j is tested by the corresponding spectral coefficient. Recall that from Criterion 6.1, $x_j/0$ and $x_j/1$ are r_α- and $r_{\alpha j}$-testable if and only if $r_{\alpha j} \neq 0$. Minimal-covering-selection techniques analogous to prime-implicant table methods can then be applied to select a spectral signature.

Following this method will not always yield a minimal signature, since we always include r_0. This is done to avoid a detailed analysis of all the internal lines and those x_i for which $r_i \neq 0$. At worst, a signature with one more than the minimal number of coefficients will be selected.

Consider the network in Fig. 6.20, which realizes

$$f(X) = x_1x_3 + \bar{x}_1\bar{x}_3 + x_2$$

with the spectrum

$$\begin{array}{cccccccc} r_0 & r_1 & r_2 & r_3 & r_{12} & r_{13} & r_{23} & r_{123} \\ 6; & 0 & -2 & 0; & 0 & 2 & 0; & 2 \end{array}$$

The appropriate covering table is Table 6.2. $\{r_0, r_{13}\}$ and $\{r_0, r_{123}\}$ are suitable signatures.

The covering table approach rapidly becomes unwieldy as the number of inputs increases. We shall thus introduce three different signatures that are more easily determined.

A basis signature[33] consists of r_0 and n other spectral coefficients whose subscripts, if written as n-bit vectors, form a basis over the space of n-bit

Table 6.2. A spectral coefficient covering table for single-input stuck-at faults for the network in Fig. 6.20.[a]

	r_1	r_2	r_3	r_{12}	r_{13}	r_{23}	r_{123}
x_1			√		√	√	√
x_3	√			√	√		√

[a] Note that x_2 is r_0-testable since $r_2 \neq 0$.

vectors. Basis signatures are of considerable theoretical importance since they consist of the coefficients used for the canonic classification of Edwards (see Section 3.5) and the linearization method due to Karpovsky.[44] They are of less practical importance since they always consist of $n + 1$ coefficients. Thus this establishes an upper bound on the size of a single stuck-at fault signature for internally unate networks.

To write a coefficient with its subscript as a binary vector, the variables involved in the coefficient are indicated by 1s in the binary vector. Hence, for a three-variable function, $r_0, r_1, r_2, r_{12}, \ldots, r_{123}$ become $r_{000}, r_{001}, r_{010}, r_{011}, \ldots, r_{111}$, respectively.

A basis signature is constructed iteratively. At each step the spectral coefficients eligible to enter the basic are examined. The selection is made by highest magnitude and within highest magnitude by the order of the coefficient, i.e., the number of 1s in its binary subscript. If two coefficients have the same magnitude and order, the one whose subscript has the smaller decimal value is chosen.

For example, consider an internally unate network realizing the function whose spectrum is given in Table 6.3. For the basis signature in addition to r_{0000}, we choose

r_{1001} highest magnitude
r_{0001} lowest order, magnitude 2
$\left.\begin{matrix} r_{0011} \\ r_{0101} \end{matrix}\right\}$ next lowest order, magnitude 2, chosen in decimal ordering.

The basis signature is $\{r_0, r_{14}, r_1, r_{12}, r_{13}\}$.

Since the basis signature always includes r_0, it covers all single stuck-at faults on internal lines. By definition, the basis signature contains at least

Table 6.3. Spectrum of $f(X) = \sum m(1, 3, 5, 7, 8, 10, 12, 15)$, with subscripts given in the normal decimal notation and the alternate binary form.

Decimal subscripts	Binary subscripts	Value	Decimal subscripts	Binary subscripts	Value
r_0	r_{0000}	8	r_4	r_{1000}	0
r_1	r_{0001}	−2 ←	r_{14}	r_{1001}	−6 ←
r_2	r_{0010}	0	r_{24}	r_{1010}	0
r_{12}	r_{0011}	2 ←	r_{124}	r_{1011}	−2
r_3	r_{0100}	0	r_{34}	r_{1100}	0
r_{13}	r_{0101}	2 ←	r_{134}	r_{1101}	−2
r_{23}	r_{0110}	0	r_{234}	r_{1110}	0
r_{123}	r_{0111}	−2	r_{1234}	r_{1111}	2

one coefficient $r_{i\alpha} \neq 0$, for each $i \in \{1, 2, \ldots, n\}$, $\alpha \subseteq \{1, 2, \ldots, n\}$. By Criterion 6.1, such an $r_{i\alpha}$ covers $x_i/0$ and $x_i/1$. A basis signature covers all single stuck-at faults in an internally unate network.

The basis signature is usually larger than necessary since, for example, if $r_i \neq 0$, r_0 covers $x_i/0$ and $x_i/1$. It is only those x_j for which $r_j = 0$ that the signature of $\{r_0\}$ need be augmented.

A minimal-covering signature[33] consists of r_0 plus a minimal set of nonzero spectral coefficients with x_i involved in at least one coefficient in the signature for each $r_i \neq 0$. The selection of a minimal-covering signature is another example of a prime-implicant-type covering problem, but it can often be solved very quickly by heuristic methods, working from the higher-order coefficients toward the lower-order coefficients because of the "subset" nature of the coefficients. In particular, $r_{12\ldots n}$ will suffice on its own if it is nonzero.

The latter case is true for our previous example (see Table 6.3), where $r_{1234} = 2$. The minimal-covering signature is $\{r_0, r_{1234}\}$.

The third alternative, a minimal-input signature,[33] does not always exist. When it does, it has the advantage of using only r_0 and certain first-order coefficients. We shall show later that these coefficients are of more practical utility than those of a higher order.

A minimal-input signature consists of r_0 together with a minimal number of first-order coefficients r_{i_k}, $i_k \in \{1, 2, \ldots, n\}$, $1 \leq i_k \leq t$, such that for every j, where $r_j = 0$, $1 \leq j \leq n$, there exists an $i \in \{i_1, i_2, \ldots, i_t\}$ such that $r_{ij} \neq 0$. Once again r_0 covers all single stuck-at faults on internal lines as well as on those x_i for which $r_i \neq 0$. The first-order coefficients are selected to satisfy Criterion 6.1 and thus cover all remaining input faults.

For our example in Table 6.3, $r_1 \neq 0$; hence $x_1/0$ and $x_1/1$ are r_0-testable. Since $r_2 = r_3 = r_4 = 0$, single stuck-at faults on x_2, x_3 and x_4 are not r_0-testable. However, r_{12}, r_{13} and r_{14} are all nonzero so the minimal-input signature is $\{r_0, r_1\}$.

A basis signature always has $n + 1$ coefficients. A minimal covering and a minimal input typically have less. The minimal-covering signature usually has the same or fewer coefficients than a minimal-input signature, but this is not always the case. In many instances, the minimal-input signature is preferred because it is easier to test. As noted earlier, the minimal-input signature does not always exist. This situation arises when there is one or more variables for which the corresponding first-order coefficient and all corresponding second-order coefficients are zero.

For some functions, very simple identification procedures suffice. Let γ_0, γ_1, γ_2 and γ_3 denote the number of true minterms in the subfunctions of

$f(X)$ found by fixing the values of x_i and x_j as follows:

x_i	x_j	Number of minterms
0	0	γ_0
0	1	γ_1
1	0	γ_2
1	1	γ_3

which yields

$$r_0 = \gamma_0 + \gamma_1 + \gamma_2 + \gamma_3$$
$$r_i = \gamma_0 + \gamma_1 - \gamma_2 - \gamma_3$$
$$r_j = \gamma_0 - \gamma_1 + \gamma_2 - \gamma_3$$
$$r_{ij} = \gamma_0 - \gamma_1 - \gamma_2 + \gamma_3$$

By Criterion 6.1, $x_j/0$ and $x_j/1$ are r_i-testable if and only if $r_{ij} \neq 0$. We need only consider those x_j for which $r_j = 0$, since all others are r_0-testable. Since $r_j = 0$, $r_{ij} = r_{ij} - r_j = 2(\gamma_3 - \gamma_2)$. Thus faults on x_j are r_i-testable if and only if $\gamma_2 \neq \gamma_3$.

Clearly, if $\frac{1}{2}(r_0 - r_i)$ is odd, $\gamma_2 \neq \gamma_3$. Hence $\{r_0, r_i\}$, such that $\frac{1}{2}(r_0 - r_i)$ is odd and $r_i \neq 0$, is a single stuck-at fault signature. r_0 covers all internal stuck-at faults and faults on those x_j, such that $r_j \neq 0$ (including x_i), while r_i covers the faults on the remaining inputs. This result confirms $\{r_0, r_1\}$ as a single stuck-at fault signature for any internally unate realization of the function defined by Table 6.3.

If $\frac{1}{2}r_0$ is odd, $\{r_0, r_i\}$, $r_i = 0$, suffices in the manner described above for all single stuck-at faults except $x_i/0$ and $x_i/1$. It is sufficient to add any r_α (or $r_{\alpha i}$) such that $r_{\alpha i} \neq 0$. While this may not lead to an optimal result, it can provide a good approximation that can be easily reduced.

Consider an internally unate realization of

$$f(X) = x_1\bar{x}_2 + x_1\bar{x}_3 + x_2x_3x_4$$

that has the spectrum

r_0	r_1	r_2	r_3	r_4	r_{12}	r_{13}	r_{14}	r_{23}	r_{24}	r_{34}	r_{123}	r_{124}	r_{134}	r_{234}	r_{1234}
8;	−6	0	0	−2;	−2	−2	0	0	2	2;	2	0	0	−2;	0

The basis signature is $\{r_0, r_1, r_4, r_{12}, r_{13}\}$. One possible minimal-covering signature is $\{r_0, r_{12}, r_{234}\}$.

Since $\frac{1}{2}(r_0 - r_1)$ is odd, $\{r_0, r_1\}$ is a minimal-input signature; $\{r_0, r_4\}$ is as well.

As a further example, reconsider

$$f(X) = x_1x_3 + \bar{x}_1\bar{x}_3 + x_2$$

which has the spectrum

$$\begin{array}{cccccccc} r_0 & r_1 & r_2 & r_3 & r_{12} & r_{13} & r_{23} & r_{123} \\ 6; & 0 & -2 & 0; & 0 & 2 & 0; & 2 \end{array}$$

Earlier we found $\{r_0, r_{13}\}$ and $\{r_0, r_{123}\}$ to be minimal signatures. These are minimal-covering signatures. The basis signature is $\{r_0, r_2, r_{13}, r_1\}$. Since $\frac{1}{2}r_0$ is odd, $\{r_0, r_1\}$ covers all single stuck-at faults except $x_1/0$ and $x_1/1$. However, $r_{13} \neq 0$, so $x_1/0$ and $x_1/1$ are r_3-testable and $\{r_0, r_1, r_3\}$ is a minimal-input signature.

The basis signature is of little practical utility and is not often computed. The best approach to determine a single stuck-at fault signature for an internally unate network appears to be to compute minimal covering and minimal-input signatures, and select the one with the fewest coefficients.

6.4.3 General Networks

We now turn to the problem of determining a spectral single stuck-at fault detection signature for a general combinational network, i.e., a combinational network subject to no gate or topology restrictions. Determining a signature for such a network is complicated by the need to determine explicitly the testability of the internal lines. The basic results were presented in Section 6.2. Here we consider a systematic approach to the problem and a number of methods for reducing the computation required.

Since it appears that the fraction of lines that are syndrome-testable is high for practical networks, we choose always to include r_0 in the spectral signature. Since unate lines are syndrome-(r_0)-testable, and unate lines are rather easily identified using Procedure 6.1, the decision to include r_0 in general greatly reduces the computation required to determine a signature. As a consequence, the problem of determining if a general combinational network is syndrome-testable for single stuck-at faults is a special case of the determination of a spectral signature, and the following discussion is applicable.

The first step in identifying a spectral signature is to apply Procedure 6.1 to identify the candidate syndrome-untestable lines, both input and internal. All other lines are unate and hence syndrome-testable. This step will of course identify unate or internally unate networks, which are handled as discussed earlier.

Two faults are functionally equivalent if they cause the network to realize the same faulty function. Functionally equivalent faults are tested by the same spectral coefficients. Functional equivalence partitions the set of stuck-at faults into equivalence classes. The testability of only one fault in each class needs to be explicitly determined.

Finding the fault equivalence classes is, in general, quite difficult. We follow the normal practice of only considering those functionally equivalent subclasses that can be determined from highly localized network information. The fault equivalencies for the basic gates are summarized in Table 6.4. In using these equivalencies, both input–output and inter–input equivalencies should be considered.

In constructing the fault equivalence classes, we are only specifically concerned with faults on the candidate syndrome-untestable lines, since a fault on a candidate syndrome-untestable line cannot be equivalent to a fault on a unate line. Multiple-input stuck-at faults are considered since these are more easily analyzed than a single internal fault equivalent.

Table 6.4. Equivalent input and output faults for single gates.

Gate	Output	Inputs
NOT	Stuck-at-0	Stuck-at-1
	Stuck-at-1	Stuck-at-0
AND	Stuck-at-0	Any number of inputs stuck-at-0
	Stuck-at-1	All inputs stuck-at-1
OR	Stuck-at-0	All inputs stuck-at-0
	Stuck-at-1	Any number of inputs stuck-at-1
NAND	Stuck-at-0	All inputs stuck-at-1
	Stuck-at-1	Any number of inputs stuck-at-0
NOR	Stuck-at-0	Any number of inputs stuck-at-1
	Stuck-at-1	All inputs stuck-at-0
Exclusive-OR	Stuck-at-0	All inputs stuck with an even number of ones
	Stuck-at-1	All inputs stuck with an odd number of ones
Exclusive-NOR	Stuck-at-0	All inputs stuck with an odd number of ones
	Stuck-at-1	All inputs stuck with an even number of ones

Once the fault-equivalence classes are found, a covering approach can be used to determine a spectral signature. A table would be constructed with one column for each spectral coefficient and one row for each fault-equivalence class. Each entry would indicate whether the corresponding coefficient tests the faults in the corresponding equivalence class. Conventional covering techniques could be applied to this table in order to select a spectral signature, with the proviso that r_0 always be chosen to ensure coverage of the faults on unate lines.

In constructing this table, one fault from each equivalence class must be selected. When possible, an input fault should be chosen, since its testability can be determined from $\mathbf{R}$ alone and no $\hat{\mathbf{R}}$ need be constructed. For an equivalence class containing only internal faults, the fault nearest the output should be chosen, since this fault will have the most easily constructed $\hat{\mathbf{R}}$.

The covering-table approach involves a great deal of computation, and quickly becomes unreasonable as the size of the network grows. The following heuristic approach is much simpler, and while it does not yield a minimal signature, it appears to give reasonable results.

As before, the candidate syndrome-untestable input and internal lines are identified using Procedure 6.1. A spectral signature covering all single-input stuck-at faults is then found in the manner described for internally unate networks. All single-input stuck-at faults on candidate lines are then recorded as being testable.

Beginning with the candidate syndrome-untestable internal lines closest to the input, we work through the network toward the output. Each candidate syndrome-untestable single-internal stuck-at fault is examined to see if it is functionally equivalent to a fault already known to be testable. If it is, it is recorded as being testable. If it is not, it must be explicitly examined.

When explicitly examining an internal fault, we wish to show if it is tested by one or more coefficients in the signature. To do this, we can make use of Criteria 6.5, 6.6 and 6.7. Criterion 6.5 is the general condition and is only used when the other criteria do not apply. Criterion 6.6 can be used to identify coefficients that test an internal fault that results in a faulty function independent of one or more inputs. Criterion 6.6 is a sufficient but not a necessary condition, so when it fails, the other criteria must be tried.

Criterion 6.7 concerns the testability of faults in disjoint subnetworks. It partitions the network into disjoint realizations of $g(X)$ and $h(X, g)$. Its advantage is that all faults in the subnetwork realizing $g(X)$ can be checked in turn using the same $\hat{\mathbf{R}}$ for $h(X, g)$, thus greatly reducing the size of the spectra that need to be computed. In addition, if $g(X)$ has a simple structure, i.e., it is unate or internally unate, then the testability of faults with respect to $g(X)$ is transparent.

When an internal fault is encountered that is not tested by the signature

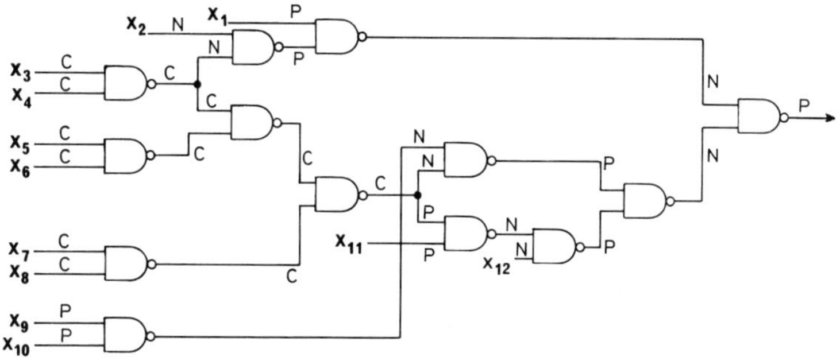

Fig. 6.21. A network used in illustrating fault equivalence. First presented by Roth and Savir.[5]

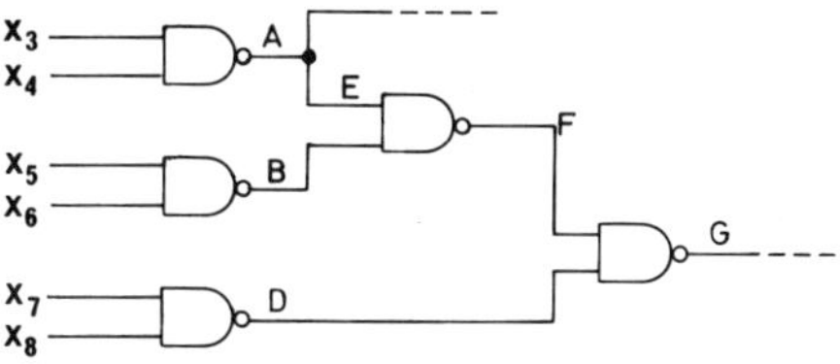

Fig. 6.22. The subnetwork of candidate syndrome-untestable lines from the network in Fig. 6.21.

constructed up to that point, a coefficient must be added to cover that fault. This may make a previous coefficient in the signature redundant. Some degree of backtracking is required at that point in order to remove the redundancy.

This approach appears to work quite well in practice. The best order in which to consider the candidate syndrome-untestable internal faults and the optimal backtracking procedure are open questions.

As an example, consider the network in Fig. 6.21. This network realizes a function with an odd number of ones, so all single- and multiple-input stuck-at faults are syndrome-testable.

Figure 6.22 shows the candidate syndrome-untestable lines from the network in Fig. 6.21. Table 6.5 shows the applicable functional equivalencies in the order considered. $E/1$ is the only candidate syndrome-untestable internal fault that is not equivalent to an input fault. Criteria 6.6 and 6.7 do not apply to this fault. $E/1$ can be shown to be syndrome-testable by constructing the appropriate $\hat{\mathbf{R}}$ and applying Criterion 6.5. The network in Fig. 6.22 is thus syndrome-testable for all single stuck-at faults.

Consider the network in Fig. 6.23 that realizes

$$f(X) = x_1x_2x_3 + x_4(\bar{x}_2 + \bar{x}_3)$$

Table 6.5. Stuck-at fault equivalences for the subnetwork shown in Fig. 6.22.

Fault	Equivalent faults
$A/0$	$x_3/1$ and $x_4/1$
$A/1$	$x_3/0$ or $x_4/0$
$B/0$	$x_5/1$ and $x_6/1$
$B/1$	$x_5/0$ or $x_6/0$
$D/0$	$x_7/1$ and $x_8/1$
$D/1$	$x_7/0$ or $x_8/0$
$E/0$	$B/0$
$E/1$	None
$F/0$	$D/0$
$F/1$	$B/0$
$G/0$	$F/1$ and $D/1$
$G/1$	$D/0$

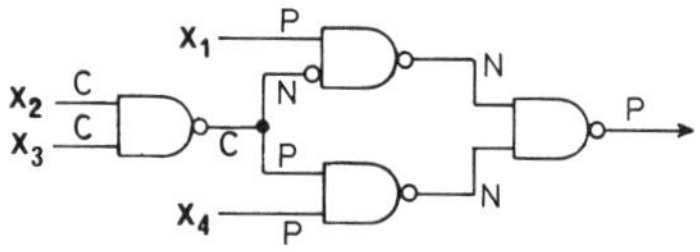

Fig. 6.23. Procedure 6.1 applied to a realization of $f(X) = x_1x_2x_3 + x_4(\bar{x}_2 + \bar{x}_3)$.

with the spectrum

r_0	r_1	r_2	r_3	r_4	r_{12}	r_{13}	r_{14}	r_{23}	r_{24}	r_{34}	r_{123}	r_{124}	r_{134}	r_{234}	r_{1234}
8;	-2	0	0	-6;	2	2	0	0	-2	-2;	-2	0	0	2;	0

Faults on x_2 or x_3 are syndrome-untestable ($r_2 = r_3 = 0$). Clearly, faults on x_2 or x_3 are both r_1- and r_4-testable. $\{r_0, r_1\}$ is one spectral signature to cover single-input stuck at faults.

Since $g = \overline{x_2x_3}$ terminates a disjoint subnetwork, $g/0$ and $g/1$ are tested by the same spectral coefficients (see Criterion 6.7). $g/1$ is functionally equivalent to $x_2/0$ (or $x_3/0$). Hence $\{r_0, r_1\}$ is a signature for all single stuck-at faults in the network in Fig. 6.23.

6.5 FURTHER CONSIDERATIONS

6.5.1 Relation between Constrained Syndrome and Spectral Coefficient Testing

Constrained syndrome-testing[29] was discussed in Section 6.3.3. At that time, we showed that any single stuck-at fault that is r_α-testable where $|\alpha|$ is minimal is also constrained syndrome-testable for an input constraint that fixes the inputs indicated by α. The values assigned to these inputs are arbitrary. In the previous section, we presented methods for deriving spectral signatures for the detection of single stuck-at faults. Here, we pursue the relation between constrained syndrome and spectral coefficient testing and, in particular, the transformation of a spectral signature to a set of constrained syndrome tests. It is assumed throughout this section that the network in question contains at least one syndrome-untestable fault.

Consider an internally unate network with a minimal input signature $\{r_0, r_{i_1}, r_{i_2}, \ldots, r_{i_p}\}$, $i_j \in \{1, 2, \ldots, n\}$, $1 \leq j \leq p$. In this case, every single-input stuck-at fault is either r_0-testable or r_{i_j}-testable for some i_j. This signature translates directly to an unconstrained syndrome test followed by a constrained test for each x_{i_j}, $1 \leq j \leq p$. In each of the constrained tests, the value assigned to the constrained input is arbitrary since all the tests clearly satisfy the minimum-cardinality condition.

The overall testing process can be improved by replacing the unconstrained syndrome test by a second constrained test in any one of the x_{i_j}. This second test in x_{i_j} must be constrained to the opposite value of the original test that constrained x_{i_j}. This replacement is possible since, if a fault is r_0-testable, it must also be r_0-testable under the constraint $x_{i_j} = 0$ or $x_{i_j} = 1$ or both. The advantage of the constrained test is that it uses one-half the input assignments (hence one-half the time) of an unconstrained test.

In Section 6.4.2 we found that $\{r_0, r_1\}$ is a minimal-input signature for an internally unate realization of

$$f(X) = x_1\bar{x}_2 + x_1\bar{x}_3 + x_2x_3x_4$$

Two constrained syndrome tests setting $x_1 = 0$ and $x_1 = 1$ are sufficient. Note that in total these two constrained tests require the same number of input assignments as the single unconstrained syndrome test, but the latter does not detect all single stuck-at faults.

Consider an internally unate network with a minimal-covering signature $\{r_0, r_{\alpha_1}, \ldots, r_{\alpha_p}\}$, $\alpha_j \subseteq \{1, 2, \ldots, n\}$, $1 \leq j \leq p$. In this form of signature, for each syndrome-untestable input x_k, $1 \leq k \leq n$, there is at least one α_j, $1 \leq j \leq p$, such that $k \in \alpha_j$. The constrained test of x_k derived from α_j fixes all the variables in $\alpha_j - \{k\}$. $|\alpha_j - \{k\}|$ is minimal, and the choice of constraint assignment is arbitrary if and only if $|\alpha_j|$ is minimal. The latter is often not the case, and the choice of input assignment must be explicitly verified using expression (6.33).

For example,

$$f(X) = x_1x_3 + \bar{x}_1\bar{x}_3 + x_2$$

has the spectrum

r_0	r_1	r_2	r_3	r_{12}	r_{13}	r_{23}	r_{123}
6;	0	−2	0;	0	2	0;	2

An internally unate realization of this function has the minimal covering signature $\{r_0, r_{13}\}$. An appropriate set of constrained tests fixes $x_1 = 0$, $x_1 = 1$ and $x_3 = 0$. Here the constraint assignments were arbitrary, and a pair of constrained tests over x_1 eliminated the need to test r_0.

Minimal-covering signatures do not always lead to simple constrained tests. Consider

$$f(X) = x_1\bar{x}_2 + \bar{x}_1x_3 + x_2\bar{x}_3 + x_4x_5 + \bar{x}_4\bar{x}_5$$

which has nonzero spectral coefficients

$$\begin{array}{cccccccc} r_0 & r_{12} & r_{13} & r_{23} & r_{45} & r_{1245} & r_{1345} & r_{2345} \\ 28; & -4 & -4 & -4 & 4; & 4 & 4 & 4 \end{array}$$

The minimal-covering signature is $\{r_0, r_{12}, r_{1345}\}$. From r_{12} we derive the constrained tests $x_1 = 0$, $x_1 = 1$ and $x_2 = 0$, which cover the faults on x_1 and x_2 as well as all syndrome (r_0)-testable faults. It remains to cover faults on x_3, x_4 and x_5.

From r_{1345}, we must constrain $\{x_1, x_4, x_5\}$, $\{x_1, x_3, x_5\}$ and $\{x_1, x_3, x_4\}$ to cover x_3, x_4 and x_5, respectively. None of these satisfy the minimal-cardinality constraint. The appropriate constraint assignment must be explicitly determined.

For example, for $\{x_1, x_4, x_5\}$ to test faults on x_3 we require a u such that

$$[r_3, r_{13}, r_{34}, r_{134}, r_{35}, r_{135}, r_{345}, r_{1345}]\mathbf{T}^3_{*u} \neq 0$$

This is most easily solved by treating the vector of coefficients as a column vector, and performing a fast transform. Any nonzero entry in the resulting column vector identifies an appropriate u. In our case,

$$[\mathbf{T}^3]\begin{bmatrix} 0 \\ -4 \\ 0 \\ 0 \\ 0 \\ 0 \\ 0 \\ 4 \end{bmatrix} = \begin{bmatrix} 0 \\ 0 \\ -8 \\ 8 \\ -8 \\ 8 \\ 0 \\ 0 \end{bmatrix}\begin{matrix} u \\ 0 \\ 1 \\ 2 \\ 3 \\ 4 \\ 5 \\ 6 \\ 7 \end{matrix}$$

The appropriate u are 2, 3, 4 or 5 corresponding to the constraints

u	x_5	x_4	x_1
2	0	1	0
3	0	1	1
4	1	0	0
5	1	0	1

To have x_1, x_3, x_5 cover faults on x_4 requires a u such that

$$[r_4, r_{14}, r_{34}, r_{134}, r_{45}, r_{145}, r_{345}, r_{1345}]\mathbf{T}^3_{*u} \neq 0$$

This leads to the constraints

u	x_5	x_3	x_1
0	0	0	0
3	0	1	1
4	1	0	0
7	1	1	1

To have x_1, x_3, x_4 test faults in x_5 requires

$$[r_5, r_{15}, r_{35}, r_{135}, r_{45}, r_{145}, r_{345}, r_{1345}]\mathbf{T}^3_{*u} \neq 0$$

which yields the constraints

u	x_4	x_3	x_1
0	0	0	0
3	0	1	1
4	1	0	0
7	1	1	1

The similarity in the last two cases arises from the fact the function is symmetric in x_4 and x_5.

One possible set of constrained tests is $\{x_1 = 0\}$, $\{x_1 = 1\}$, $\{x_2 = 0\}$, $\{x_1 = x_5 = 0, x_4 = 1\}$, $\{x_1 = x_3 = x_5 = 0\}$ and $\{x_1 = x_3 = x_4 = 0\}$.

The function being considered, however, has a minimal-input signature $\{r_0, r_1, r_2, r_4, r_5\}$, which yields the input constraints $x_1 = 0$, $x_1 = 1$, $x_2 = 0$, $x_4 = 0$ and $x_5 = 0$.

The minimal-covering signature led to six tests with a total of 60 input assignments examined. The minimal-input signature led to five tests with a total of 80 input assignments examined. Clearly there is a time-versus-setup cost tradeoff. The optimal choice will depend strongly on the testing environment and the hardware required to implement the constrained test.

For general networks, a set of constrained tests can be derived from a spectral signature in a similar fashion. The problem is that, in general, each coefficient tests a number of faults. This coefficient can be mapped to a single constrained test only when that test is consistent over those faults, i.e., the chosen assignment is applicable to each of the faults. This problem is avoided if every fault is tested by an r_α such that $|\alpha|$ is minimal, since in this case the choice of constrained assignments is completely arbitrary. Gaining this flexibility will on occasion result in additional tests.

The network in Fig. 6.23 was shown to have the spectral signature $\{r_0, r_1\}$. Clearly the minimal-cardinality requirement is satisfied, and $x_1 = 0$ and $x_1 = 1$ are an appropriate set of constraints.

6.5.2 Fault Equivalence

Functional fault equivalence was introduced in Section 6.4.3, where it was used to reduce the number of faults that had to be explicitly considered. Here we examine a similar notion in the spectral domain.

Two faults are said to be r_α-testable equivalent if they both result in some faulty function $\overset{*}{f}(X)$ such that $\overset{*}{r}_\alpha \neq r_\alpha$. There is no requirement that the two faults yield the same faulty function. Functionally equivalent faults are r_α-testable equivalent for any α. All faults on unate lines of a network are r_0-testable equivalent.

Unfortunately, a further characterization of r_α-testable equivalent faults appears to be inherently complex in that, in general, it requires knowledge of the total network. If this is indeed the case, it will be of little practical advantage, since determining equivalency would be no simpler than determing testability.

To illustrate this, consider the problem of characterizing when a line in a network can be syndrome-testable for a stuck-at-1 fault, and syndrome-untestable for a stuck-at-0 fault. The line g in Fig. 6.24 is such a line. Here $g/0$ is syndrome-untestable while $g/1$ is syndrome-testable. This network realizes a function with an even number of ones. It is easy to show that this situation can exist for a function with an odd number of ones by adding in one as-yet-uncovered minterm via an additional final OR gate.

Consider the network in Fig. 6.25a. $g/0$ and $g/1$ are clearly syndrome-testable, since g is a unate (positive) line. Figure 6.25b is the same network with a disjoint spectral translation. Here $g/0$ is syndrome-untestable while $g/1$ is syndrome-testable.

This example is discouraging, since the two networks are identical in the immediate vicinity of g. Spectral conditions for a line to be testable for only one of the two possible stuck-at faults can be constructed, but they are in fact more complex than Criterion 6.5, since they make use of the spectra of both $g(X)$ and $h(X, g)$.

The problem is clearly associated with the reconvergent fan-out paths, but it appears the source can be far removed from the effect. Further research in this area is needed since any results would be useful in reducing the cost of determining the testability of a network.

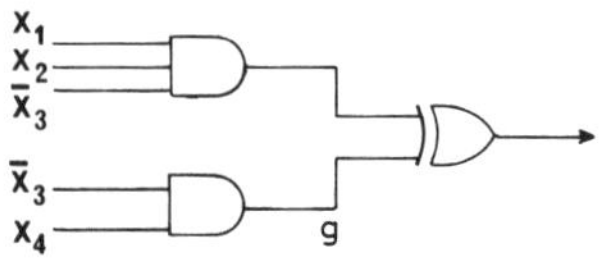

Fig. 6.24. A network for which g/0 is syndrome-untestable and for which g/1 is not.

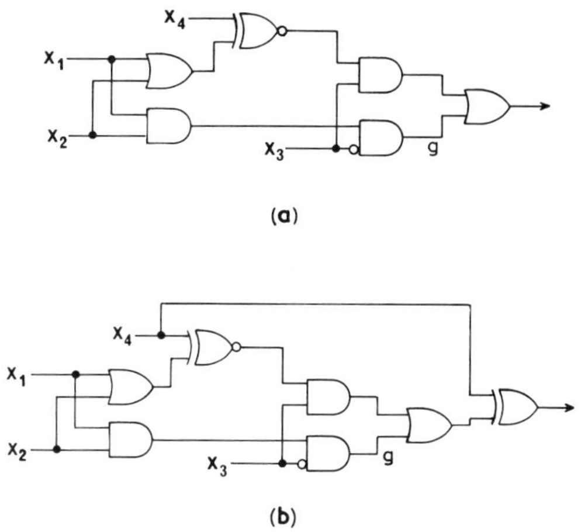

Fig. 6.25. (a) A network for which g/0 and g/1 are syndrome-testable (g is a unate line). (b) A very similar network for which g/0 is syndrome-untestable while g/1 is.

6.5.3 Implementation of Tests

Implementation of the test procedures described above depends on the environment in which the test is to be performed. The approach is quite different for a separate test unit and for on-chip self-testing, for instance. The following discussion will concentrate on the latter case.

The goal of on-chip self-testing is to achieve a testable design with a minimum of additional hardware and with no undue degradation of the performance of the original circuitry. Figure 6.26 shows the logical structure of syndrome-testing a single-output combinational function. The input source must generate each of the possible input assignments in turn. The syndrome accumulator counts the number of ones in the resulting data steam. Both functions can be performed by simple counters, but, as we shall show shortly, a more economical input source can often be achieved.

A major factor in on-chip self-testing is the number of additional pins required to accomplish the test. Rather than bringing the entire syndrome off the chip, a scheme such as that shown in Fig. 6.27 can be used to reduce the

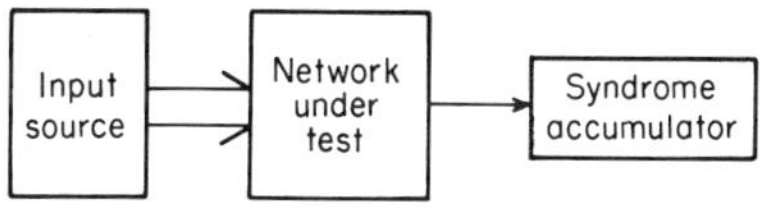

Fig. 6.26. The logical structure of syndrome testing.

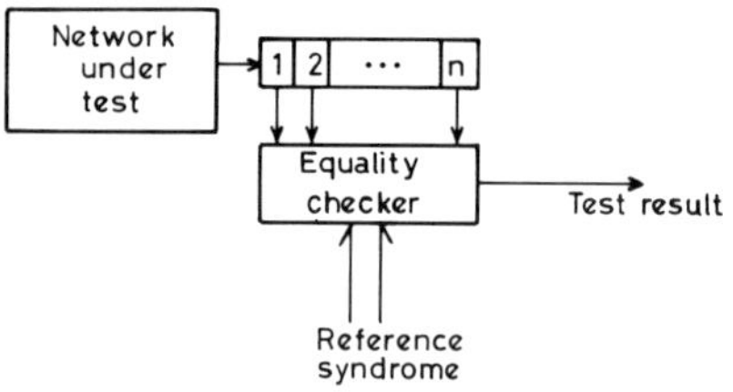

Fig. 6.27. Use of a reference syndrome and equality checker to reduce the test result to a single pin.

syndrome to a single test-result pin. The reference syndrome is fixed in the chip design. Another alternative would be to use a down-counter that initializes to the correct syndrome at the beginning of the test and decrements the count each time a one occurs in the test response. At the end of the test, a value of zero in the counter represents a successful test.

At the relatively slow test rate of 1 MHz, a 20-input combinational network can be syndrome-tested in just over 1 sec. Larger networks must be partitioned by a scheme such as that shown in Fig. 6.28, which was originally suggested by Savir.[28] The "extra pins" must be tri-stated so they can be used as inputs or outputs depending on which block is being tested. The term pin is perhaps misleading, since in many design approaches these lines would not be brought off the chip. In situations where there are several test points to be syndrome-tested, a multiplexer arrangement can be introduced so that only one syndrome accumulator is required. This accumulator must be capable of measuring a variety of syndrome values.

The vast majority of practical testing situations involve sequential networks. Unfortunately, sequential networks are inherently more difficult

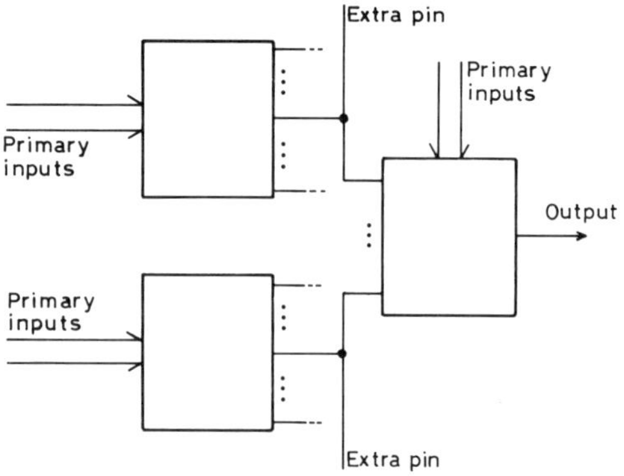

Fig. 6.28. A partitioning scheme to permit the syndrome testing of a combinational network with more than 20 inputs.

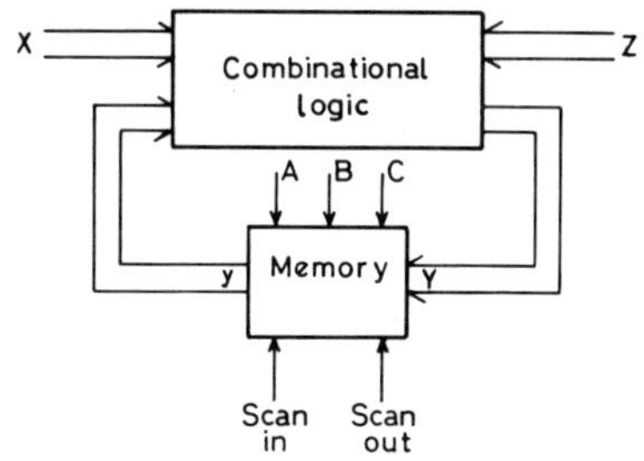

Fig. 6.29. The logical structure of level-sensitive-scan-design (LSSD).

to test than combinational networks.[14] A variety of techniques have been developed to allow a sequential network to be tested in a purely combina-. tional fashion (see Parker and Williams[14] for a comprehensive survey). Here we describe one approach, level-sensitive-scan-design (LSSD).[15–19] and show how it can be used as a basis for the testing techniques described in preceding sections.

The block structure of LSSD is shown in Fig. 6.29. The memory consists of some number of rather complex data latches—Fig. 6.30. During normal operation, the latch functions as a simple level–sensitive flip-flop with input Y_i, output y_i and clock A. During testing the latch can be loaded from the scan-in pin using clock B. Clock C allows the value of y_i to be latched at the scan-out pin. A series of these latches can be formed into a shift register as shown in Fig. 6.31a. An alternating sequence of B and C clocks shifts data through this register.

This design approach has a number of advantages. During normal operation, the sequential network is clocked using A with no degradation in performance. During testing, the sequential network can be single-stepped using A. The values of Y_i and y_i can be scanned out and in, respectively. When the latches are formed into a shift register, only four additional pins, B, C,

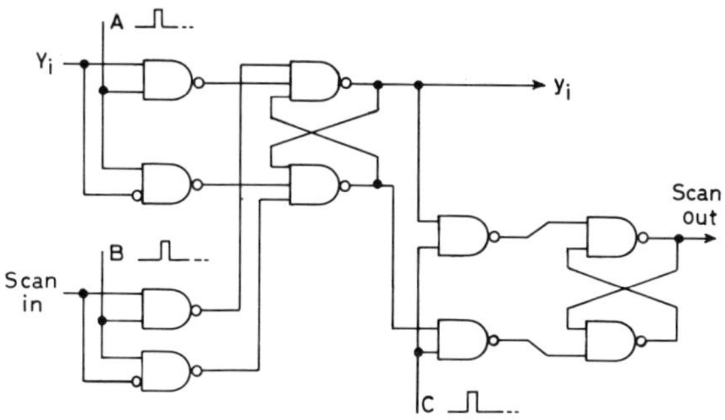

Fig. 6.30. A single LSSD memory data latch.

scan-in and scan-out are required. The principal disadvantage is the complexity of the data latches, which varies from 4 to 23%[14] depending upon the degree to which the additional hardware is incorporated into the functional unit.

LSSD has a number of additional characteristics that reduce the chances of nonfunctional faults related to certain AC parameters. In addition, it tends to partition the logic into compact blocks. Access to the shift register of memory latches is sufficient to implement the traditional test set approach.

Implementation of syndrome, constrained syndrome or spectral coefficient testing in an LSSD network requires an efficient method for generating the required input conditions. One could shift them in the same way test vectors are shifted-in. This would require some external mechanism for generating the data pattern.

A simpler approach is to add Exclusive-OR gates turning the LSSD shift register into a linear-feedback-shift-register (LFSR).[35-39] With the proper feedback, via the added Exclusive-OR gates, an LFSR can be made to generate the input assignments required in syndrome, constrained syndrome, and spectral coefficient testing. The problem of how to choose the appropriate feedback has been discussed in detail by Barzilai *et al.*[40] Here we simply outline the results without proof. Familiarity with the Galois field of two elements,[41] $GF(2)$, and its properties are assumed. In such a field, arithmetic operations are all performed mod 2.

A shift register, Fig. 6.31a, is converted to an LFSR by the addition of Exclusive-OR gates that feed the contents of certain cells back into the input, (see Fig. 6.31b for an example). If the LFSR in Fig. 6.31b is initialized to zeroes and then "primed" with a single one, it will generate the eight assignments to three variables in a pseudorandom fashion. See Table 6.6 for details.

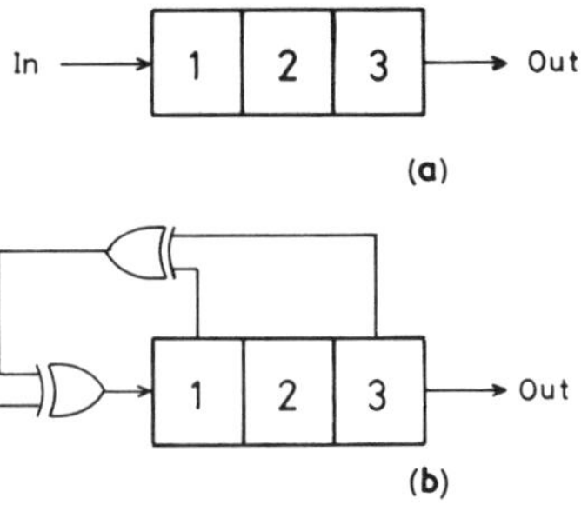

Fig. 6.31. (a) A three-bit shift register. Shifting is performed one cell per cycle from left (1) to right (3). (b) The register in (a) modified to include linear feedback, thus becoming a linear feedback shift register.

In general, let

$$P(x) = a_0 \oplus a_1 x \oplus a_2 x^2 \oplus \cdots \oplus a_n x^n;$$
$$a_n = 1; \quad a_j \in \{0, 1\}, \quad 0 \le j < n \tag{6.38}$$

be a polynomial over $GF(2)$. $P(x)$ is irreducible if it cannot be factored (over $GF(2)$), and primitive if $N = 2^n - 1$ is the smallest integer such that $x^n - 1$ is divisible by $P(x)$.

Given an irreducible primitive polynomial with coefficient a_i, $0 \le i \le n$, the recurrence relation

$$y_{m+n} = a_0 y_m \oplus a_1 y_{m+1} \oplus \cdots \oplus a_{n-1} y_{m+n-1} \tag{6.39}$$

with initial values $y_0 = y_1 = \cdots = y_{n-1} = 0$, $y_n = 1$ generates a length $2^n + n - 1$ string, all of whose length n contiguous substrings are distinct.[42] An LFSR based on such a polynomial will thus generate all 2^n *n-bit* patterns.

For example, the LFSR in Fig. 6.31b is based on the irreducible primitive polynomial

$$P(x) = x^3 \oplus x^2 \oplus 1$$

Suppose we are given a particular shift register and are required to test exhaustively a single contiguous block of cells. If this block is first in the

Table 6.6. Use of the LFSR in Fig. 6.31b to generate the eight possible three-bit data patterns.[a]

		Register contents		
Cycle	Input	1	2	3
0	1	0	0	0
1	0	1	0	0
2	0	1	1	0
3	0	1	1	1
4	0	0	1	1
5	0	1	0	1
6	0	0	1	0
7	0	0	0	1

[a] The register contents must be initialized to zeroes. The single 1 input is required to initiate the cycle. Note that when continuously clocked, the LFSR has a cycle length of 7. The initial pattern 000 is not regenerated.

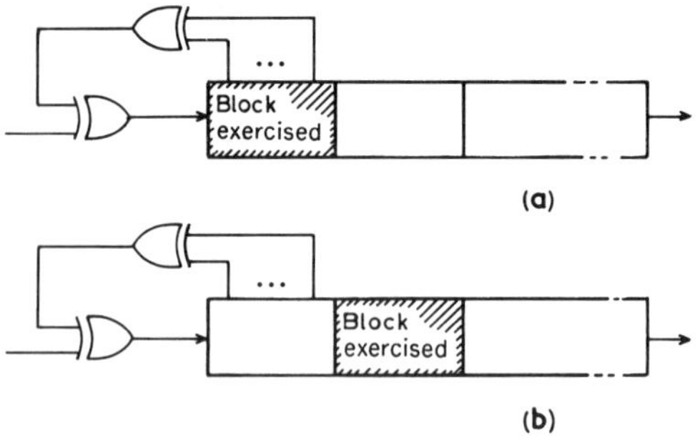

Fig. 6.32. (a) Using an LFSR to exercise exhaustively a left-most contiguous block of cells. The actual feedback is derived from a primitive polynomial over *GF*(2) of the correct degree. (b) The same scheme will work for a contiguous block elsewhere in the register, but extra cycles are required.

register (Fig. 6.32a), one need only select an appropriate irreducible primitive polynomial and convert this block to an LFSR (Fig. 6.32b). If the block lies elsewhere in the register, the same approach will work but extra cycles are required to allow the input assignments to be moved to the appropriate section of the shift register.

Exhaustive testing of a single discontiguous block of cells is accomplished in the same way (Fig. 6.33). In this case, the following theorem due to Barzilai *et al.*[40] must be satisfied.

Theorem 6.1. The irreducible primitive polynomial $P(x)$ exhaustively exercises cell positions $i_1, i_2, \ldots, i_p$ if and only if the p residue classes $\{x^{i_j} \bmod P(x)\}$ are linearly independent over $GF(2)$.

For a given shift register and single discontiguous set of cells, it is a simple matter to identify candidate primitive polynomials and to test each in turn until one that satisfies the above theorem is found.

Barzilai *et al.*[40] have presented a solution for the case of several designated subsets, either contiguous or discontiguous. The interested reader should consult the reference. A certain difficulty arises in applying the multiple-block

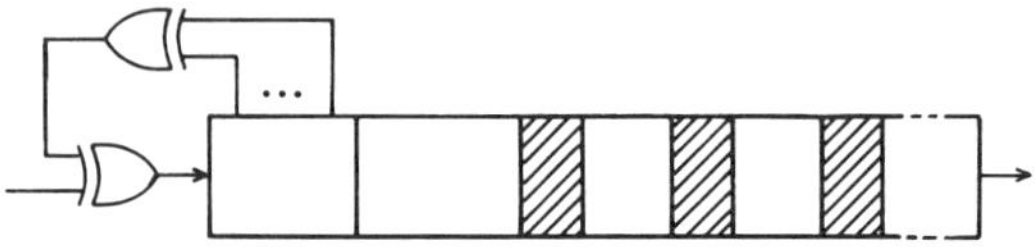

Fig. 6.33. Using an LFSR to exercise exhaustively a discontinuous set of cells in the shift register.

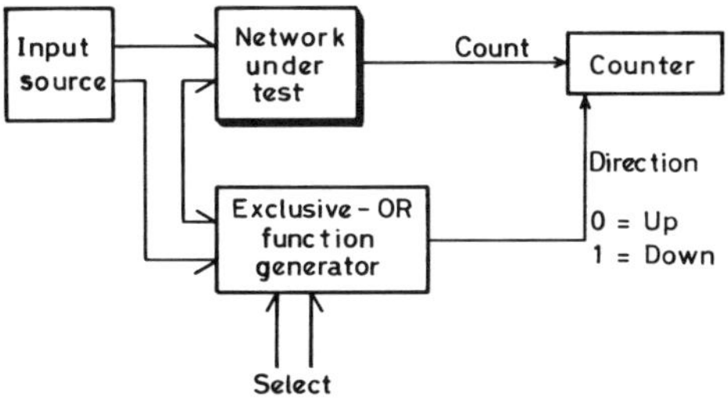

Fig. 6.34. A structure for generating the spectral coefficients in $\{0, 1\}$ coding.

solution to the testing methods described here. If the blocks are of unequal size, all but the largest will see certain input assignments more than once. While this can be accounted for in specific instances, a general theory is evasive.

As noted earlier, a simple counter is sufficient to accumulate a syndrome value. Spectral-coefficient testing requires the appropriate Exclusive-OR function be generated. The general structure is shown in Fig. 6.34.

Figure 6.35 shows a slightly simpler alternative. This network determines the number of input assignments for which the function realized by the network under test and the chosen Exclusive-OR function disagree. Since the number of agreements plus the number of disagreements equals 2^n, the value of the spectral coefficient, which in $\{+1, -1\}$ coding is the number of agreements minus the number of disagreements, is easily determined.

A general Exclusive-OR function generator can be implemented by a tree of Exclusive-OR gates with a preceding level of AND gates to select the active input variables. The case of $n = 4$ is shown in Fig. 6.36. Note that if only r_0 and certain first-order coefficients are required, a simple multiplexer can be used in lieu of the Exclusive-OR function generator.

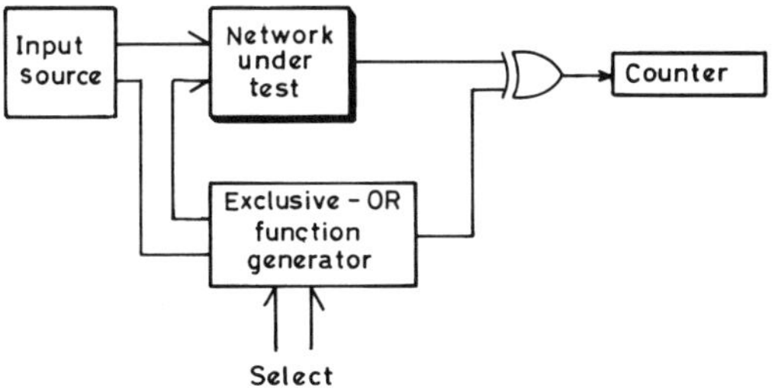

Fig. 6.35. A simpler test structure than the one shown in Fig. 6.34. Here the true value of r_0 is produced, but for all other r_α the value produced is $r_\alpha + 2^{n-1}$, where n is the number of inputs to the network under test.

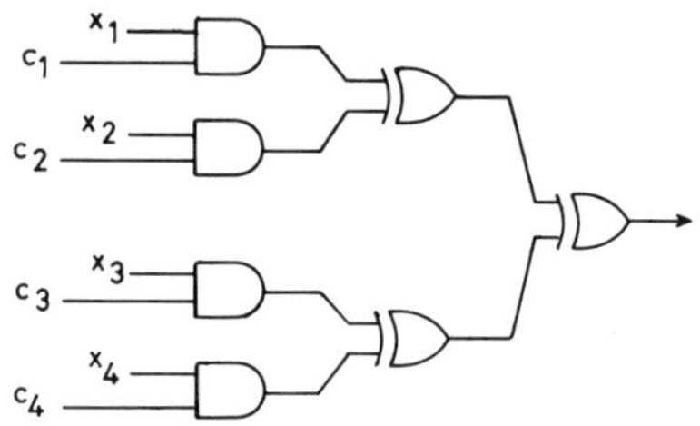

Fig. 6.36. A four-input Exclusive-OR function generator. x_i is selected as an active input by setting $c_i = 1$; otherwise, x_i is inactive.

6.6 CHAPTER SUMMARY

In this chapter we have examined the use of the spectral coefficients in the detection of faults in combinational networks. We have shown that the spectral domain is useful for determining the syndrome- or constrained-syndrome-testability of a network, and that the spectral coefficients are themselves useful fault detectors.

The use of the spectral coefficients for fault detection in the manner described here was first suggested by Bennetts and Hurst.[43] Certain preliminary results have appeared in Susskind[22] and Miller and Muzio.[33–34] We have shown that the spectral approach encompasses Savir's notion of syndrome-testing[28–31] as well as the functional attribute tests of Tzidon *et al.*[21]

No mention has been made of the important work of Karpovsky[44–48] concerning the application of spectral techniques to problems in fault detection and analysis. This work is not at the gate level and thus cannot be readily compared to the work presented here.

The thrust of the presentation has been to demonstrate the applicability of the spectrum in fault detection. For that reason, a number of interesting related concepts have been omitted, since they do not lend themselves to spectral analysis. For example, we have not considered the use of an LFSR as the data compression scheme. This idea is receiving considerable attention because of the ease of its implementation. Detailed studies of the effectiveness of such schemes are required and one such study has been given by Carter.[49]

The chapter has been largely a theoretical study with only brief concern for the practical implementation of the results. More research is required concerning models other than the simplistic single stuck-at fault model, the practical computation of spectral signatures, and the physical implementation of the techniques presented in different design environments.

References

1. Brewer, M. A., and Friedman, A. D., "Diagnosis and Reliable Design of Digital Systems." Computer Science Press, Potomac, Maryland, 1976.
2. Hayes, J. P., On realizations of Boolean functions requiring a minimal or near-minimal number of tests, *IEEE Trans. Comput.* **C-20,** 1506–1513 (1971).
3. Kohavi, I., and Kohavi, Z., Detection of multiple faults in combinational logic networks, *IEEE Trans. Comput.* **C-21,** 556–568 (1972).
4. Goel, P., An explicit enumeration algorithm to generate tests for combinational logic circuits, *IEEE Trans. Comput.* **C-30,** 215–222 (1981).
5. Savir, J., and Roth, J. P., Testing for and distinguishing between failures, *Proc. 12th Symp. Fault-Tolerant Comput.* 165–172 (1982).
6. Karpovsky, M. G., and Su, S. Y. H., Detection and location of input and feedback bridging faults in combinational networks, *IEEE Trans. Comput.* **C-29,** 523–527 (1980).
7. Xu Shiyi, Shanghai University of Science and Technology and SUNY, Binghamton, New York, private communication with the authors.
8. Roth, J. P., Methods of testing for shorts, *IBM Tech. Disclosure Bull.*, 3108–3109 (1967).
9. Friedman, A. D., Diagnosis of short-circuit faults in combinational circuits, *IEEE Trans. Comput.* **C-23,** 746–752 (1974).
10. Iosupowicz, A., Optimal detection of bridging faults and stuck-at faults in two-level logic, *IEEE Trans. Comput.* **C-27,** 452–455 (1978).
11. Carr, W. N., and Mize, J. P., "MOS/LSI Design and Applications," pp. 229–258. Texas Instruments Electronics Series. McGraw-Hill, New York, 1972.
12. Fleisher, H., and Maissel, L. I., An introduction to array logic, *IBM J. Res. Devel.* **19,** 98–109 (1975).
13. "Signetics Field Programmable Logic Arrays." Signetics, Sunnyvale, California, 1976.
14. Williams, T. W., and Parker, K. P., Design for testability—A survey, *IEEE Trans. Comput.* **C-31,** 2–15 (1982).
15. DasGupta, S., Eichelberger, E. B., and Williams, T. W., LSI chip design for testability, *Digest 1978 Internat. Solid-State Circuits Conf.*, 216–217 (1978).
16. Eichelberger, E. B., and Williams, T. W., A logic design structure for LSI testability, *J. Design Automat. Fault-Tolerant Comput.* **2,** 165–178 (1978).
17. Eichelberger, E. B., and Williams, T. W., A logic design structure for LSI testing, *Proc. 14th Design Automat. Conf.*, 462–468 (1977).
18. Eichelberger, E. B., Muehldorf, E. J., Walter, R. G., and Williams, T. W., A logic design structure for testing internal arrays, *Proc. 3rd USA–Japan Comput. Conf.*, 266–272 (1978).
19. Arzoumanian, Y., and Waicukouski, J., Fault diagnosis in LSSD environment, *Proc. 1981 Internat. Test Conf.*, 86–88 (1981).
20. Roth, J. P., "Computer Logic Testing and Verification." Computer Science Press, Potomac, Maryland, 1980.
21. Tzidon, A., Berger, I., and Yoelli, M., A practical approach to fault detection in combinational networks, *IEEE Trans. Comput.* **C-27,** 968–971 (1978).
22. Susskind, A. K., Testing by verifying Walsh coefficients, *IEEE Trans. Comput.* **C-32,** 198–201 (1983).
23. Carter, W. C., The ubiquitous parity bit, *Proc. 12th Internat. Symp. Fault-Tolerant Comput.*, 289–296 (1982).
24. Galiay, J., Crouzet, Y., and Vergniault, M., Physical versus logical fault models for MOS LSI circuits; impact on their testability, *IEEE Trans. Comput.* **C-29,** 527–531 (1980).

25. Ostapko, D. L., and Hong, S. J., Fault analysis and test generation for programmable logic arrays, *Proc. 8th Internat. Symp. Fault-Tolerant Comput.*, 326–331 (1978).
26. Cha, C. W., A testing strategy for PLAs, *Proc. 15th Design Automat. Conf.*, 326–331 (1978).
27. Smith, J. E., Detection of faults in programmable logic arrays, *IEEE Trans. Comput.* **C-28,** 845–853 (1979).
28. Savir, J., Syndrome testable design of combinational circuits, *IEEE Trans. Comput.* **C-29,** 442–451 (1980). (Correction, *ibid.*, 1012.)
29. Savir, J., Syndrome testing of "syndrome untestable" combinational circuits, *IEEE Trans. Comput.* **C-30,** 606–608 (1981).
30. Barzilai, Z., Savir, J., Markowsky, G., and Smith, M. G., The weighted syndrome sums approach to VLSI testing, *IEEE Trans. Comput.* **C-30,** 996–1000 (1981).
31. Savir, J., Syndrome testable design of combinational circuits, *Proc. 9th Internat. Symp. Fault-Tolerant Comput.*, 137–140 (1979).
32. Markowsky, G., Syndrome-testability can be achieved by circuit modification, *IEEE Trans. Comput.* **C-30,** 604–606 (1981).
33. Muzio, J. C., and Miller, D. M., Spectral signatures for fault testing in combinational circuits, Technical Report, Tektronix, Inc., Beaverton, Oregon, 1981.
34. Muzio, J. C., and Miller, D. M., Spectral methods for fault detection in combinational circuits, Technical Report, Tektronix, Inc., Beaverton, Oregon, 1981.
35. Golomb, S. W., "Shift Register Sequences." Holden-Day, San Francisco, California, 1967.
36. David, R., Feedback shift register testing, *Proc. 8th Internat. Symp. Fault-Tolerant Comput.*, 103–107 (1978).
37. Frohwerk, R. A., Signature analysis: A new digital field service method, *Hewlett-Packard J.*, 2–8 (May 1977).
38. Smith, J. E., Measures of the effectiveness of fault signature analysis, *IEEE Trans. Comput.* **C-29,** 510–514 (1980).
39. Benowitz, N., An advanced fault isolation system for digital logic, *IEEE Trans. Comput.* **C-24,** 489–497 (1975).
40. Barzilai, Z., Coppersmith, D., and Rosenberg, A. L., Exhaustive generation of bit patterns with applications to VLSI self-testing, *IEEE Trans. Comput.* **C-32,** 190–193 (1983).
41. Van der Waerden, B. L., "Modern Algebra," Vol. I, Vol. II. Frederick Unger, New York, 1949, 1950.
42. Peterson, W. W., "Error Correcting Codes." MIT Press, Cambridge, Massachusetts, 1961.
43. Bennetts, R. G., and Hurst, S. L., Rademacher–Walsh spectral transform: A new tool for problems in digital fault diagnosis?, *IEE Comput. Digital Tech.* **1,** 38–44 (1978).
44. Karpovsky, M. G., "Finite Orthogonal Series in the Design of Digital Devices." Wiley, New York, 1976.
45. Karpovsky, M. G., Error detection in digital devices and computer programs with the aid of linear recurrent equations over finite commutative groups, *IEEE Trans. Comput.* **C-26,** 208–218 (1977).
46. Karpovsky, M. G., Error detection for polynomial computations, *Comput. Digital Tech.* **2,** 50–56 (1979).
47. Karpovsky, M. G., Detection and location of errors by linear inequality checks, *IEE Proc.* **129,** 86–92 (1982).
48. Karpovsky, M. G., Testing for numerical computations, *IEE Proc.* **127,** 69–76 (1980).
49. Carter, J. L., The Theory of signature testing for VLSI, unpublished IBM report (1982).

7

Conclusions

In appreciating the material that we have attempted to cover in the preceding chapters of this text, the reader should, it is hoped, have acquired a working knowledge of the techniques and general state of development of spectral methods applied to the digital logic field. The end-of-chapter references and the end of-volume bibliography cover the principal sources of original material and will provide a more rigorous and in-depth treatment for those readers who require it.

As we noted in an early section of Chapter 4, we have not in this text pursued to any extent the direct use of the spectral coefficients themselves for the realisation of logic functions, even though this is a fundamental possibility. Such techniques, however, were implied in the majority of early pioneering works in this spectral area, resulting in synthesising networks such as the one illustrated in Fig. 4.1, but hardware and component limitations are still not conducive to serious consideration of such possibilities. Instead, our whole emphasis here has been upon the individual and collective meaning of the spectral coefficient values for a given function, and we have used their particular information content extensively to give us insight and guidance into the structure of functions. In Chapter 1 we noted that the problem with our conventional Boolean data was that the information content was discrete rather than global; with the completion of this text it will be apparent how useful global parameters, each of which tells us something but usually not everything about the complete function, can be. The circuit realisations of our design procedures, however, remain normal digital logic gates, although not necessarily confined to the most common AND, OR, NAND, NOR family.

Whilst the use of the spectral domain for the classification of Boolean logic functions is of undisputed significance, resulting in the most compact and simple means of function classification, the use of spectral data in other areas such as we have considered is not without continuing dispute, development, and discussion. In network synthesis much remains to be done in designing easily testable networks, for example. Similarly, in the testing and fault diagnosis of existing networks, the use of spectral coefficient signatures is not yet of proven commercial advantage. Indeed, these two considerations should be considered parts of the same problem of "good" digital network design, although how and when a cohesive design philosophy can be achieved remains open. Nevertheless we remain optimistic that the power and many unique properties of the spectral domain will provide working tools for future digital logic engineering.

Finally, this whole subject area of the application of orthogonal transforms to digital logic data has much in common with many other fields of engineering and research that involve discrete transforms. We have not consciously emphasised any parallels between our work and other fields, but the discerning reader may have appreciated that work in pattern recognition, orthogonal encoding of information, and many other areas have similar mathematical involvements. A significant publication that attempts a broad overview of these areas is Beauchamp.[1]

Many research and development aspects remain to be investigated, and questions concerning the viability of designing or testing in the spectral domain versus designing or testing in the conventional functional domain answered. The increasing use of CAD software for digital design purposes, especially for stand-alone designers' work stations, emphasises the need for computationally attractive design methodologies. Hence a great deal of research work remains to be done in the digital logic field. We therefore trust that this text may be useful to research workers in this field, as well as to the many others who are concerned with digital logic theory and practice.

Reference

1. Beauchamp, K. G., "Applications of Walsh and Related Functions," Academic Press, London, 1984.

A

Comments upon Geometrical Constructions

Geometric constructions have been used in places in the main body of this work, in order to illustrate the logical structure or processing under consideration.

For immediate reference we detail below the principal constructions that have been used, and particularly define the identification order of the entries on each format.

A.1 KARNAUGH MAPS

The principle of Karnaugh maps is well established and needs no further comment here. However, it will be recalled that we have defined the mathematical significance of the x_i input variables of a function $f(X)$ with x_1 as

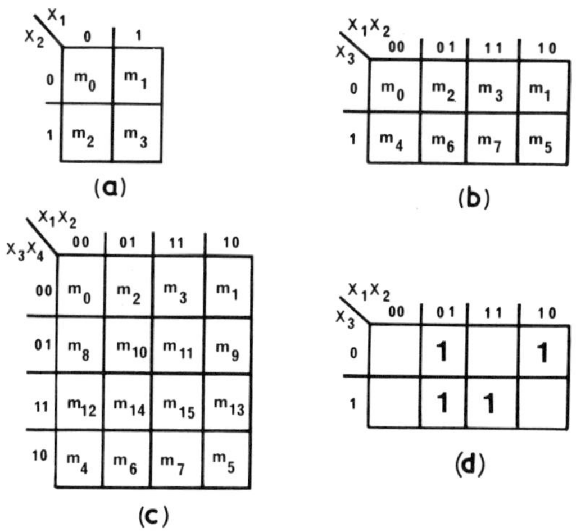

Fig. A.1. Karnaugh map formats, showing the minterm identifications m_j, $j = 0$ to $2^n - 1$, as used in this text. (a) $n = 2$; (b) $n = 3$; (c) $n = 4$; (d) example $f(X) = \bar{x}_1x_2 + x_2x_3 + x_1\bar{x}_2\bar{x}_3 = \sum 1, 2, 6, 7$. (Note that x_1 is taken as the least-significant input variable, x_n as the most-significant.)

the least-significant digit and x_n as the most-significant digit. This definition therefore defines the minterm squares of the Karnaugh maps in decimal form as shown in Fig. A.1.

A.2 REED–MULLER COEFFICIENT MAPS

The Reed–Muller coefficient map is a more recent introduction than the Karnaugh map[1], and may require the following brief comments.

Recall that the positive, canonic Reed–Muller expansion for any function $f(X)$ is

$$f(X) = a_0 \oplus a_1x_1 \oplus a_2x_2 \oplus \cdots \oplus a_nx_n \oplus a_{n+1}x_1x_2 \oplus \cdots \oplus a_{2^n-1}x_1x_2\cdots x_n$$

where the 2^n coefficients a_α, $\alpha = 0$ to $2^n - 1$, each take the value 0 or 1 as appropriate to realise $f(X)$. If we consider the decimal equivalent of each product term $x_ix_j \ldots$, again with x_1 as the least-significant digit, then we may replace each a_α coefficient with b_β, where β takes the same numerical identification number as the decimal equivalent of the product term it qualifies. Hence we have

$$f(X) = b_0 \oplus b_1x_1 \oplus b_2x_2 \oplus b_3x_1x_2 \oplus \cdots \oplus b_{2^n-1}x_1x_2\cdots x_n$$

For $n = 3$, we have

$$f(X) = b_0 \oplus b_1x_1 \oplus b_2x_2 \oplus b_3x_1x_2 \oplus b_4x_3 \oplus b_5x_1x_3 \oplus b_6x_2x_3 \oplus b_7x_1x_2x_3$$

The Reed–Muller map, therefore, is a map construction to show geometrically the pattern of the 0 and 1 values of the latter coefficients for any given function $f(X)$. As with the Karnaugh map, it may be more convenient to indicate the one-valued entries only, blank map entries implying the zero-valued case.

Figure A.2 shows the published constructions for $n = 2$, 3 and 4. The following points should be noted:

(i) The 0, 1 entries in this map construction do not represent the 0, 1 output values of $f(X)$ as in a Karnaugh map, but are the appropriate zero- and non-zero-coefficient values of the positive canonic Reed–Muller expansion realising $f(X)$.

(ii) The individual b_β coefficients are immediately identified by the x_i axes identifications. For example, in the three-variable case, coefficient b_0 is outside any x_i area of the map, b_1 lies inside the x_1 boundary but outside the x_2 and x_3 boundaries, ..., up to b_7 lying within x_1, x_2 and x_3 boundaries. An identical situation holds for any n.

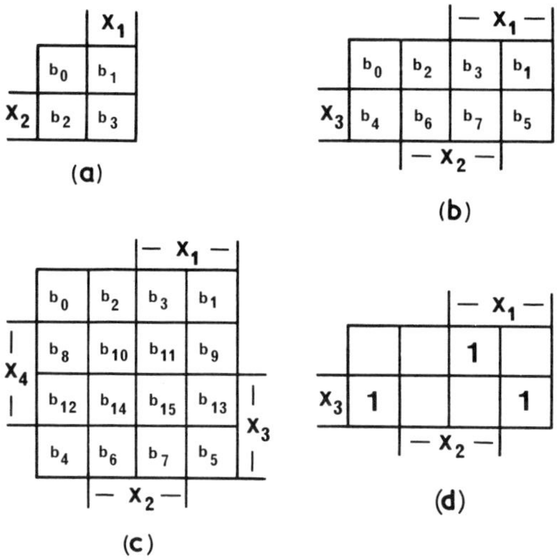

Fig. A.2. The Reed–Muller map formats, showing the positive canonic Reed–Muller coefficient identification b_β, $\beta = 0$ to $2^n - 1$, in the expansion $f(X) = b_0 \oplus b_1x_1 \oplus b_2x_2 \oplus \cdots \oplus b_{2^n-1}x_1x_2\cdots x_n$. (a) $n = 2$; (b) $n = 3$; (c) $n = 4$; (d) example $f(X) = x_1x_2 + \bar{x}_1x_3 = x_3 \oplus x_1x_2 \oplus x_1x_3$.

The use of this map construction for the manipulation and minimisation of generalised Reed–Muller expansions not confined to the positive, canonic case may be found in the original published paper.

A.3. SPECTRAL COEFFICIENT MAPS

The spectral coefficient map is also a recent introduction,[2] and may require a brief comment.

Like the Reed–Muller coefficient map, the spectral coefficient map does not plot the 0, 1 output value of a function $f(X)$, but instead shows the spectral coefficients of $f(X)$. The map entries, therefore, are the positive and negative integer values of the 2^n spectral coefficients of $f(X)$.

Figure A.3 shows the published constructions for $n = 2$, 3 and 4. Again, the spectral coefficients associated with a particular x_i input variable will be found within the appropriate x_i area of the map construction, while outside each x_i are are those coefficients not dependent upon this particular x_i.

It will be noted that we may employ this map construction to display either **R** spectral coefficients obtained from the $\{0, 1\}$ coding of the output vector of $f(X)$, or the **S** spectral coefficients obtained from the $\{+1, -1\}$ recoding of $f(X)$. In general **S** coefficients will be encountered, although it should be appreciated that these may be termed **R** rather than **S** in other publications.

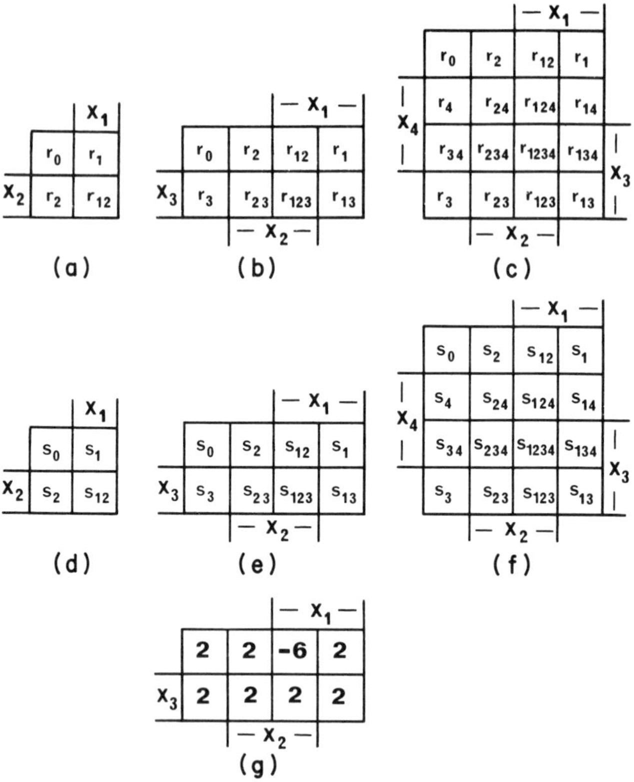

Fig. A.3. The spectral coefficient map formats for $n = 2$, 3 and 4. (a)–(c) r_i identification; (d)–(f) s_i identification; (g) example $f(X) = x_1x_2 + \bar{x}_1\bar{x}_2x_3$; spectrum $\mathbf{S} = 2; 2\,2\,2; -6\,2\,2; 2$.

This map construction gives a powerful insight into the manipulation of spectral coefficients corresponding to given operations in the functional domain. Further details may be found in the original published paper.

A.4 SPECTRUM AMPLITUDE CHARTS

The spectral coefficients for any given $f(X)$ may be plotted so as to indicate their sign and magnitude pattern by the straightforward bar chart construction shown in Fig. A.4. Again, either the **R** or the **S** values may be plotted. Note that since the linear relationships between **R** and **S** are

$$r_0 = \tfrac{1}{2}\{2^n - s_0\}, \qquad r_\alpha = -\tfrac{1}{2}s_\alpha, \quad \alpha \neq 0$$

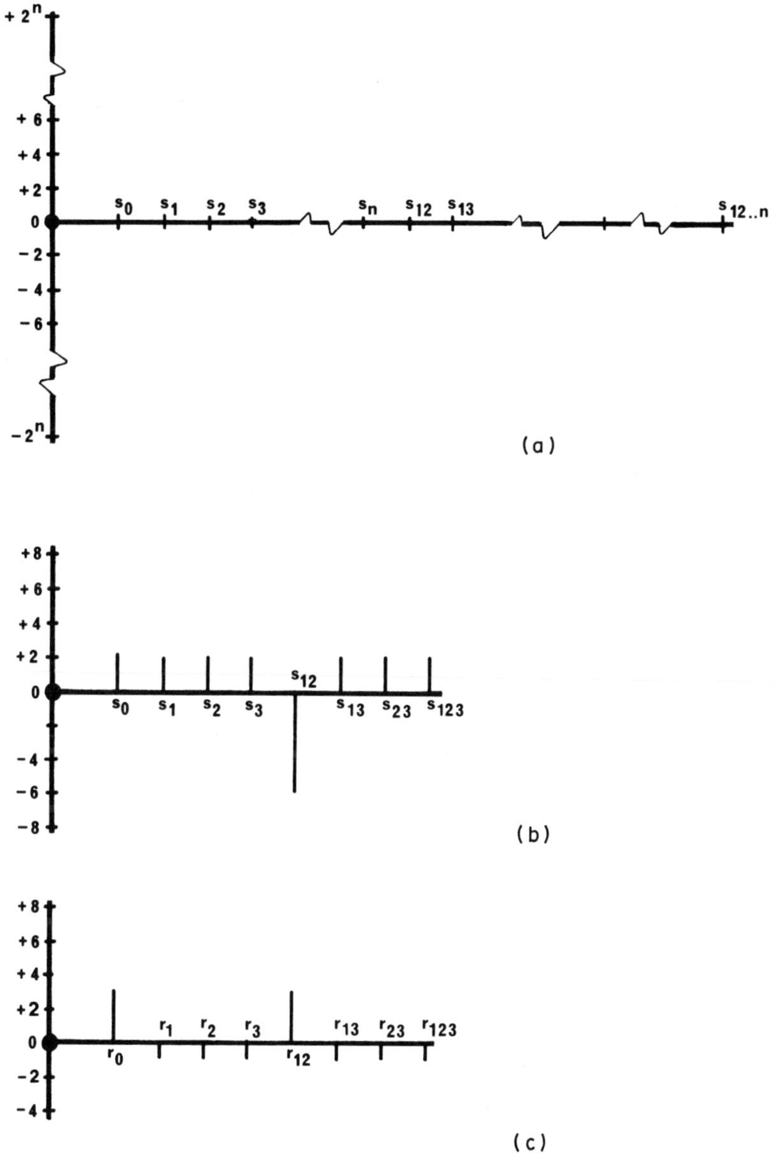

Fig. A.4. Spectrum amplitude charts for plotting either the r_i or the s_i coefficients of $f(X)$. (a) General format for **S**; format for **R** similar except for reduced range of the negative y axis; (b) example $f(X) = x_1x_2 + \bar{x}_1\bar{x}_2x_3$; **S** $=$ 2; 2 2 2; -6 2 2; 2; (c) same example, **R** $=$ 3; -1 -1 -1; 3 -1 -1; -1.

the range of possible **S** coefficient values is greater than the range of possible **R** coefficient values.

This construction needs no further comment; the parallel between this format and that for showing the harmonic amplitude components of a Fourier series of an a.c. waveform will be obvious.

References

1. Wu, X., Chen, X. and Hurst, S. L., Mapping of Reed–Muller coefficients and the minimisation of Exclusive-OR switching functions, *Proc. IEE* **E129,** 15–20 (1982).
2. Chen, X., The mapping of spectral coefficients and its application, *Comput. Electr. Engrg.* **9,** 167–180 (1982).

B

Canonic Spectral Classification of All Binary Functions of $n \leq 5$ under Spectral Invariance Operations

The classification of binary functions of $\leq n$ variables by means of their spectral coefficients was introduced in Chapter 3. It may be recalled that the spectral coefficients of any given function $f(X)$ may be sign-changed and reordered by the operations

(i) negation of the input variables;
(ii) permutation of the input variables;
(iii) negation of the output function;
(iv) input spectral translation, involving the replacement of an input variable x_i with $x_i \oplus x_j$, $i \neq j$, in the function domain; and
(v) output (or disjoint) spectral translation, involving the modification of the function output $f(X)$ to $\{f(X) \oplus x_i\}$ in the function domain.

These five operations each leave the magnitudes of the spectral coefficients invariant. They may be applied equally to the coefficients of **R** or **S**.

Table 3.6 gave the positive canonic classification for functions of $n \leq 4$ under these five invariant operations; references 4 and 21 of Chapter 3 give corresponding $n \leq 5$ classification tabulations, consisting in total of 48 canonic entries. However, in these $n \leq 5$ tabulations, two entries under the full five invariance operations gave identical spectral coefficient values; these were identified as canonic functions 45a and 45b in reference 4, and as functions 45 and 46 in reference 21.

Table B.1 gives the $n \leq 5$ canonic classification omitting the invariance operation (v), hence expanding each entry of the previous classification tabulations into more than one entry. We give the canonic spectral coefficients of **S**, as used in Table 3.6. Conversion to the canonic spectral coefficients of **R** are directly available from the positive canonic relationships

$$|r_0| = \tfrac{1}{2}|2^n - s_0|, \qquad |r_i| = \tfrac{1}{2}|s_i|, \quad i = 1 \text{ to } n$$

The class numbers in parentheses are the class identifications used in reference 4 of Chapter 3. Entries marked * are the canonic classification entries of reference 4 under the final invariance operation (v).

The functions 1–84 have linearly separable core functions, see Fig. 3.6. For additional information concerning the function groups 183, 184 (45a) and 185, 186 (45b), see reference 4 of Chapter 3, pp. 540, 541.

Table B.1

$n \leq 5$ Canonic classification function ref. no.	Zero- and first-order coefficients $s_0, s_1, \ldots, s_5$ only							Summary of complete canonic spectrum
1 (1)	0	32	0	0	0	0		1 × 32, 31 × 0
2 (1)	32	0	0	0	0	0	*	
3 (2)	2	30	2	2	2	2		1 × 30, 31 × 2
4 (2)	30	2	2	2	2	2	*	
5 (3)	0	28	4	4	4	4		1 × 28, 15 × 4,
6 (3)	4	28	4	4	4	0		16 × 0
7 (3)	28	4	4	4	4	0	*	
8 (4)	2	26	6	6	6	2		1 × 26, 7 × 6,
9 (4)	6	26	6	6	2	2		24 × 2
10 (4)	26	6	6	6	2	2	*	
11 (5)	0	24	8	8	4	4		1 × 24, 3 × 8,
12 (5)	4	24	8	8	4	4		16 × 4, 12 × 0
13 (5)	8	24	8	4	4	4		
14 (5)	24	8	8	4	4	4	*	
15 (6)	0	24	8	8	8	0		1 × 24, 7 × 8,
16 (6)	8	24	8	8	0	0		24 × 0
17 (6)	24	8	8	8	0	0	*	
18 (7)	2	22	10	10	6	2		1 × 22, 3 × 10,
19 (7)	6	22	10	10	2	2		4 × 6, 24 × 2
20 (7)	10	22	10	6	2	2		
21 (7)	22	10	10	6	2	2	*	
22 (8)	6	22	10	6	6	6		1 × 22, 1 × 10,
23 (8)	2	22	10	6	6	6		10 × 6, 20 × 2
24 (8)	10	22	6	6	6	6		
25 (8)	22	10	6	6	6	6	*	

Table B.1 (*Continued*)

$n \leq 5$ Canonic classification function ref. no.	Zero- and first-order coefficients $s_0, s_1, \ldots, s_5$ only						Summary of complete canonic spectrum
26 (9)	0	20	12	12	4	4	1 × 20, 3 × 12,
27 (9)	4	20	12	12	4	0	12 × 4, 16 × 0
28 (9)	12	20	12	4	4	0	
29 (9)	20	12	12	4	4	0 *	
30 (10)	0	20	12	8	8	4	1 × 20, 1 × 12,
31 (10)	4	20	12	8	4	4	4 × 8, 14 × 4,
32 (10)	4	20	12	8	8	4	12 × 0
33 (10)	8	20	12	8	4	4	
34 (10)	12	20	8	8	4	4	
35 (10)	20	12	8	8	4	4 *	
36 (11)	0	20	8	8	8	8	1 × 20, 6 × 8,
37 (11)	4	20	8	8	8	8	15 × 4, 10 × 0
38 (11)	8	20	8	8	8	8	
39 (11)	20	8	8	8	8	8 *	
40 (12)	2	18	14	14	2	2	1 × 18, 3 × 14,
41 (12)	14	18	14	2	2	2	28 × 2
42 (12)	18	14	14	2	2	2 *	
43 (13)	2	18	14	10	6	6	1 × 18, 1 × 14,
44 (13)	2	18	14	10	6	2	2 × 10, 6 × 6,
45 (13)	6	18	14	10	6	2	22 × 2
46 (13)	10	18	14	6	6	2	
47 (13)	14	18	10	6	6	2	
48 (13)	18	14	10	6	6	2 *	
49 (14)	2	18	10	10	10	6	1 × 18, 3 × 10,
50 (14)	2	18	10	10	6	6	9 × 6, 19 × 2
51 (14)	6	18	10	10	10	6	
52 (14)	6	18	10	10	6	6	
53 (14)	10	18	10	10	6	6	
54 (14)	18	10	10	10	6	6 *	
55 (15)	0	16	16	16	0	0	4 × 16, 28 × 0
56 (15)	16	16	16	0	0	0 *	
57 (16)	0	16	16	12	4	4	2 × 16, 2 × 12,
58 (16)	4	16	16	12	4	4	14 × 4, 14 × 0
59 (16)	12	16	16	4	4	4	
60 (16)	16	16	12	4	4	4 *	
61 (17)	0	16	16	8	8	8	2 × 16, 8 × 8,
62 (17)	0	16	16	8	8	0	22 × 0
63 (17)	8	16	16	8	8	0	
64 (17)	16	16	8	8	8	0 *	

(*Continued*)

Table B.1 *(Continued)*

$n \leq 5$ Canonic classification function ref. no.	Zero- and first-order coefficients $s_0, s_1, \ldots, s_5$ only							Summary of complete canonic spectrum
65 (18)	0	16	12	12	8	8		1 × 16, 2 × 12,
66 (18)	0	16	12	12	8	4		4 × 8, 14 × 4,
67 (18)	4	16	12	12	8	8		11 × 0
68 (18)	4	16	12	12	8	4		
69 (18)	8	16	12	12	8	4		
70 (18)	8	16	12	8	8	4		
71 (18)	12	16	12	8	8	4		
72 (18)	16	12	12	8	8	4	*	
73 (19)	2	14	14	14	6	6		3 × 14, 1 × 10,
74 (19)	6	14	14	14	6	6		7 × 6, 21 × 2
75 (19)	10	14	14	6	6	6		
76 (19)	14	14	14	6	6	6	*	
77 (20)	2	14	14	10	10	10		2 × 14, 4 × 10,
78 (20)	2	14	14	10	10	2		4 × 6, 22 × 2
79 (20)	6	14	14	10	10	2		
80 (20)	10	14	14	10	10	2		
81 (20)	14	14	10	10	10	2	*	
82 (21)	0	12	12	12	12	12		6 × 12, 10 × 4,
83 (21)	4	12	12	12	12	0		16 × 0
84 (21)	12	12	12	12	12	0	*	
85 (22)	4	20	12	4	4	4		1 × 20, 1 × 12,
86 (22)	12	20	4	4	4	4		30 × 4
87 (22)	20	12	4	4	4	4	*	
88 (23)	2	18	14	6	6	6		1 × 18, 1 × 14,
89 (23)	6	18	14	6	6	6		12 × 6, 18 × 2
90 (23)	14	18	6	6	6	6		
91 (23)	18	14	6	6	6	6	*	
92 (24)	2	18	10	6	6	6		1 × 18, 1 × 10,
93 (24)	6	18	10	6	6	6		15 × 6, 15 × 2
94 (24)	10	18	6	6	6	6		
95 (24)	18	10	6	6	6	6	*	
96 (25)	0	16	16	8	8	4		2 × 16, 4 × 8,
97 (25)	0	16	16	8	4	4		16 × 4, 10 × 0
98 (25)	4	16	16	8	8	4		
99 (25)	8	16	16	8	4	4		
100 (25)	16	16	8	8	4	4	*	
101 (26)	0	16	12	8	8	8		1 × 16, 1 × 12,
102 (26)	4	16	12	8	8	8		6 × 8, 15 × 4,
103 (26)	8	16	12	8	8	8		9 × 0
104 (26)	12	16	8	8	8	8		
105 (26)	16	12	8	8	8	8	*	

Table B.1 (*Continued*)

$n \leq 5$ Canonic classification function ref. no.	Zero- and first-order coefficients $s_0, s_1, \ldots, s_5$ only						Summary of complete canonic spectrum
106 (27)	0	16	8	8	8	8	1 × 16, 12 × 8,
107 (27)	8	16	8	8	8	8	19 × 0
108 (27)	16	8	8	8	8	8 *	
109 (28)	0	16	8	8	8	4	1 × 16, 8 × 8,
110 (28)	4	16	8	8	8	8	16 × 4, 7 × 0
111 (28)	8	16	8	8	8	4	
112 (28)	16	8	8	8	8	4 *	
113 (29)	2	14	14	10	10	6	2 × 14, 2 × 10,
114 (29)	2	14	14	10	6	6	10 × 6, 18 × 2
115 (29)	6	14	14	10	10	6	
116 (29)	6	14	14	10	6	6	
117 (29)	10	14	14	10	6	6	
118 (29)	14	14	10	10	6	6 *	
119 (30)	2	14	10	10	10	10	1 × 14, 5 × 10,
120 (30)	2	14	10	10	10	6	7 × 6, 19 × 2
121 (30)	6	14	10	10	10	10	
122 (30)	6	14	10	10	10	6	
123 (30)	10	14	10	10	10	6	
124 (30)	14	10	10	10	10	6 *	
125 (31)	2	14	10	10	10	6	1 × 14, 3 × 10,
126 (31)	2	14	10	10	6	6	13 × 6, 15 × 2
127 (31)	6	14	10	10	10	6	
128 (31)	6	14	10	10	6	6	
129 (31)	10	14	10	10	6	6	
130 (31)	14	10	10	10	6	6 *	
131 (32)	2	14	10	10	6	6	1 × 14, 3 × 10,
132 (32)	6	14	10	10	6	6	13 × 6, 15 × 2
133 (32)	10	14	10	6	6	6	
134 (32)	14	10	10	6	6	6 *	
135 (33)	0	12	12	12	12	8	4 × 12, 4 × 8,
136 (33)	0	12	12	12	12	4	12 × 4, 12 × 0
137 (33)	4	12	12	12	12	8	
138 (33)	4	12	12	12	12	4	
139 (33)	8	12	12	12	12	4	
140 (33)	12	12	12	12	8	4 *	
141 (34)	4	12	12	12	12	4	4 × 12, 28 × 4
142 (34)	4	12	12	12	4	4	
143 (34)	12	12	12	12	4	4 *	
144 (35)	0	12	12	12	8	8	4 × 12, 4 × 8,
145 (35)	4	12	12	12	8	8	12 × 4, 12 × 0

(*Continued*)

Table B.1 (*Continued*)

$n \le 5$ Canonic classification function ref. no.	Zero- and first-order coefficients $s_0, s_1, \ldots, s_5$ only							Summary of complete canonic spectrum
146 (35)	8	12	12	12	8	8		
147 (35)	12	12	12	8	8	8	*	
148 (36)	0	12	12	12	8	8		3 × 12, 6 × 8,
149 (36)	4	12	12	12	8	8		13 × 4, 10 × 0
150 (36)	4	12	12	8	8	8		
151 (36)	8	12	12	12	8	8		
152 (36)	12	12	12	8	8	8	*	
153 (37)	4	12	12	12	4	4		4 × 12, 28 × 4
154 (37)	12	12	12	4	4	4	*	
155 (38)	0	12	12	8	8	8		2 × 12, 8 × 8,
156 (38)	4	12	12	8	8	8		14 × 4, 8 × 0
157 (38)	8	12	12	8	8	8		
158 (38)	12	12	8	8	8	8	*	
159 (39)	0	12	12	8	8	8		2 × 12, 8 × 8,
160 (39)	4	12	12	8	8	8		14 × 4, 8 × 0
161 (39)	4	12	12	8	8	4		
162 (39)	8	12	12	8	8	4		
163 (39)	12	12	8	8	8	4	*	
164 (40)	0	12	8	8	8	8		1 × 12, 10 × 8,
165 (40)	4	12	8	8	8	8		15 × 4, 6 × 0
166 (40)	8	12	8	8	8	8		
167 (40)	12	8	8	8	8	8	*	
168 (41)	2	10	10	10	10	10		6 × 10, 10 × 6,
169 (41)	6	10	10	10	10	10		16 × 2
170 (41)	10	10	10	10	10	10	*	
171 (42)	2	10	10	10	10	10		6 × 10, 10 × 6,
172 (42)	2	10	10	10	10	6		16 × 2
173 (42)	6	10	10	10	10	10		
174 (42)	6	10	10	10	10	6		
175 (42)	10	10	10	10	10	6	*	
176 (43)	2	10	10	10	10	10		6 × 10, 10 × 6,
177 (43)	6	10	10	10	10	2		16 × 2
178 (43)	10	10	10	10	10	2	*	
179 (44)	2	10	10	10	10	6		4 × 10, 16 × 6,
180 (44)	6	10	10	10	10	6		12 × 2
181 (44)	6	10	10	10	6	6		
182 (44)	10	10	10	10	6	6	*	
183 (45a)[a]	0	8	8	8	8	8		16 × 8, 16 × 0
184 (45a)	8	8	8	8	8	8	*	

Table B.1 *(Continued)*

$n \leq 5$ Canonic classification function ref. no.	Zero- and first-order coefficients $s_0, s_1, \ldots, s_5$ only							Summary of complete canonic spectrum
185 (45b)[a]	0	8	8	8	8	8		16 × 8, 16 × 0
186 (45b)	8	8	8	8	8	8	*	
187 (46)[a]	0	8	8	8	8	4		12 × 8, 16 × 4, 4 × 0
188 (46)	4	8	8	8	8	8		
189 (46)	8	8	8	8	8	4	*	
190 (47)[a]	0	8	8	8	8	8		16 × 8, 16 × 0
191 (47)	8	8	8	8	8	0	*	

[a] Functions (45a), (45b), (46) and (47) of reference 4 (Chapter 3) are listed as (45), (46), (47) and (48), respectively, in reference 21 (Chapter 3).

C

The Properties of Vectors and Matrices over the Complex Field

For reference purposes, we list below the salient properties of vectors and matrices, such as may be encountered in the spectral area of digital logic. It should be noted that these mathematical properties hold over the complex field, and hence are applicable to any-valued logic. For the binary case ($p = 2$), the entries of all vectors and matrices are confined to the two values $\{+1, -1\}$ unless otherwise stated.

Also note that the symbols used here represent numbers or vectors only, and do not have the same logical meaning as in the main body of this book.

C.1 COMPLEX CONJUGATE

The complex conjugate (or merely conjugate) $\bar{v}$ of any complex number v, where

$$v = y + jz$$

y, z = any positive or negative real numbers, $j = \sqrt{-1}$, is given by

$$\bar{v} = y - jz$$

For example, given $v = 2.3 + j0.7$, $\bar{v} = 2.3 - j0.7$.

Where the value of v is any real positive or real negative number, the complex conjugate $\bar{v} \equiv v$. This clearly holds for the $\{+1, -1\}$ binary logic ($p = 2$) case. For the ternary logic ($p = 3$) case, however, we have

$$v = +1, \quad \text{or} \quad -0.866 + j0.5, \quad \text{or} \quad -0.866 - j0.5$$

or in terms of the complex 120° unit operator a

$$v = 1, \qquad a, \qquad a^2$$

giving the respective complex conjugate values

$$\begin{aligned}\bar{v} &= 1, \quad -0.866 - j0.5, \quad -0.866 + j0.5\\ &= 1, \quad a^2, \quad a\end{aligned}$$

and similar values for all higher-valued cases.

C.2 COMPLEX CONJUGATE OF ANY ROW OR COLUMN VECTOR

The complex conjugate of any row or column vector **V** is given by $\bar{\mathbf{V}}$, where all entries in $\bar{\mathbf{V}}$ are the complex conjugates of the respective entries in **V**. For example, given

$$\mathbf{V} = \begin{bmatrix} 1.2 + j0 \\ 3.7 - j1.5 \\ 0.6 + j0.2 \\ -1.9 + j0 \end{bmatrix}, \qquad \bar{\mathbf{V}} = \begin{bmatrix} 1.2 - j0 \\ 3.7 + j1.5 \\ 0.6 - j0.2 \\ -1.9 - j0 \end{bmatrix}$$

Where the row (column) entries are all real positive or real negative numbers, then clearly $\bar{\mathbf{V}} \equiv \mathbf{V}$.

C.3 INNER PRODUCT OF ANY TWO VECTORS

Given any two row or column vectors,

$$\mathbf{V} = v_1, v_2, \ldots, v_N, \qquad \mathbf{W} = w_1, w_2, \ldots, w_N$$

where b_i, w_i, $i = 1$ to N, are the complex number values in the vectors, then the inner product of vectors **V** and **W** is defined as

$$\begin{aligned}\mathbf{V}\cdot\mathbf{W} &= \bar{v}_1 w_1 + \bar{v}_2 w_2 + \cdots + \bar{v}_N w_N\\ &= \sum_{i=1}^{N} \bar{v}_i w_i\end{aligned}$$

This may equally be written as

$$\mathbf{V}\cdot\mathbf{W} = [\bar{\mathbf{V}}]^t\mathbf{W}$$

Where the row (column) entries are all real positive or real negative numbers, the inner product, therefore, is merely $\sum_{i=1}^{N} v_i w_i$.

C.4 ORTHOGONALITY

Two row (column) vectors **V**, **W** are orthogonal if the inner vector product (see C.3) is zero, giving

$$\mathbf{V}\cdot\mathbf{W} = \mathbf{W}\cdot\mathbf{V} = 0$$

that is, $\sum_{i=1}^{N} \bar{v}_i w_i = \sum_{i=1}^{N} \bar{w}_i v_i = 0$. Where the row and column entries of **V**, **W** are all real positive or real negative numbers, orthogonality is therefore then shown by

$$\sum_{i=1}^{N} v_i w_i = 0$$

When each of m vectors **V**, **W**, ... is orthogonal to all other $m - 1$ vectors, we say that the vectors are mutually orthogonal.

Finally, if m vectors are mutual orthogonal, and if each vector **V**, **W**, ... has unit length, then vectors **V**, **W**, ... are said to form a normal orthogonal basis. Recall that the length (or absolute value or norm) $|v|$ of any vector **V** is given by

$$|v| = \sum_{i=1}^{N} (v_i \bar{v}_i)^{1/2} = K$$

with $K = 1.0$ for the unit length case.

C.5 COMPLEX CONJUGATE OF ANY MATRIX M

The complex conjugate $\bar{\mathbf{M}}$ of any matrix **M** is defined as

$$\bar{\mathbf{M}} = \begin{bmatrix} \bar{\mathbf{A}} \\ \bar{\mathbf{B}} \\ \vdots \end{bmatrix}$$

where **A**, **B**, ... are the row vectors constituting **M**; thus all entries in $\bar{\mathbf{M}}$ are the complex conjugates of the corresponding entries in **M**.

Where all row and column entries of **M** are real, then $\bar{\mathbf{M}} \equiv \mathbf{M}$.

C.6 TRANSPOSED CONJUGATE (TRANJUGATE) OF ANY MATRIX M

The transposed conjugate or tranjugate $\mathbf{M}^*$ of any matrix **M** is defined as

$$\mathbf{M}^* = [\bar{\mathbf{M}}]^t$$

that is, the transpose of the complex conjugate of **M**.

Where all row and column entries of **M** are real, then $\mathbf{M}^* \equiv \mathbf{M}^t$.

C.7 HERMITIAN SQUARE MATRIX

A matrix **M** is said to be Hermitian when

$$\mathbf{M}^* = \mathbf{M}$$

that is, the transpose of the complex conjugate of **M** is equal to **M**. Note that a Hermitian matrix must be a square matrix in order that the transpose does not modify its order, and, also, all its diagonal entries must be real.

Where all row and column entries are real, we then have $\mathbf{M}^* \equiv \mathbf{M}^t = \mathbf{M}$, that is, **M** is symmetric. Examples of complex and real Hermitian matrices are

$$\begin{bmatrix} 2.5 & 3.1 + j2 & 1.5 - j6 \\ 3.1 - j2 & -6.1 & -7.8 + j4 \\ 1.5 + j6 & -7.8 - j4 & 1.7 \end{bmatrix}, \quad \begin{bmatrix} 1.5 & -2.0 & 4.3 \\ -2.0 & 11.7 & -3.7 \\ 4.3 & -3.7 & 2.8 \end{bmatrix}$$

C.8 UNITARY SQUARE MATRIX

A square matrix **M** is said to be unitary when

$$\mathbf{M}\mathbf{M}^* = \mathbf{M}^*\mathbf{M} = \mathbf{I}$$

where **I** is the identity (or unit) matrix. Hence it follows that the inverse of a unitary matrix is given by

$$\mathbf{M}^{-1} = \mathbf{M}^*$$

Where all row and column entries are real, then $\mathbf{M}\mathbf{M}^* \equiv \mathbf{M}\mathbf{M}^t = \mathbf{I}$, whence $\mathbf{M}^{-1} = \mathbf{M}^t$.

C.9 ORTHOGONAL MATRIX

An orthogonal matrix **M** is a matrix whose row vectors are mutually orthogonal. Hence from the preceding definitions (C.4 and C.6), it follows that

$$\mathbf{M}\mathbf{M}^* = K[\mathbf{I}]$$

where K is a scalar. Note that a unitary matrix is a particular case where **M** is square and $K = 1.0$.

Where all row and column entries are real, then $\mathbf{M}\mathbf{M}^* \equiv \mathbf{M}\mathbf{M}^t = K[\mathbf{I}]$.

C.10 INVERSE OF ORTHOGONAL MATRIX

The inverse $\mathbf{M}^{-1}$ of any orthogonal matrix **M** follows from C.9., namely,

$$\mathbf{M}^{-1} = \frac{1}{K}[\mathbf{M}^*]$$

where $\mathbf{M}^*$ is the transposed conjugate of **M**.

Where all row and column entries of **M** are real, then

$$\mathbf{M}^{-1} = \frac{1}{K}[\mathbf{M}^t]$$

In addition, when **M** is real and symmetric, then $\mathbf{M}^t = \mathbf{M}$, giving

$$\mathbf{M}^{-1} = \frac{1}{K}[\mathbf{M}]$$

and if in addition **M** is unitary, then

$$\mathbf{M}^{-1} = \mathbf{M}$$

References

For further details of matrix algebra see

Birkhoff, G., and MacLane, S., "A Survey of Modern Algebra." Macmillan, New York, 1977.

Ayres, F., "Theory and Problems of Matrices." Shaum's Outline Series, McGraw-Hill, New York, 1962.

Hohn, F. E., "Elementary Matrix Algebra." Macmillan, New York, 1964.

Bibliography

This bibliography is divided into two sections, the first textbooks and other published works containing information of relevance to the subject area of this book, and the second more specific published papers and other documentation.

Most of these references have been listed in individual chapters, but are repeated here for completeness.

TEXTBOOKS AND OTHER PUBLISHED WORKS

Abramowitz, M., and Stegun, I. A., "Handbook of Mathematical Functions." Dover, New York, 1964.

Ahmed, N., and Rao, K. R., "Orthogonal Transforms for Digital Signal Processing." Springer-Verlag, New York, 1975.

Ayres, F., "Theory and Problems of Matrices." Schaum's Outline Series, McGraw-Hill, New York, 1962.

Baumslag, B., and Chandler, B., "Theory and Problems of Group Theory." Schaum's Outline Series, McGraw-Hill, New York, 1968.

Beauchamp, K. G., "Walsh Functions and Their Applications." Academic Press, London, 1975.

Beauchamp, K. G., "Applications of Walsh and Related Functions." Academic Press, London, 1984.

Bennetts, R. G., "Introduction to Digital Board Testing." Crane-Russak, New York and Edward Arnold, London, 1982.

Birkhoff, G., and Maclane, S., "A Survey of Modern Algebra," 4th ed. Macmillan, London and New York, 1977.

Breuer, M. A., and Friedman, A. D., "Diagnosis and Reliable Design of Digital Systems." Computer Science Press, Potomac, Maryland, 1975.

Carr, W. N., and Mize, J. P., "MOS/LSI Design and Applications." Texas Instruments Electronics Series. McGraw-Hill, New York, 1972.

Clare, C. R., "Designing Logic Systems using State Machines." McGraw-Hill, New York, 1973.

Curtis, H. A., "A New Approach to the Design of Switching Circuits." D. van Nostrand, London, 1962.

Davio, M., Deschamps, J. P., and Thayse, A., "Discrete and Switching Functions." McGraw-Hill, New York, 1978.

Davio, M., Deschamps, J. P., and Thayse, A., "Digital Systems with Algorithm Implementation." Wiley (Interscience), New York, 1983.

Dertouzos, M. L., "Threshold Logic: A Synthesis Approach." MIT Press, Cambridge, Massachusetts, 1965.

Dietmeyer, D. L., "Logical Design of Digital Systems." Allyn and Bacon, Boston, Massachusetts, 1971.

Edwards, C. R., Matrix methods in combinational logic design, Ph.D. Thesis, University of Bath, United Kingdom, 1973.

Eris, E., Minterm-interchange applications to digital circuit design, Ph.D. Thesis, University of Bath, United Kingdom, 1979.

Friedman, A. D., and Menon, P. R., "Theory and Design of Switching Circuits." Computer Science Press, Potomac, Maryland, 1975.

Golomb, S. W., "Shift Register Sequences." Holden-Day, San Francisco, California, 1967.

Hall, M., "The Theory of Groups." Macmillan, New York, 1959.

Hammermesh, M., "Group Theory and Its Applications." Addison Wesley, Reading, Massachusetts, 1962.

Haring, D. R., "Sequential Circuit Synthesis: State Assignment Aspects." MIT Press, Cambridge, Massachusetts, 1966.

Harmuth, H. F., "Transmission of Information by Orthogonal Functions." Springer-Verlag, New York, 1969.

Hennie, F. C., "Iterative Arrays of Logic Circuits." MIT Press, Cambridge, Massachusetts, 1961.

Ho, J. C. H., Application of spectral techniques to fault detection and isolation, M.Sc. Thesis, University of Manitoba, Canada, 1983.

Hohn, F. E., "Elementary Matrix Algebra." Macmillan, London and New York, 1964.

Hurst, S. L., "The Logical Processing of Digital Signals." Crane-Russak, New York, and Edward Arnold, London, 1978.

Hurst, S. L., "Threshold Logic: An Engineering Survey." Mills and Boon, London, 1971 (translated: "Schwellwertlogik." UTB, Heidelberg, 1974).

Hurst, S. L., "Custom-Specific Integrated Circuits: Design and Fabrication." Dekker, New York, 1984.

Karpovsky, M. G., "Finite Orthogonal Series in the Design of Digital Devices." Wiley, New York, 1976.

Korn, G. A., and Korn, T. M., "Mathematical Handbook for Scientists and Engineers." McGraw-Hill, New York, 1968.

Lechner, R. J., Harmonic analysis of switching functions, *in* "Recent Developments in Switching Theory," (A. Mukhopadhyay, ed.). Academic Press, New York, 1971.

Lee, S. C., "Digital Circuits and Logic Design." Prentice-Hall, Englewood Cliffs, New Jersey, 1976.

Lee, S. C., "Modern Switching Theory and Logical Design." Prentice Hall, Englewood Cliffs, New Jersey, 1978.

Lewin, D., "Computer-Aided Design of Digital Systems." Crane-Russak, New York and Edward Arnold, London, 1977.

Lewis, P. M., and Coates, C. L., "Threshold Logic." Wiley, New York, 1967.

Lloyd, A. M., Orthogonal transforms in digital logic design, Ph.D. Thesis, University of Bath, United Kingdom, 1980.

Lui, P. K., The application of spectral techniques to the detection of single and multiple stuck-at faults in irredundant computational networks, M.Sc. Thesis, University of Manitoba, Canada, 1983.

Mano, Morris M., "Digital Logic and Computer Design." Prentice-Hall, Englewood Cliffs, New Jersey, 1979.

Miller, D. M., A spectral estimate of function complexity, *in* NATO Final Report 1623, "An Integrated Theory of Digital Logic Employing Rademacher–Walsh Transforms and Two-Place Decompositions," University of Bath, United Kingdom, 1980.

Mukhopadhyay, A. (Ed.), "Recent Developments in Switching Theory." Academic Press, New York, 1971.

Muroga, S., "Threshold Logic and Its Application." Wiley (Interscience), New York, 1971.

Muroga, S., "Logic Design and Switching Theory." Wiley (Interscience), New York, 1978.

Peterson, W. W., "Error-Correcting Codes." MIT Press, Cambridge, Massachusetts, 1961.

Rescher, N., "Many-Valued Logic." McGraw-Hill, New York, 1969.

Rine, D. C. (Ed.), "Computer Science and Multiple-Valued Logic." North-Holland, Amsterdam, 1977.

Roth, J. P., "Computer Logic, Testing and Verification." Computer Science Press, Potomac, Maryland, 1980.

Schrieber, H., and Sandy, G. F. (Eds.), "Applications of Walsh Functions and Sequency Theory." IEEE, New York, 1974.

Sheng, C. L., "Threshold Logic." Academic Press, New York, 1969.

Tokmen, V. H., An investigation into the properties of multi-valued spectral logic, Ph.D. Thesis, University of Bath, United Kingdom, 1980.

Wallis, J. S., "Hadamard Matrices." Lecture Notes No. 292. Springer-Verlag, New York, 1972.

Van der Waerden, B. L., "Modern Algebra," Vol. I, Vol. II' Frederick Unger, New York, 1949, 1950.

OTHER PUBLISHED PAPERS

Almaini, A. E. A., and Woodward, M. E., An approach to the control variable selection problem for universal logic modules, *Digital Process.* **3,** 189–206 (1977).

Andrews, H. C., and Caspari, K. C., A generalized technique for spectral analysis, *Trans. IEEE* **C-19,** 16–25 (1970).

Arzoumanian, Y., and Waicukouski, J., Fault diagnosis in LSSD environment, *Proc. 1981 Internat. Test Conf.*, 86–88 (1981).

Barzilai, Z., Savir, J., Markowsky, C., and Smith, M. G., The weighted syndrome sums approach to VLSI testing, *IEEE Trans. Comput.*, **C-30,** 996–1000 (1981).

Barzilai, Z., Coppersmith, D., and Rosenberg, A. L., Exhaustive generation of bit patterns with applications to VLSI self-testing, *IEEE Trans. Comput.* **C-32,** 190–193 (1983).

Bennetts, R. G., and Hurst, S. L., Rademacher–Walsh spectral transform: a new tool for problems in digital fault diagnosis?, *IEE Comput. Digital Tech.* **1,** 38–44 (1978).

Benowitz, N., An advanced fault isolation system for digital logic, *IEEE Trans. Comput.* **C-24,** 489–497 (1975).

Besslich, P. W., On the Walsh–Hadamard transform and prime implicant extraction, *Trans. IEEE* **EMC-20,** 516–519 (1978).

Besslich, P. W., Determination of the irredundant forms of a Boolean function using Walsh–Hadamard analysis and dyadic groups, *IEE Comput. Digital Tech.* **1,** 143–151 (1978).

Besslich, P. W., Computer aided design of logic circuits using transform methods, *Proc. IEE Internat. Conf. Comp. Aided Design Manufacture*, 75–79 (July 1979).

Besslich, P. W., Fast transform procedure for the generation of near-minimal covers of Boolean functions, *Proc. IEE* **128,** 250–254 (1981).

Braeckelmann, W., Custom-made integrated circuits, *Proc. Internat. Conf. New Trends Integrated Circuits, Paris*, 99–107 (1981).

Brown, R. D., A recursive algorithm for sequency-ordered fast Walsh transforms, *IEEE Trans. Comput.* **C-26,** 819–822 (1977).

Carter, J. L., The theory of signature testing for VLSI, Technical Report, IBM Watson Research Center, Yorktown Heights, New York, 1982.

Carter, W. C., The ubiquitous parity bit, *Proc. 12th Internat. Symp. Fault-Tolerant Comput.*, 289–296 (1982).

Cha, C. W., A testing strategy for PLAs, *Proc. 15th Design Automat. Conf.*, 326–331 (1978).

Chen, X., The mapping of spectral coefficients and its applications, *Comput. Electr. Engrg.* **9,** 167–180 (1982).

Chen, X., and Hurst, S. L., A consideration of the number of input terminals on universal logic gates and their realizations, *Internat. J. Electron.* **50,** 1–13 (1981).

Chen, X., and Hurst, S. L., A comparison of universal-logic-module realizations and their application in the synthesis of combinatorial and sequential logic networks, *Trans. IEEE* **C-31,** 140–147 (1982).

Chen, X., and Wu, X., Derivation of universal logic modules for $n \geq 3$ by algebraic means, *Proc. IEE* **E128,** 205–211 (1981).

Chow, C. K., On the characterization of threshold functions, SCTLD, pp. 34–38. IEEE Special Publication No. S134, September, 1961.

Chrestenson, H. E., A class of generalized Walsh functions, *Pacific J. Math.* **5,** 17–31 (1955).

Clarke, C. K. P., Hadamard transformation: Walsh spectral analysis of television signals, British Broadcasting Corp. Research Dept. Eng. Division Report No. BBC/RD/1975/26, 1975.

Coleman, R. P., Orthogonal functions for the logical design of switching circuits, *IEEE Trans. Comput.* **EC-10,** 379–383 (1961).

Cooley, J. W., and Tukey, J. W., An algorithm for the machine calculation of complex Fourier series, *Math. Comput.* **19,** 297–301 (1965).

Das, S. R., and Sheng, C. L., On detecting total or partial symmetry of switching functions, *Trans. IEEE* **C-20,** 352–355 (1971).

DasGupta, S., Eichelberger, E. B., and Williams, T. W., LSI chip design for testability, *Digest 1978 Internat. Solid-State Circuits Conf.*, 216–217 (1978).

David, R., Feedback shift register testing, *Proc. 8th Internat. Symp. Fault-Tolerant Comput.*, 103–107 (1978).

Davies, A. C., On the definition and generation of Walsh functions, *Trans. IEEE* **C-21,** 187–189 (1972).

Durrani, T. S., and Nightingale, J. M., Sequential generation of binary orthogonal functions, *IEE Electron. Lett.* **13,** 377–380 (1971).

Edwards, C. R., Characterisation of threshold functions under the Walsh transform and linear translation, *IEE Electron. Lett.* **11,** 563–565 (1975).

Edwards, C. R., The application of the Rademacher–Walsh transform to Boolean function classification and threshold logic synthesis, *Trans. IEEE* **C-24,** 48–62 (1975).

Edwards, C. R., The design of easily tested circuits using mapping and spectral techniques, *Radio Electron. Eng.* **47,** 321–342 (1977).

Edwards, C. R., I^2L threshold circuits for binary/quaternary encoding and decoding, *Internat. J. Electron.* **44,** 445–448 (1978).

Edwards, C. R., A special class of universal logic gate and its evaluation under the Walsh transform, *Internat. J. Electron.* **44,** 49–59 (1978).

Edwards, C. R., and Hurst, S. L., Preliminary considerations of combinatorial and sequential digital systems under symmetry methods, *Internat. J. Electron.* **40,** 499–507 (1976).

Edwards, C. R., and Hurst, S. L., A digital synthesis procedure under function symmetries and mapping methods, *Trans. IEEE Comput.* **C-27,** 985–997 (1978).

Eichelberger, E. B., and Williams, T. W., A logic design structure for LSI testing, *Proc. 14th Design Automat. Conf.*, 462–468 (1977).

Eichelberger, E. B., and Williams, T. W., A logic design structure for LSI testability, *J. Design Automat. Fault-Tolerant Computing* **2,** 165–178 (1978).

Eichelberger, E. B., Muehldorf, E. J., Walter, R. G., and Williams, T. W., A logic design structure for testing internal arrays, *Proc. 3rd USA–Japan Comput. Conf.*, 266–272 (1978).

Eris, E., Relationships between Rademacher–Walsh spectra of Boolean functions, *IEE Comput. Digital Tech.* **1,** 45–48 (1978).

Fine, N. J., On the Walsh functions, *Trans. Am. Math. Soc.* **65,** 372–414 (1949).

Fino, B. J., and Algazi, V. R., Unified matrix treatment of the fast Walsh–Hadamard transform, *IEEE Trans. Comput.* **C-25,** 1142–1145 (1976).

Fleisher, H., and Maissel, L. I., An introduction to array logic, *IBM J. Res. Develop.* **19,** 98–109 (1975).

Friedman, A. D., Diagnosis of short-circuit faults in combinational circuits, *IEEE Trans. Comput.* **C-23,** 746–752 (1974).

Frohwerk, R. A., Signature analysis: A new digital field service method, *Hewlett-Packard J.* 2–8 (May 1977).

Gabelman, I. J., The synthesis of Boolean functions using a single threshold element, *Trans. IRE* **EC-11,** 639–642 (1967).

Galiay, J., Crouzet, Y., and Vergniault, M. Physical versus logical fault models for MOS LSI circuits; impact on their testability, *IEEE Trans. Comput.* **C-29,** 527–531 (1980).

Geadah, Y. A., and Corinthios, M. J. G., Natural dyadic and sequency order algorithms and processors for the Walsh–Hadamard transform, *Trans. IEEE* **C-26,** 435–442 (1977).

Goel, P., An explicit enumeration algorithm to generate tests for combinational logic circuits, *IEEE Trans. Comput.* **C-30,** 215–222 (1981).

Golomb, S. W., On the classification of Boolean functions, *Trans. IRE* **CT-6** (Suppl.), 176–186 (May 1959).

Goto, E., and Takahashi, H., Some theorems in threshold logic for enumerating Boolean functions, *Proc. IFIP Congress, North-Holland, Amsterdam*, 747–752 (1962).

Haar, A., Zur Theorie der orthogonalen Funktionensysteme, *Math. Ann.* **69,** 331–371 (1914).

Hartman, R. F., Design and market potential for gate arrays, *Lambda* **1,** No. 3, 55–59 (1980).

Hayes, J. P., On realization of Boolean functions requiring a minimal or near-minimal number of tests, *IEEE Trans. Comput.* **C-20,** 1506–1513 (1971).

Hellerman, L., A catalogue of three-variable OR-invert and AND-invert logic circuits, *Trans. IEEE* **EC-12,** 198–223 (1963).

Huffman, G. D., Gate array logic, *EDN* **26,** No. 19, 86–96 (1981).

Hurst, S. L., The application of Chow parameters and Rademacher–Walsh matrices in the synthesis of binary functions, *Comput. J.* **16,** 165–173 (1973).

Hurst, S. L., Detection of symmetries in combinatorial functions by spectral means, *IEE J. Electron. Circuits Systems* **1,** 173–180 (1977).

Hurst, S. L., An engineering consideration of spectral transforms for ternary logic synthesis, *Comput. J.* **22,** 173–183 (1979).

Hurst, S. L., Custom lsi design: The universal-logic-module approach, *Proc. IEEE Internat. Conf. Circuits Comput., New York*, (ICCC 80), 1116–1119 (1980).

Hurst, S. L., The Haar transform in digital network synthesis, *Proc. IEEE 11th Internat. Symp. Multiple-Valued Logic*, 10–18 (1981).

Iosupowicz, A., Optimal detection of bridging faults and stuck-at faults in two-level logic, *IEEE Trans. Comput.* **C-27,** 452–455 (1978).

Karp, R., Functional decomposition and switching circuit design, *SIAM J.* **11,** 291–335 (1963).

Karpovsky, M. G., Error detection in digital devices and computer programs with the aid of linear recurrent equations over finite commutative groups, *IEEE Trans. Comput.* **C-26,** 208–218 (1977).

Karpovsky, M. G., Error detection for polynomial computations, *Comput. Digital Tech.* **2,** 50–56 (1979).

Karpovsky, M. G., Testing for numerical computations, *Proc. IEE* **127,** 69–76 (1980).

Karpovsky, M. G., Detection and location of errors by linear inequality checks, *Proc. IEE* **129,** 86–92 (1982).

Karpovsky, M. G., and Su, S. Y. H., Detection and location of input and feedback bridging faults in combinational networks, *IEEE Trans. Comput.* **C-29,** 523–527 (1980).

Kitahasi, T., and Tanaka, A., Orthogonal expansion of many-valued logical functions and its application to their realization with single-threshold element, *Trans. IEEE* **C-21,** 211–218 (1972).

Kohavi, I., and Kohavi, Z., Detection of multiple faults in combinational logic networks, *IEEE Trans. Comput.* **C-21,** 556–568 (1972).

Kremer, H., Algorithms for the Haar functions and the fast Haar transform, *Proc. Symp. Theory Appl. Walsh Functions, Hatfield,* United Kingdom (1971).

Langheld, E., and Hurst, S. L., Die spektrale Darstellung binärer Logikfunctionen, *Elektronik* No. 13, 61–66; No. 14, 69–74 (1981).

Liebler, M. E., and Roesser, R. P., Multiple real-valued Walsh functions, *Proc. IEEE Theory Appl. Multiple-Valued Logic Drsign,* pp. 84–102 (1971).

Lloyd, A. M., Spectral addition techniques for the synthesis of multivariable logic networks, *IEE Comput. Digital Tech.* **1,** 152–164 (1978).

Lloyd, A. M., A consideration of orthogonal matrices, other than the Rademacher–Walsh types, for the synthesis of digital networks, *Internat. J. Electron.* **47,** 205–212 (1979).

Lloyd, A. M., Design of multiplexer universal-logic-module networks using spectral techniques, *Proc. IEE* **E127,** 31–36 (1980).

Lotfi, Z. M., and Tosser, A. F., Graphical exhaustive analysis of minimum MUX synthesis of switching functions, *Comput. Elect. Eng.* **7,** 235–242 (1980).

McClusky, E. J., Detection of group invariance or total symmetry of a Boolean function, *Bell Syst. Tech. J.* **35,** 1445–1453 (Nov. 1956).

Marcus, M. P., The detection and identification of symmetric switching functions with the use of tables of combinations, *Trans. IRE* **E-C5,** 237–239 (1956).

Markowsky, G., Syndrome-testability can be achieved by circuit modification, *IEEE Trans. Comput.* **C-30,** 604–606 (1981).

Miller, D. M., Spectral symmetry tests, *Proc. 11th Internat. Symp. Multiple-Valued Logic,* 130–134 (1981).

Miller, D. M., and Muzio, J. C., The distribution of symmetry information in the spectrum of a Boolean function, *Electron. Lett.* **15,** 816–817 (1979).

Miller, D. M., and Muzio, J. C., Detection of symmetries in totally specified or partially specified combinational functions, *IEE J. Comput. Digital Tech.* **2,** 203–209 (1979).

Miller, D. M., Muzio, J. C., and Hurst, S. L., Spectral method of Boolean function complexity, *IEE Electron. Lett.* **18,** 572–574 (1982).

Minnick, R. C., Linear input logic, *Trans. IRE,* **EC-10,** 6–16 (1961).

Moraga, C., Ternary spectral logic, Report No. AIUD/23/76, 1976, University of Dortmund, Dortmund, Federal Republic of Germany.

Moraga, C., Ternary spectral logic, *Proc. IEEE 7th Internat. Symp. Multiple-Valued Logic,* 7–12 (1977).

Moraga, C., Complex spectral logic, *Proc. IEEE 8th Internat. Symp. Multiple-Valued Logic,* 149–156 (1978).

Moraga, C., Spectral logic design, Report No. AIUD 57/78, 1978, University of Dortmund, Dortmund, Federal Republic of Germany.

Moraga, C., Introducing disjoint spectral translation in multiple-valued logic design, *Electron. Lett.* **14,** 241–243 (1978).

Moraga, C., Characterisation of ternary threshold functions using a partial spectrum, *Electron. Lett.* **15,** 803–805 (1979).

Moraga, C., Spectral characterisation of ternary threshold functions, *Electron. Lett.* **15,** 712–713 (1979).

Moraga, C., On a property of the Chrestenson spectrum, Report No. AIUD/MVL/8102, 1981, University of Dortmund, Dortmund, Federal Republic of Germany.

Mukhopadhyay, A., Detection of total or partial symmetry of a switching function with the use of decomposition charts, *Trans. IEEE* **EC-12,** 553–557 (1963).

Muroga, S., Tsuboi, T., and Baugh, C. A., Enumeration of threshold functions of eight variables, *Trans. IEEE* **C-19,** 818–825 (1970).

Muzio, J. C., Non-orthogonal transforms for logical design, Technical Report CS.77006-R, Virginia Polytechnic Institute and State University, February 1977.

Muzio, J. C., The evaluation of large magnitude orthogonal spectral coefficients, University of Manitoba Scientific Report No. 87, 1977.

Muzio, J. C., Concerning transforms for three-valued systems, Technical Report CS.7700/R, Virginia Polytechnic Institute and State University, 1977.

Muzio, J. C., The decomposition of Rademacher–Walsh spectra, Technical Report CS7707-R, Virginia Polytechnic Institute and State University, February 1977.

Muzio, J. C., Evaluation of the spectra of sum and product functions, *IEE J. Comput. Digital Tech.*, **1,** 113–118 (1978).

Muzio, J. C., Concerning low-order spectral coefficients, *IEE Comput. Digital Tech.* **2,** 179–202 (1979).

Muzio, J. C., Composite spectra and the analysis of switching circuits, *Trans. IEEE* **C-29,** 750–753 (1980).

Muzio, J. C., and Hurst, S. L., The computation of complete and reduced sets of orthogonal spectral coefficients for logic design and pattern recognition purposes, *Comput. Electric. Engrg.* **5,** 231–249 (1978).

Muzio, J. C., and Miller, D. M., Spectral methods for fault detection in combinational circuits, Technical Report, Tektronix, Inc., Beaverton, Oregon, 1981.

Muzio, J. C., and Miller, D. M., Spectral signatures for fault testing in combinational circuits, Technical Report, Tektronix, Inc., Beaverton, Oregon, 1981.

Muzio, J. C., Miller, D. M., and Hurst, S. L., Multi-variable symmetries and their detection, *IEE-E Proc.*, **130,** 141–148 (1983).

New, A. M., The statistical efficiency of universal-logic elements in the realisation of logic functions, *Proc. IEE* **E129,** 93–99 (1982).

Ostapko, D. L., and Hong, S. J., Fault analysis and test generation for programmable logic arrays, *Proc. 8th Internat. Symp. Fault-Tolerant Comput.*, 326–331 (1978).

Paley, R. E. A. C., A remarkable series of orthogonal functions, *Proc. London Math. Soc.* **34**(2), 241–279 (1932).

Pichler, F. R., Walsh functions and linear system theory, *Proc. 1970 Symp. Appl. Walsh Functions, Washington, D.C.* 175–182 (1970).

Pichler, F. R., On discrete dyadic systems, *Proc. Symp. Theory Appl. Walsh Functions, Hatfield,* United Kingdom, 1–17 (1971).

Picton, P. D., Clock-steering synthesis using spectral techniques, *IEE Electron. Lett.* **16,** 409–411 (1980).

Posa, J. G., Gate arrays, *Electronics* **53,** 45–158 (September 1980).

Pradhan, D. K., Hsiao, M. Y., Patel, A. M., and Su, S. Y., Shift registers designed for on-line fault detection, *Proc. IEEE 8th Internat. Conf. Fault-Tolerant Computing*, 173–184 (1978).

Pratt, W. K., Andrews, H. C., and Kane, J., Hadamard transform image processing, *Proc. IEEE* **57,** 58–68 (1969).

Preparata, F. P., On the design of universal Boolean functions, *Proc. IEEE* **C-20,** 418–423 (1971).

Rademacher, H., Einige Sätze über Reihen von allgemeinen orthogonal funktionen, *Math. Ann.* **87**, 112–138 (1922).

Rao, K. R., Narasimham, M. A., and Revuluri, K., A family of discrete Haar transforms, *Comput. Electrical Engrg.* **2,** 367–368 (1975).

Roth, J. P., Method of testing for shorts, *IBM Tech. Disclosure Bull.* 3108–3109 (1967).

Savir, J., Syndrome testable design of combinational circuits, *Proc. 9th Internat. Symp. Fault-Tolerant Computing*, 137–140 (1979).

Savir, J., Syndrome testable design of combinational circuits, *IEEE Trans. Co.nput.* **C-29,** 442–451 (1980) (correction, p. 1012).

Savir, J., Syndrome testing of "syndrome untestable" combinational circuits, *IEEE Trans. Comput.* **C-30,** 606–608 (1981).

Savir, J., and Roth, J. P., Testing for and distinguishing between failures, *Proc. 12th Symp. Fault-Tolerant Computing*, 165–172 (1982).

Selfridge, R. E., Generalized Walsh transforms, *Pacific J. Math.* **5,** 451–480 (1955).

Shanks, J., Computation of the fast Walsh–Fourier transform, *Trans. IEEE* **EC-18,** 457–459 (1969).

Shanon, C. E., A symbolic analysis of relay and switching circuits, *Trans. IEEE* **57,** 713–723 (1938).

Shannon, C. E., The synthesis of two-terminal switching circuits, *Bell System Tech. J.* **28,** 59–98 (1949).

Smith, J. E., Detection of faults in programmable logic arrays, *IEEE Trans. Comput.* **C-28,** 845–853 (1979).

Smith, J. E., Measures of the effectiveness of fault signature analysis, *IEEE Trans. Comput.* **C-29,** 510–514 (1980).

Staff of the Harvard Computational Laboratory, Synthesis of Electronic Computing Systems, Harvard University Press, Cambridge, Massachusetts, 1957.

Stankovic, M. M., Tosik, Z. J., and Nikolic, S. I., Synthesis of Maitra cascades by means of spectral coefficients, *Proc. IEE* **E130,** 101–108 (1983).

Susskind, A. K., Testing by verifying Walsh coefficients, *IEEE Trans. Comput.* **C-32,** 198–201 (1983).

Tabloski, T. F., and Moule, F. J., A numerical expansion technique and its application to minimal multiplexer circuits, *Trans. IEEE* **C-25,** 684–702 (1976).

Tanaka, S., and Tahara, M., Functional completeness and polypheks in 3-valued logic, *Trans. IECE, Japan* **53C,** 111–117 (1970).

Toda, I., On the number of types of self-dual logic functions, *Trans. IRE* **EC-11,** 282–284 (1962).

Tokmen, V. H., Some properties of the spectra of ternary logic functions," *Proc. IEEE 9th Internat. Symp. Multiple-Valued Logic*, 88–93 (1979).

Tokmen, V. H., Evaluation of the spectrum of multiple-valued logic networks, *Comput. Electric. Engrg.* **6,** 233–237 (1979).

Tokmen, V. H., Disjoint decomposability of multi-valued functions by spectral means, *Proc. IEEE 10th Internat. Symp. Multiple-Valued Logic*, 88–93 (1980).

Tull, M. P., and Lee, S. C., A new method of realizing parallel processing machines using multiple-valued logic, *Proc. IEEE 10th Internat. Symp. Multiple-Valued Logic*, 36–44 (1980).

Tzafestas, S., Frangakis, G., and Pimenidis, T., Global Walsh function generators, *Electron. Engrg.* **48,** 45–49 (1976).

Tzidon, A., Berger, I., and Yoelli, M., A practical approach to fault detection in combinational networks, *IEEE Trans. Comput.* **C-27,** 968–971 (1978).

Ulman, J. L., Computation of the Hadamard transform and the R transform in ordered form, *Trans. IEEE* **EC-19,** 359–360 (1970).

Voith, R. P., ULM implicants for minimisation of universal logic module circuits, *Trans. IEEE,* **C-26,** 417–424 (1977).

Walsh, J. L., A closed set of orthogonal functions, *Amer. J. Math.* **45,** 5–24 (1923).

Wendling, S., Gagneux, G., and Stamon, G., Use of the Haar transform and some of its properties in character recognition, *Proc. Internat. Joint Conf. Pattern Recognition*, Colorado, USA, 844–848 (1976).

Wheaton, L. B., and Wayne-Current, K., A threshold logic modulo-four multiplier circuit for residue number system nonrecursive digital filters, *Proc. IEEE 11th Internat. Symp. Multiple-Valued Logic*, 48–53 (1981).

Whiteman, A. L., An infinite family of Hadamard matrices of Williamson type, *J. Combinatorial Theory Ser. A* **14,** 334–340 (1973).

Williams, T. W., and Parker, K. P., Design for testability—a survey, *IEEE Trans. Comput.* **C-31,** 2–15 (1982).

Winder, R. O., Fundamentals of threshold logic, Air Force Cambridge Research Lab. Report No. 1, Contract AFCRC-68-0066, January 1968.

Winder, R. O., Threshold functions through $n = 7$, Scientific Report No. 7, Air Force Cambridge Research Lab., Contract AF19(604)-8423, October 1969.

Wu, X., Chen, X. and Hurst, S. L., Mapping of Reed–Muller coefficients and the minimisation of Exclusive-OR switching functions, *Proc. IEE-E* **129,** 15–20 (1982).

Yablonski, S. W., On algorithmic obstacles to the synthesis of minimal contact networks, *Problemy Kibernet.* **No. 2,** 75–121 (1955) (in Russian).

Yajima, S., and Ibaraki, T., Realization of arbitrary logic functions by completely monotonic functions and its application to threshold logic, *Trans. IEEE* **C-17,** 328–351 (1968).

Yanagita, M., *et al.*, "Synthesis methods for ternary logic functions, based upon NAND-type polyphecks," *Proc. IEEE Internat. Symp. Multiple-Valued Logic, Japan*, 172–174 (1983).

Yau, S. S., and Tang, C. K., Universal logic modules and their application, *Trans. IEEE* **C-19,** 141–149 (1970).

Index

P

Q

R

S

T

U

W